Statistical Analysis of Profile Monitoring

WILEY SERIES IN PROBABILITY AND STATISTICS

Established by WALTER A. SHEWHART and SAMUEL S. WILKS

A complete list of the titles in this series appears at the end of this volume.

Statistical Analysis of Profile Monitoring

RASSOUL NOOROSSANA

Industrial Engineering Department
Iran University of Science and Technology
Tehran, Iran

ABBAS SAGHAEI

Industrial Engineering Department
Islamic Azad University – Science and Research Branch
Tehran, Iran

AMIRHOSSEIN AMIRI

Industrial Engineering Department
Shahed University
Tehran, Iran

WILEY

A JOHN WILEY & SONS, INC., PUBLICATION

Published by John Wiley & Sons, Inc., Hoboken, New Jersey.
Published simultaneously in Canada.

Library of Congress Cataloging-in-Publication Data

Noorossana, Rassoul, 1959–
Statistical analysis of profile monitoring / Rassoul Noorossana, Abbas Saghaei, Amirhossein Amiri.
p. cm.
Includes bibliographical references and index.
ISBN 978-0-470-90322-3 (cloth)
1. Process control–Statistical methods. 2. Quality control–Statistical methods. I. Saghaei, Abbas, 1972– II. Amiri, Amirhossein, 1979– III. Title.
TS156.8.N656 2011
658.5–dc22

2011002197

oBook ISBN: 9781118071984
ePDF ISBN: 9781118071960
ePub ISBN: 9781118071977

10 9 8 7 6 5 4 3 2 1

Contents

Preface

The aim of this book is to summarize major achievements in statistical profile monitoring methods. Statistical profile monitoring can be considered as a potential subarea of statistical quality control that has recently attracted attention of many researchers and practitioners. One major reason behind this attractiveness is the wide range of applications that one can identify for the concept of profile monitoring in different service and manufacturing settings. It should not be too long before the concept, methods, and issues related to statistical profile monitoring and its related analyses are introduced in different engineering and statistical textbooks and software packages.

It is well known that, in standard statistical process control applications, one is traditionally concerned with monitoring performance of a process or product using measurements on a single quality characteristic or a vector of quality characteristics at a given time or space. However, in many applications of statistical process control, quality of a process or product is best characterized and summarized by a functional relationship. In the literature of profile monitoring, this functional relationship is usually referred to as profile, signature, or waveform. Fortunately, advances in technology have made it possible for process or statistical engineers and practitioners to collect a large number of process or product measurements to reconstruct this functional relationship with the aim of understanding and evaluating its stability over time using statistical methods.

This book addresses the fundamental concepts, methods, and issues related to statistical profile monitoring. The book begins with an introduction to the concept of profile monitoring and its applications in practice, and then throughout the remaining nine chapters, issues related to simple linear profiles, complex nonlinear profiles, and roundness profiles or profiles associated with geometric specifications are discussed.

This book can be used as a major reference or textbook for researchers, engineers, and statisticians who are interested in advanced topics in statistical process control. In addition, this book can serve as a major reference for senior undergraduate students who are familiar with the basic concepts and methods of statistical process control or graduate students in advanced quality control course.

RASSOUL NOOROSSANA, PHD
Professor of Applied Statistics

ABBAS SAGHAEI, PHD
Associate Professor of Industrial Engineering

AMIRHOSSEINE AMIRI, PHD
Assistant Professor of Industrial Engineering

Contributors

Amirhossein Amiri is Assistant Professor at Shahed University. He holds BS, MS, and PhD degrees in industrial engineering from Khajeh Nasir University of Technology, Iran University of Science and Technology, and Tarbiat Modares University, respectively. He is a member of the Iranian Statistical Association. His research interests are statistical quality control, profile monitoring, and six sigma.

Jeffrey B. Birch is Professor of Statistics and Director of Graduate Programs at Virginia Polytechnic Institute, where he has been a faculty member since 1977. Dr. Birch received his MS (1970) and PhD (1977) degrees in biostatistics from the University of Washington. His area of primary research includes most aspects of regression analysis including robust regression, nonparametric and semiparametric regression, and profile monitoring. Dr. Birch is a former Associate Editor of Biometrics (1989–1993).

Eric Chicken is Associate Professor in Department of Statistics at Florida State University. He holds a BS in mathematics from University of Wisconsin-Madison and a PhD in statistics from Purdue University. His main area of research interest is nonparametric statistics, including methods involving nonparametric regression and wavelet analysis.

Bianca Maria Colosimo is Associate Professor in Department of Mechanical Engineering at Politecnico di Milano (Italy), where she received her PhD degree in industrial engineering in 2001. Her research interest is mainly in the area of quality monitoring and process optimization with special attention to discrete part manufacturing. She is a senior member of American Society for Quality (ASQ) and a member of the Editorial Board of *Journal of Quality Technology*. She is also serving as a referee for several other international scientific journals.

Majid Eyvazian is a lecturer at Iran University of Science and Technology. He holds BS, MS, and PhD degrees in industrial engineering from Iran University of Science and Technology. His primary research interests include applied probability and engineering statistics and its applications in statistical quality control. He has been teaching probability theory and engineering statistics in Industrial Engineering

Department at Iran University of Science and Technology and Khajeh Nasir Toosi University of Technology.

Longcheen Huwang received his PhD in statistics from Cornell University in 1991. He is currently Full Professor of the Institute of Statistics, National Tsing Hua University, Taiwan. His research interests include quality improvement, statistical inference, and experimental design.

Willis A. Jensen is Associate at W.L. Gore & Associates, Inc., where he provides statistical support and training for Medical Products Division in Flagstaff, AZ. He earned his BS and MS degrees in statistics from Brigham Young University and PhD in statistics from Virginia Tech. He is a member of American Statistical Association (ASA) and a senior member of ASQ. His research interests include profile monitoring, statistical process control, and other problems in industrial statistics. He is a member of Editorial Board for Quality Engineering and his publications have appeared in *Quality and Reliability Engineering International* and *Journal of Quality Technology*.

Reza Baradaran Kazemzadeh is Associate Professor in Tarbiat Modares University. He received his BS in operations research and MS in industrial engineering. He holds, PhD in technology management from University of Bradford. His research interests are quality engineering and management, profile monitoring, applied multivariate statistics, and operation management.

Mahmoud A. Mahmoud is Associate Professor of Statistics at Cairo University, Faculty of Economics and Political Science, Department of Statistics. He holds his BS (1992) and MS (1997) in statistics from Cairo University, and PhD (2004) in statistics from Virginia Tech. His primary area of interest is statistical quality control and improvement, regression analysis, and multivariate analysis.

Rassoul Noorossana is Professor of Statistics at Iran University of Science and Technology. He received his BS in engineering from Louisiana State University in 1983, and his MS and PhD in engineering management and statistics from University of Louisiana in 1986 and 1990, respectively. His primary research interests include statistical process control, process optimization, and statistical analysis. He is the editor of *Journal of Industrial Engineering International* and serves on the editorial review board of many journals. He is a senior member of ASQ, Iranian Society for Quality, Iranian Statistical Association, and Industrial Engineering Society. He is ASQ certified six sigma black belt and a quality engineer.

Massimo Pacella is Assistant Professor at University of Salento. He received his MS in computer science engineering from University of Lecce and PhD in manufacturing and production systems from Polytechnic of Milan. He is a member of Italian Association for Manufacturing. His research interests are in statistical quality control, profile monitoring, and applications of artificial intelligence in industrial engineering.

Abbas Saghaei is Associate Professor of Industrial Engineering at Islamic Azad University, Science and Research Branch. He received his PhD in industrial engineering from Iran University of Science and Technology in 2004. He also holds BS and MS degrees in industrial engineering. His primary research interests include statistical process control, six sigma, statistical learning, and process optimization. He is a senior member of ASQ, board member of Iranian Quality Management Society. He is ASQ certified six sigma black belt, a quality engineer, and reliability engineer.

Quirico Semeraro is Professor of Manufacturing and Dean of IV College of Engineering at Politecnico di Milano, Italy, where he is a faculty member since 1982. His area of primary research includes several aspects of manufacturing system optimization.

James D. Williams is Vice President, Chief Analyst in Decision Analytics and Research Organization of JP Morgan Chase Consumer Bank. He earned his BA degree in mathematics education, MS in statistics from Brigham Young University, and PhD in statistics from Virginia Tech. He is chair of ASA Quality & Productivity Section for 2011. His research interests include profile monitoring, multivariate SPC, nonlinear regression, robust regression, and econometrics. His publications have appeared in *Journal of Quality Technology*, *Communications in Statistics – Theory and Methods*, *Quality and Reliability International*, and *Journal of Agricultural, Biological, and Environmental Statistics*.

Arthur Yeh is Professor of Statistics and Chair of Department of Applied Statistics and Operations Research at Bowling Green State University. Over the years, Dr. Yeh has conducted and published research in several areas of industrial statistics, including, among others, optimal experimental designs, computer experiments, univariate and multivariate control charts, multivariate process capability indices, univariate and multivariate run-by-run process control, and profile monitoring. He has also worked as a consultant for various local and international companies in both traditional and modern high-tech manufacturing environments. He currently serves as Associate Editor for *The American Statistician*. He has also served in the past as President of the Northwest Ohio Chapter of the ASA, and Chair of the Toledo Section of the ASQ. He is a senior member of ASQ and an elected member of ISI, in addition to being a member of ASA, IMS, ICSA, and INFORMS.

Changliang Zou is Assistant Professor in Department of Statistics at Nankai University. He obtained his BS, MS, and PhD degrees in statistics from School of Mathematical Sciences, Nankai University, in 2003, 2006, and 2008, respectively. He worked as Postdoctoral Researcher at Quality Lab, Department of IELM, Hong Kong University of Science and Technology. His research interest includes statistical process control and quality improvement, nonparametric regression and dimension reduction. Special focuses are profile monitoring, multivariate statistical process control, semiparametric modeling and nonparametric lack of fit tests. He has authored more than 30 refereed journal publications in *JASA*, *Annals of Statistics*, *Technometrics*, *Journal of Quality Technology*, *Statistica Sinica*, *IIE Transactions*, *Annals of Operations Research*, and other technical journals.

CHAPTER 1

Introduction to Profile Monitoring

Abbas Saghaei

Industrial Engineering Department, Islamic Azad University – Science and Research Branch, Tehran, Iran

Rassoul Noorossana

Industrial Engineering Department, Iran University of Science and Technology, Tehran, Iran

INTRODUCTION

Quality can play an important role in the success and prosperity of many manufacturing and service organizations. A company that can fulfill customers' needs on time, with competitive cost and superior quality, can easily dominate its competitors. Hence, it is logical for organizations to view quality as business strategy. International Organization for Standardization (ISO) provides a comprehensive definition of quality in its ISO 9001:2008 quality management systems. According to this standard, quality is defined as "the degree to which a set of inherent characteristics fulfills requirements." However, Montgomery (2009) and others define quality as inversely proportional to variability. This modern definition of quality implies that variability reduction in the key quality characteristics should be of prime concern to practitioners.

Different quality improvement and variability reduction tools and methods exist that one can employ in practice to improve process performance. Statistical process control (SPC), a subarea of statistical quality control (SQC), is one of the improvement methods that can be effective in practice. SPC consists of a set of powerful tools that helps practitioners to improve quality of products and services by achieving process stability and reduction of process variability. SPC includes seven major problem-solving tools, which can be employed to improve quality. These tools, which often are referred to as "the magnificent seven", are as follows:

Statistical Analysis of Profile Monitoring, First Edition. Edited by Rassoul Noorossana, Abbas Saghaei and Amirhossein Amiri.

1. Histogram or stem-and-leaf plot
2. Check sheet
3. Pareto chart
4. Cause-and-effect diagram
5. Defect concentration diagram
6. Scatter diagram
7. Control chart

Among these seven tools, control chart is often viewed as a featured tool of SPC. Since its introduction by Walter A. Shewhart in 1924, control charts have been applied to processes in different manufacturing and service industries. Control chart is a helpful tool that plots measurements of a quality characteristic against time or sample number with the aim of distinguishing random, common, or chance causes of variation from the assignable causes of variation. Chance causes of variation are inherent natural variability of the process and are a cumulative effect of many inevitable small causes. Montgomery (2009) refers to this natural variability as "background noise." A process that operates only in the presence of chance causes or background noise is said to be statistically in-control. On the other hand, variability arising from other sources of variation such as materials, personnel, machines, environment, measurement system, and methods when compared to chance causes of variation are larger and will eventually move process to an unacceptable level of performance with respect to the quality characteristic of interest. Montgomery (2009) and others refer to these sources of variability as assignable or special causes of variation. According to Deming (1982), special causes of variation refer to "something special, not part of the system of common causes." A process that operates in the presence of assignable causes is said to be statistically out-of-control. Figure 1.1 illustrates the chance and assignable causes of variation in a process at different times. Except the first case where process operates in-control, the other cases indicate presence of assignable cause(s) leading to an out-of-control condition. Presence of an assignable cause will be eventually detected by a control chart when an unusual point or pattern appears on a control chart.

A typical Shewhart control chart is shown in Figure 1.2. A Shewhart control chart consists of a center line and symmetric upper and lower control limits. The center line is the center of gravity for the observations or the place where most of the observation should fall if process operates only in the presence of chance causes of variation. The upper and lower control limits that show the acceptable region for the sample statistic are determined using statistical considerations.

A fundamental assumption in any Shewhart control chart is that the plotted statistic should be computed on the basis of independently and identically distributed random variables. Departure from these premises may significantly affect the performance of control charts. Control charts, based on the type of quality characteristic, can be divided into two general categories of variable and attribute control charts. The quality characteristics used in the variable control charts are measured on continuous scale. Length, temperature, and weight are examples of measurements made in continuous

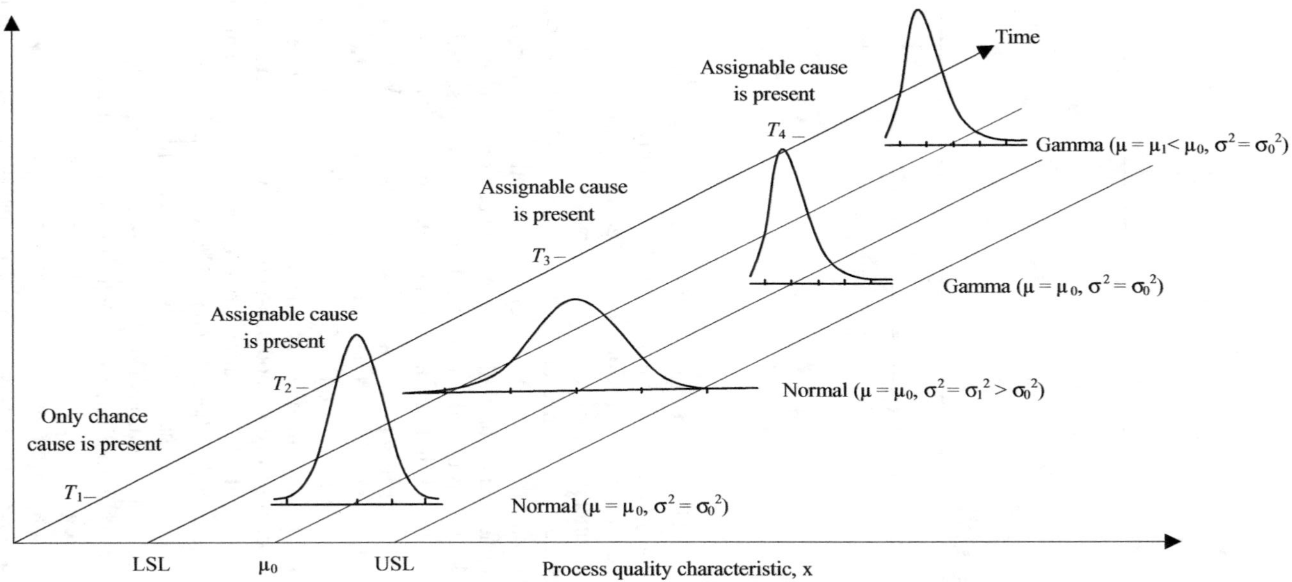

Figure 1.1 A process in the presence of chance and assignable causes of variation.

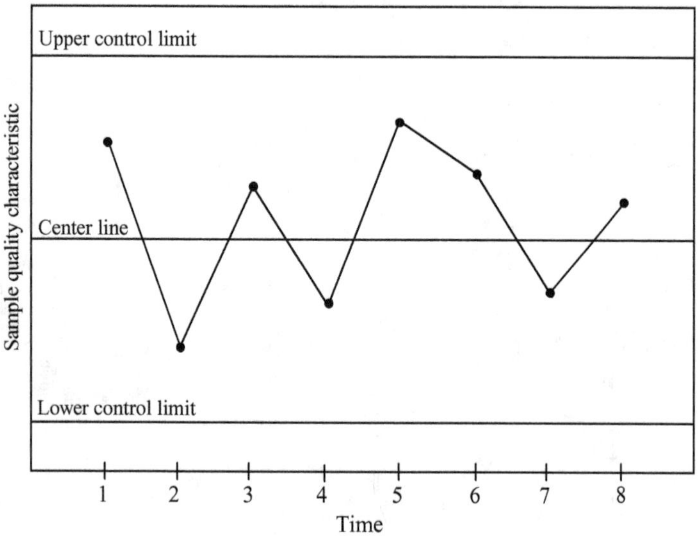

Figure 1.2 A typical control chart.

or measurable scale. However, attribute control charts are based on quality characteristics, which can only take certain integer values or can only be expressed in discrete or countable scale. Number of conforming products in a shipment, surface defects on a product, and patients arriving at an emergency room of a hospital with a trauma during a day are examples of measurements made in discrete or countable scale. A concise classification for univariate control charts based on continuous and discrete scales of measurement and correlation status between observations is provided by Montgomery (2009). This classification of control charts is presented in Figure 1.3.

In control charting, it is important to distinguish between Phase I or retrospective phase and Phase II or prospective phase analyses. According to Woodall et al. (2004) and others, in Phase I analysis of control charting, a set of historical process data is used to study process variation and evaluate its stability over time. In phase I, after identifying and eliminating anomalous observations and verifying process stability, process performance is modeled and unknown parameters are estimated. Retrospective analysis of Phase I allows one to construct trial control limits and determine if the process has been in-control when historical set of observations were collected. In Phase II analysis, one is concerned with process monitoring and detecting out-of-control conditions using online data to quickly identify shifts in the process from the trial control limits constructed in Phase I to determine if the process is under statistical control.

In standard SPC applications, one is traditionally concerned with monitoring performance of a process or product considering measurements on a single quality characteristic or a vector of quality characteristics at a given time or space. However, advances in technology have allowed engineers and practitioners to collect a large

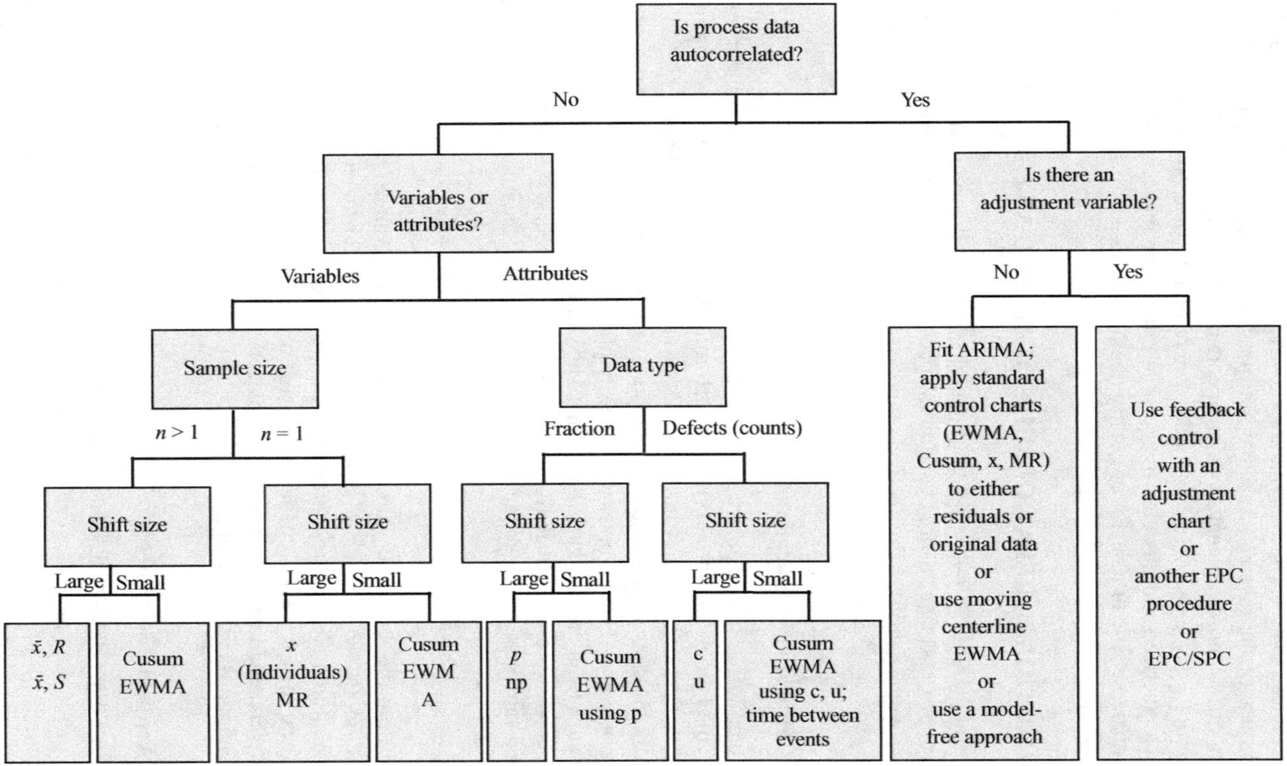

Figure 1.3 Classification of univariate control charts. (Adapted from Montgomery 2009.)

number of process or product measurements to reconstruct the entire functional relationship for the process or product performance. This functional relationship is usually referred to as a profile, signature, or waveform. For each profile it is assumed that n values of the response variable are measured along with the corresponding values of one or more explanatory or independent variables.

Section 1.1 presents several examples where quality of a process or product is better characterized and modeled by a profile rather than measurements on a single quality characteristic or a vector of quality characteristics.

1.1 FUNCTIONAL RELATIONSHIPS QUALIFIED AS PROFILES

Profiles can be used in many different manufacturing and service areas to evaluate product or process performance over time or space. In this section, we discuss practical situations where a profile can effectively represent or characterize a product or process performance.

1.1.1 Calibration Applications

Profile monitoring has extensive applications in calibration of measurement instruments. This is to ascertain their proper performance over time, determine optimum calibration frequency, and avoid overcalibration. Croarkin and Varner (1982) proposed a monitoring scheme initially developed to address calibration issues in optical imaging systems. Their proposed scheme requires plotting deviations of the measured values from the standard values on a Shewhart control chart for lower, middle, and upper values of the standards. In the calibration process, it is assumed that the measured values are related to the standard values through the following relationship:

$$y_{ij} = f(x_i) + \varepsilon_{ij}, \quad i = 1, 2 \ldots n, \quad j = 1, 2 \ldots, \tag{1.1}$$

where y_{ij} are the measured values, x_i are the standard values, n is the number of observations in the jth random sample, and ε_{ij} are the error terms assumed to be independent and identically distributed (i.i.d.) normal random variables with mean zero and variance σ^2. This scheme is now part of the ISO 11095, "Linear Calibration Using Reference Material." Figure 1.4 depicts the relationship between the measured values and the standard amounts.

1.1.2 Artificial Sweetener

Kang and Albin (2000) discussed the case of aspartame, an artificial sweetener, where the amount of aspartame that can be dissolved per liter of water (y_i) is a function of temperature (x_i). Figure 1.5 shows the milligrams of aspartame dissolved per liter of water for several samples. This figure indicates that as temperature increases, the amount of aspartame dissolved per liter of water increases up to a certain level and then drops. This pattern appears form sample to sample and according to the profile

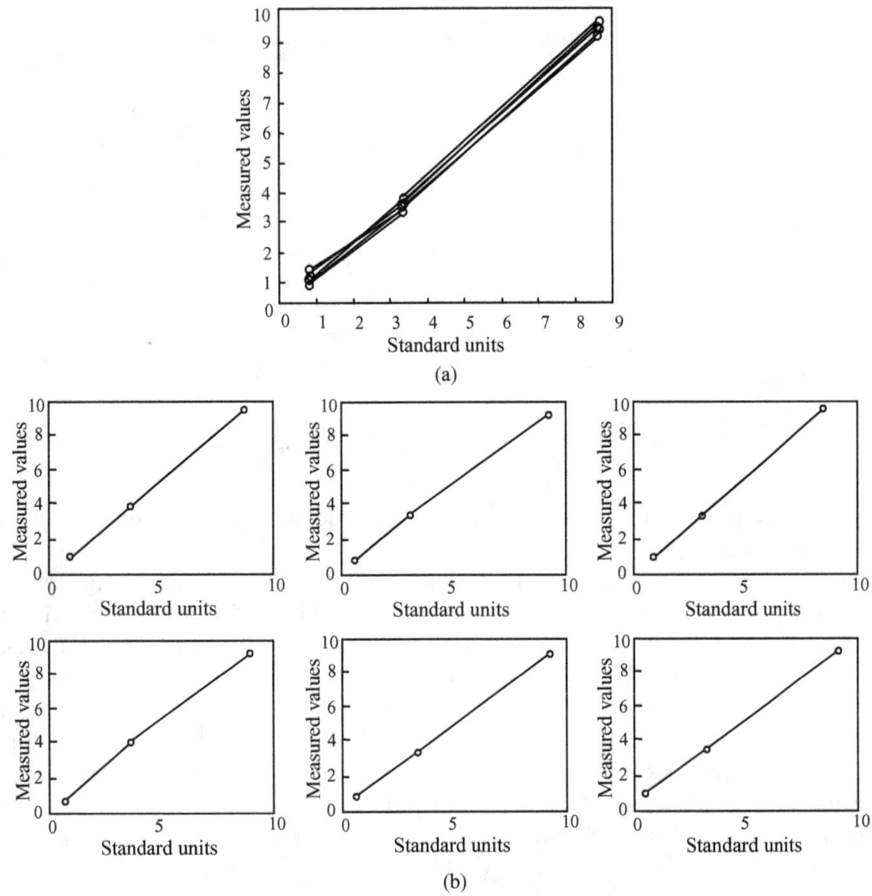

Figure 1.4 Plot of the line width reference standards (in upper, middle, and lower ends of measurement range) versus the measured values in (a) all samples in one figure (b) samples in separate figures.

of the process at different sampling period one needs to decide about the status of the process.

1.1.3 Mass Flow Controller

Kang and Albin (2000) considers mass flow controller (MFC) as an example where monitoring a profile is preferred technically over monitoring a single measurement over time. MFC is a device that controls flow of gases in a gas chamber during the semiconductors manufacturing operation, where photoresist is etched away and the required patterns for the layer of chips is created. This device includes four main components: (1) a bypass, (2) a sensor, (3) an electronic board, and (4) a regulating valve. The measuring side contains the bypass, sensors, and one part of electronic

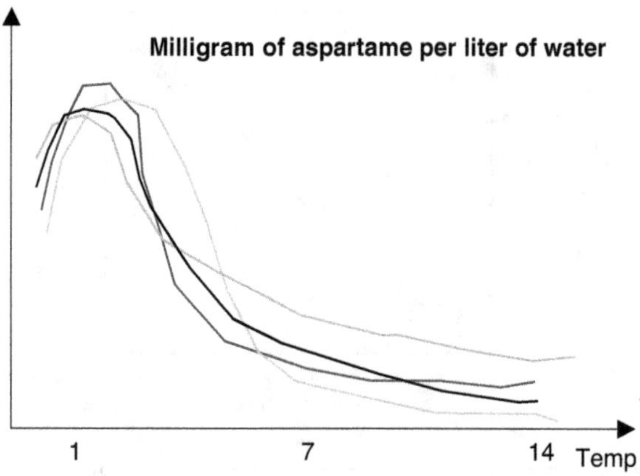

Figure 1.5 Aspartame profiles. (Adapted from Kang and Albin 2000.)

board. The other elements form the controlling side. A schematic view of MFC is shown in Figure 1.6.

Since MFC plays an important role in this semiconductor manufacturing process, performance of this device should be evaluated constantly. The common practice in evaluating performance of the MFC device is to break into the gas lines and recalibrate the device at regular intervals, which takes approximately 4 hours. According to Sheriff (1995), "a $1500 MFC device may cost more than $250,000 in production downtime during its six or seven-year life time." Hence, an SPC scheme that helps to eliminate unnecessary recalibrations of the device by differentiating assignable causes from random causes could lead to significant process improvement and annual savings. Kang and Albin (2000) provide an effective statistical process monitoring

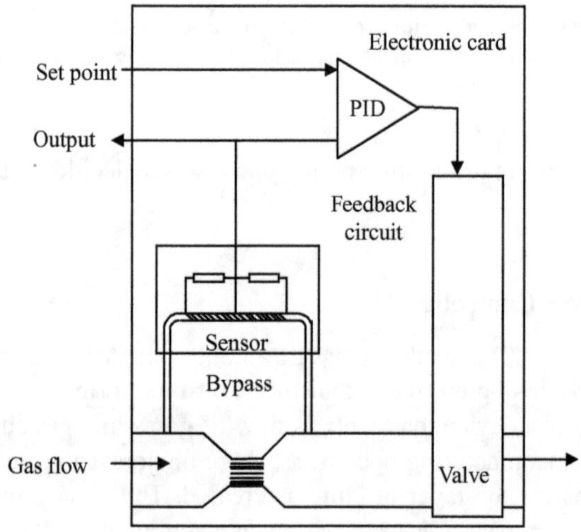

Figure 1.6 Schematic of MFC.

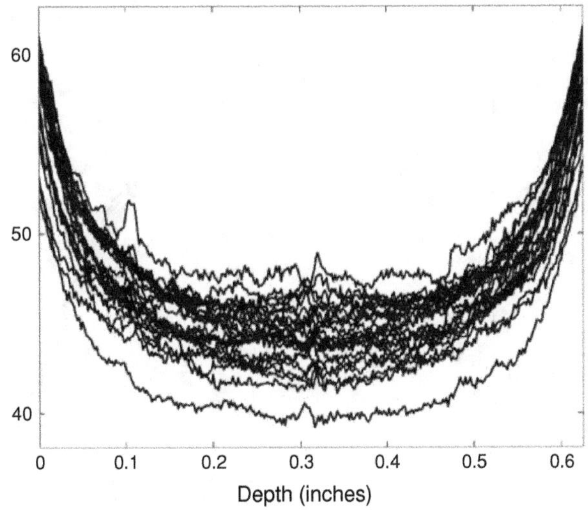

Figure 1.7 Vertical density profiles. (Adapted from Walker and Write 2002.)

scheme, which compares performance of the device represented by a linear profile to the theoretical linear profile that exists between the response and the explanatory variables. Figure 1.6 depicts the linear relationship between the response or measure pressure and the explanatory variable or flow rate.

According to Kang and Albin (2000), removal of the etcher and evaluation of the MFC performance through collection of the required observations only takes around 20–30 minutes, which is a significant reduction in time. Section 1.1.4 briefly presents several examples where profiles, signatures, or waveform signals can be used to well model performance of a process or product.

1.1.4 Vertical Density

Walker and Write (2002) considered vertical density data from engineered wood board that contain particleboard and medium density fiberboard. The vertical density is measured at certain depth across the thickness of the wood board. The measurements on a sample form the vertical density profile, which is of a bathtub shape. In other words, the wood density is higher on the top and bottom surfaces and drops as we get close to the middle section. Each vertical density profile consists of $n = 314$ measurements taken 0.002 in apart. Figure 1.7 illustrates the vertical density measurements (y_i) as a function of depth (x_i) for several wood boards.

1.1.5 Engine Torque

Amiri et al. (2010) described an example in automotive industry where profiles of polynomial form were applicable. The relationship between the torque produced by

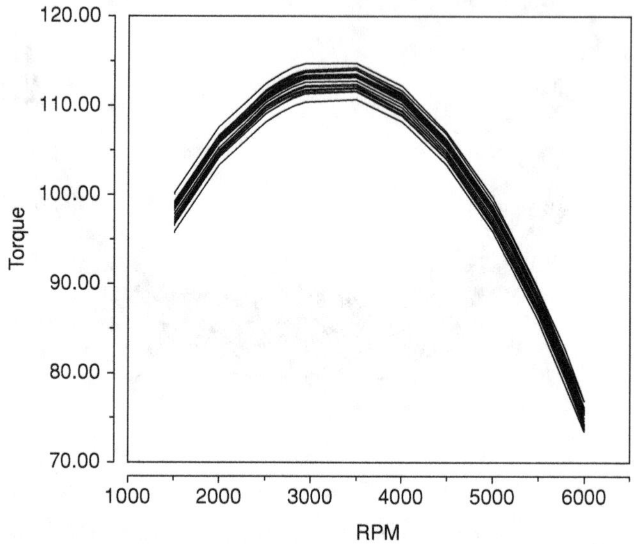

Figure 1.8 Engine torque profiles. (Adapted from Amiri et al. 2009.)

an automobile engine and the engine speed in revolution per minute is an important quality characteristic that needs to be monitored in order to distinguish conforming engines from nonconforming ones. Conforming engines ought to yield polynomial profiles very close to the theoretical engine profile, which is well described by a second-order polynomial model. Figure 1.8 illustrates the engine torque data for 26 engines as a function of engine speed in revolution per minute. For each engine, the torque value is measured at 14 different engine speeds.

1.1.6 Stamping Force

Jin and Shi (1999) studied process performance in a stamping operation. In this operation, stamping tonnage sensors are usually used to measure the stamping force for each stamped part. Figure 1.9 shows the total tonnage or stamping force (y_i), which is the sum of the outputs of all tonnage sensors mounted on the press, as a function of the crank angle (x_i). A complete stamping cycle covers crank angles from 0° to 360°, which can be divided into different segments according to different forming stages of a stamping process. This complex "waveform signals" as the authors refers to or profiles can be used to identify any potential process failures in different forming stages.

1.1.7 Location Chart

Boeing (1998) proposed location control chart for situations when several measurements of the same variable are made on each manufactured part. Although one can include this situation in the profile framework and analyze the data accordingly, but

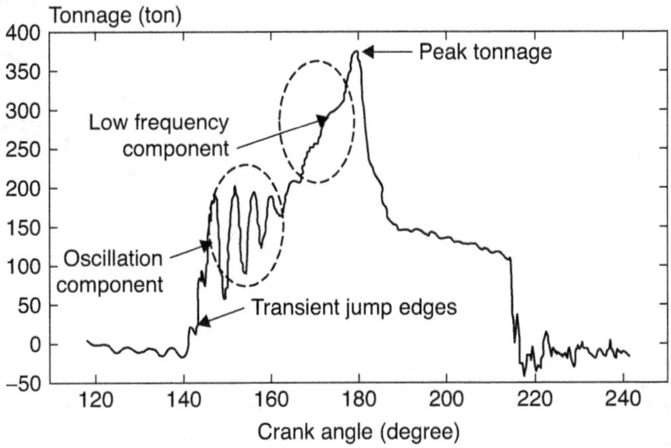

Figure 1.9 Stamping force profiles. (Adapted from Jin and Shi 1999.)

the solution proposed by Boeing (1998) is based on control limits for each location. These control limits depend only on the responses measured at that location and the multivariate structure of the data is obviously ignored. Figure 1.10 shows a location chart where the response variable is the upper flange angle measured at $n = 15$ different locations for 13 parts.

1.1.8 Acceptance Sampling Application

Tsong et al. (1997) developed an acceptance-sampling rule based on dissolution profiles of pharmaceutical products such as tablets that require long dissolution time. The acceptance rule requires fitting a dissolution profile to tablets of the approved

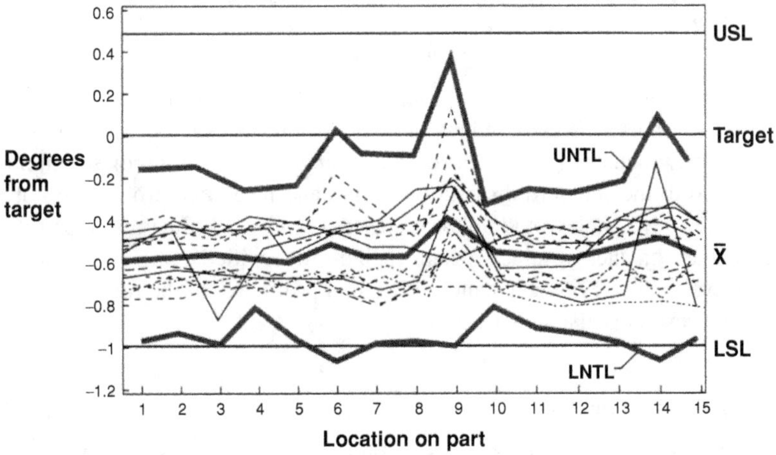

Figure 1.10 An illustration of location chart. (Adapted from Boeing 1998.)

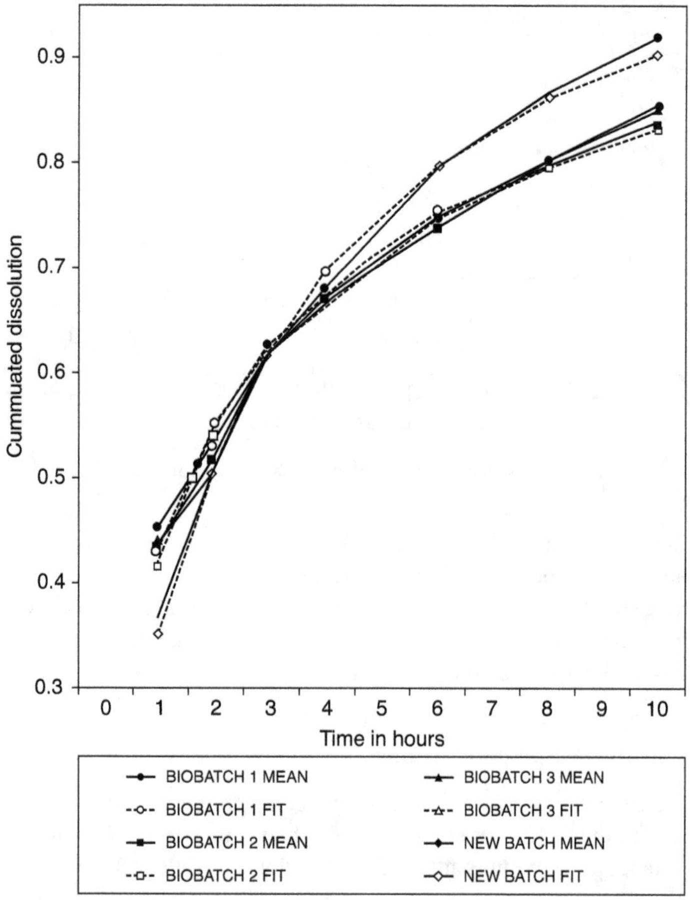

Figure 1.11 Dissolution profiles for three standard batches and a new batch of tablets. (Adapted from Tsong et al. 1997.)

standard batches and estimating the parameters of the profile. Once the standard profile is estimated, one can use batch release testing to decide whether to release or abandon a new batch. In the batch release testing, the new batch dissolution profile is compared to the standard dissolution profile and if the two profiles do not match well then the new batch would not be released. Figure 1.11 illustrates the profiles for three standard batches of tablets where each profile is fitted using dissolution results from twelve tablets. For each standard batch, a profile using the mean values of the dissolution results is also constructed.

1.1.9 Geometric Profile

Geometrical profiles such as roundness, flatness, and cylindricity can be considered as a natural extension of two-dimensional profiles. Gardner et al. (1997) used spatial

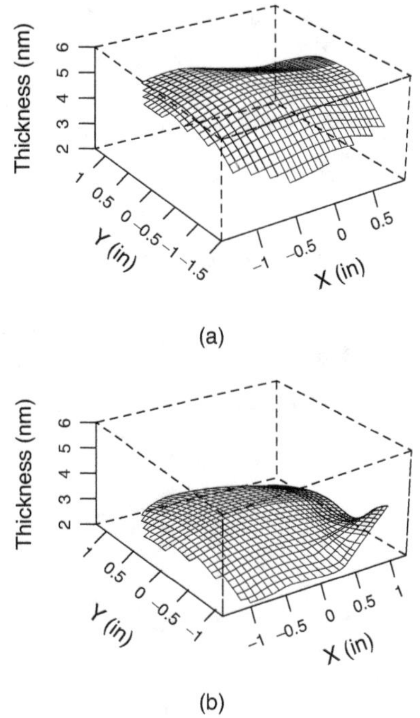

Figure 1.12 Spatial profiles for wafer surfaces (a) with no equipment faults and (b) with known equipment faults. (Adapted from Gardner et al. 1997.)

information to model spatial signatures or profiles of wafer surfaces. Their methodology consists of modeling and comparing observed wafer surfaces to a baseline wafer surface to detect and diagnose various types of equipment faults. The response variable in their study was defined as the gate oxide thickness and the explanatory variables were defined as the x and y distances from the center of the wafer. Figures 1.12a, b illustrate the spatial profiles for gate oxide thickness produced under a fault-free condition and known equipment failures, respectively.

1.2 FUNCTIONAL RELATIONSHIPS NOT QUALIFIED AS PROFILES

In most practical applications of SPC, a univariate quality characteristic or a vector of quality characteristics is used to evaluate performance of a process or product. However, there exist situations where quality of a process or product can be well characterized by a profile. Although these situations appear to be increasingly common in practice, but there exist occasions where a functional relationship looks very much like a profile but in fact it is not really a profile. Woodall (2007) believes that most of situations where functional relationship cannot be considered as profile involve applications where times series data are collected on individuals while time

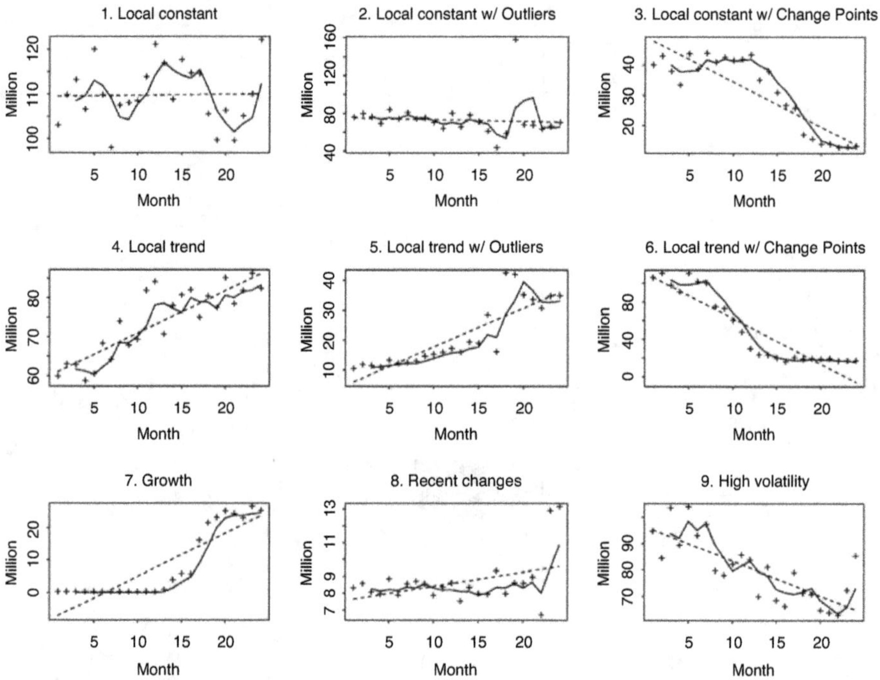

Figure 1.13 Customer telephone usage patterns (cross) with moving average smoothing (solid line) and global linear regression fit (dashed line). (Adapted from Jiang et al. 2007.)

index remains the same. As an example of profile that involves time series data with differing time index, he refers the force of an automobile air bag on the passenger since its deployment as a function of time.

Jiang et al. (2007) presents the concept of business activity monitoring (BAM) and points out that real-time BAM begins with profile monitoring. They discuss the case of monthly customer telephone usage in a telecommunications company from October 2001 to September 2003 and consider it as an example of profile. Figure 1.13 illustrates these patterns of monthly telephone usage for nine different customers. Although the shape of the time series data may look like a profile but certainly under the definition provided by Woodall (2007), their time series data on customer telephone usage would not be qualified as profile.

Brown et al. (2004) consider the case where data collected on cross-section of paper during manufacture are broken into different segments, and a stochastic model or a profile is fitted to each segment. Then estimated model parameters are used as inputs into multivariate quality control charts to monitor the manufacturing process. It is obvious that this is the case of multiple time series observations that are interpreted as successive profiles, and since only one time index exists then this example again would not be qualified as profile.

1.3 STRUCTURE OF THIS BOOK

The purpose of this section is to provide a broad overview on the structure of the book to help reader to better understand the structure of the book and what one should expect in each chapter.

1.3.1 Chapter 2

In many calibration applications, the functional relationship between the measured and real values is given by a simple linear regression model. Obviously, in these situations simple linear profile monitoring techniques could be used to establish a control scheme. Chapter 2 is devoted to Phase I and Phase II control charting methods for monitoring simple linear profiles. The calibration problem in the optical imaging systems can be considered as a representative example of such profiles. Another helpful example on simple linear profile is introduced by Mestek et al. (1994). The main focus of this example is to study the calibration curves in the photometric determination of Fe^{3+} with sulfosalicylic acid. Detailed description about this example is also provided in this chapter.

1.3.2 Chapter 3

In certain cases, quality of a process or product can be well characterized by a multiple linear or polynomial profile. In Chapter 3, several approaches for monitoring such profiles are discussed. Zou et al. (2007) present a motivating example in semiconductor manufacturing. In this example, the quality characteristic to be monitored is the profile of trenches associated with deep reactive ion etching process, which is a key process in microelectromechanical systems fabrication operation. In this process, a trench with vertical and smooth sidewalls can be regarded to have an ideal profile. However, negative and positive profiles that are associated with overetching and underetching, respectively, are considered unacceptable. Figure 1.14 illustrates various types of trenches.

It is obvious that the entire profile cannot be represented by a polynomial or a general linear profile. Since information contained in the left and right corners, which are symmetric, would allow one to evaluate process performance; hence, it would suffice to consider only the profile associated with one side of the trench to establish a control scheme. If the corner is rotated 45° counterclockwise then a quadratic

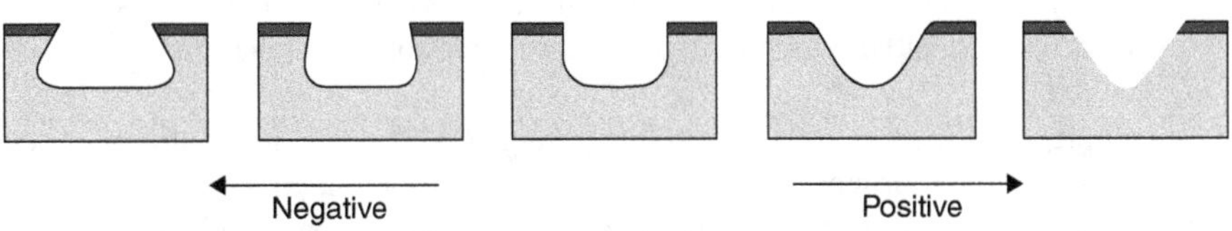

Figure 1.14 Negative, positive, and desired profiles. (Adapted from Zou et al. 2007.)

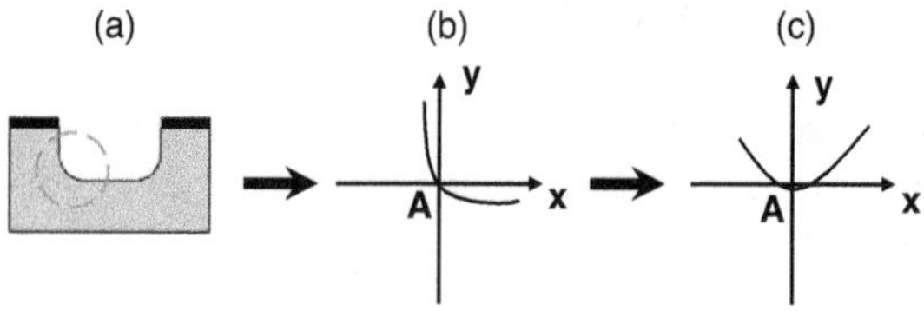

Figure 1.15 Modeling the left-corner profile of a trench. (Adapted from Zou et al. 2007.)

polynomial model can be used to model the resulting profile. Figure 1.15 shows the profile for the shape of the sidewalls of the trench.

1.3.3 Chapter 4

Profile monitoring applications are not restricted to regression models with continuous response variables. In many practical applications, quality of a process or product can be well characterized by a binary response variable in the form of conforming or nonconforming product. For instance, Yeh et al. (2009) discuss the case of compressive strength of an alloy fastener originally discussed by Montgomery et al. (2006). In this example, the compressive strength of alloy fastener used in aircraft manufacturing is considered as a critical-to-quality characteristic, which is tested at particular load to examine whether it will fail or stand the load. Under this condition, the profile of interest consists of a response, which is a binary variable and an explanatory variable (load strength), which is continuous in nature. They use logistic regression model to develop control charting techniques suitable for monitoring profiles with binary responses. Chapter 4 focuses on profile monitoring techniques suitable for binary response profiles. Healthcare, public health surveillance, and hospitality industries are among the industries where profiles with binary or categorical response variables are encountered easily. As an example, in healthcare and public health surveillance, one considers the mortality rate after a cardiac surgery. It is important to note that that mortality rate depends on many variables, including age, diabetic status, and Parsonnet score.

1.3.4 Chapter 5

There are many instances where performance of a process or product can be well modeled by a nonlinear relationship. To handle monitoring issues related to such situations, one can use either parametric nonlinear regression or nonparametric smoothing methods. The purpose of parametric nonlinear profile monitoring is to reduce the complex nonlinear profile into a few parameters through the nonlinear model estimation and then form a control chart scheme on the estimated parameters of each individual profile. The parametric nonlinear regression approach and its associated control schemes are the main subject of this chapter. A vertical density data of the wood board presented in Figure 1.7 of Section 1.2.3 is a typical example of a nonlinear profile.

1.3.5 Chapter 6

A main assumption in profile monitoring using parametric models is that the data follows a particular assumed parametric form. It is obvious that there exists many occasions where this assumption fails to hold. In other words, the observed profile f cannot readily admit to a parameterization using a parametric functional relationship. Under such conditions, a more general form of the profile that allows for a nonparametric representation of the relationship would be appropriate. The vertical density profile data from Walker and Wright (2002) is considered in this chapter to evaluate performance of the nonlinear profile monitoring methods.

1.3.6 Chapter 7

Parker et al. (2001) provide a calibration example respecting the force balance in the wind tunnel experiments at NASA Langley Research Center. Three orthogonal force components representing response variables and three orthogonal torque components respecting explanatory variables are measured simultaneously. Figure 1.16 illustrates these components. A close look of the variables involved indicates that this is the case of a multivariate multiple linear profile because of the correlation structure that exists between the response variables. It is obvious that if the correlation structure between the response variables is ignored by assuming separate profiles then misleading results should be expected. This chapter provides various approaches for Phase I and Phase II monitoring of multivariate multiple and multivariate simple linear profiles.

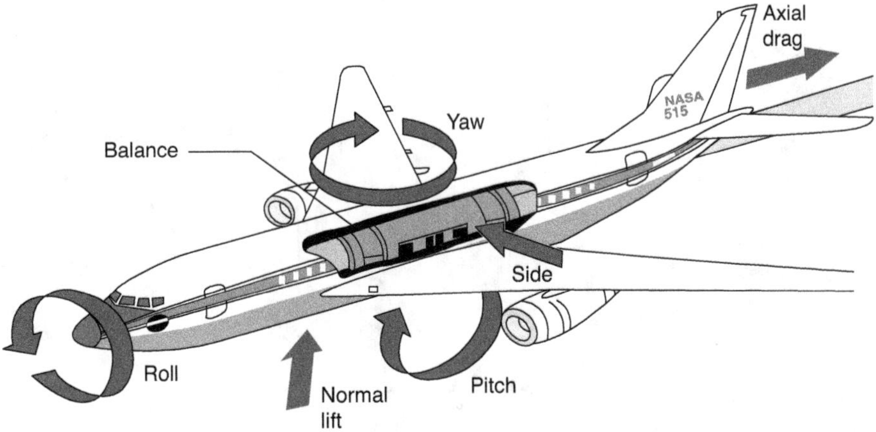

Figure 1.16 A schematic view of forces and moments. (Adapted from Parker et al. 2001.)

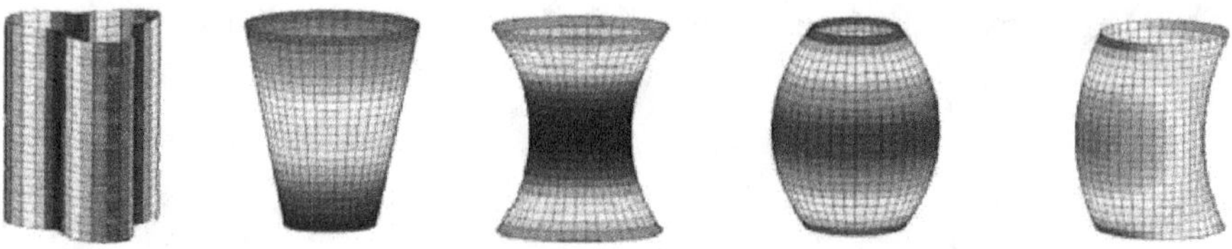

Figure 1.17 Various types of nonconforming shapes for a cylinder.

1.3.7 Chapter 8

In the SPC-related literature, there are various cases in which we are interested in monitoring geometric specifications of products. Roundness, flatness, straightness, or cylindricity of products could be examples of such quality characteristics. It is obvious that under such circumstances, one would be interested in detecting any departure from the baseline shape of the product. Figure 1.17 shows various types of nonconforming shapes for a machined cylinder discussed by Henke et al. (1999) and Zhang et al. (2005). Detailed discussions on the methods for monitoring quality characteristics related to the geometric specifications of manufactured products are presented in this chapter.

1.3.8 Chapter 9

Staudhammer et al. (2007) consider a case where thickness of lumber is a function of the distance on the board. The relationship between the lumber thickness and the distance is presented in Figure 1.18. A close analysis of the patterns on this figure indicates that observations within each profile are not independent and as

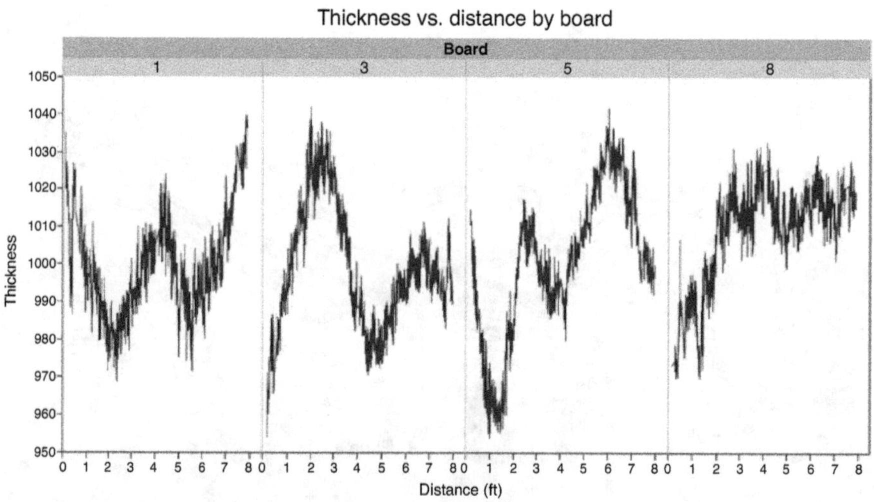

Figure 1.18 Plot of thickness versus distance.

a result a fundamental assumption of profile monitoring approaches is violated. Violation of this assumption if not addressed properly would significantly affect the performance of the approach used to monitor processes that inherently generate autocorrelated observations. This chapter addresses various issues related to both within- and between-profile autocorrelations and several approaches for monitoring profiles when the independence assumption is violated are discussed.

1.3.9 Chapter 10

There are many instances where a parametric linear or nonlinear model may not well represent a profile. In practice, process engineers may want to avoid spending too much time on fitting a parametric model to a set of data representing a complicated profile. In such cases, if an appropriate right parametric nonlinear regression model is not selected to represent the complex profile then misleading results should be expected. Zou et al. (2008) refer to the asymptotic normal distribution of the estimators that may significantly affect the in-control and out-of-control performance of parametric-based control schemes, slow convergence rate of the iterative procedures, and having a similar form for the in-control and out-of-control models as potential problems that one may encounter while working with nonlinear regression models. In this chapter, assuming that a profile can be well represented by a regression function, several nonparametric methods such as spline and wavelet methods are discussed, which help to address issues when data does not follow a particular parametric form.

REFERENCES

Amiri, A., Jensen, W. A., and Kazemzadeh, R. B. (2010) A case study on monitoring polynomial profiles in the automotive industry. *Quality and Reliability Engineering International*, **26** (5), 509–520.

Boeing Commercial Airplane Group, Materiel Division, Procurement Quality Assurance Department (1998) *Advanced Quality System Tools, AQS D1-9000-1*. The Boeing Company, Seattle, WA.

Brown, P. E., Diggle, P. J., and Henderson, R. (2004) A model-based approach to quality control of paper production. *Applied Stochastic Models in Business and Industry*, **20**, 173–184.

Croarkin, C. and Varner, R. (1982) Measurement assurance for dimensional measurements on integrated-circuit photomasks. *NBS Technical Note 1164*, US Department of Commerce, Washington, DC, USA.

Deming, W. E. (1982) *Out of The Crisis*, MIT Press, Cambridge.

Gardner, M. M., Lu, J. -C., Gyurcsik, R. S., Wortman, J. J., Hornung, B. E., Heinisch, H. H., Rying, E. A., Rao, S., Davis, J. C., and Mozumder, P. K. (1997) Equipment fault detection using spatial signatures. *IEEE Transactions on Components, Packaging, and Manufacturing Technology—Part C*, **20**, 295–304.

Henke, R. P., Summerhays, K. D., Baldwin, J. M., Cassou, R. M., and Brown, C. W. (1999) Methods for evaluation of systematic geometric deviations in machined parts and their relationships to process variables. *Precision Engineering*, **23**, 273–292.

Jiang, W., Au, T., and Tsui, K. L. (2007) A statistical process control approach to business activity monitoring. *IIE Transactions*, **29**, 235–249.

Jin, J. and Shi, J. (1999) Feature-preserving data compression of stamping tonnage information using wavelets. *Technometrics*, **41**, 327–339.

Kang, L. and Albin, S. L. (2000) On-line monitoring when the process yields a linear profile. *Journal of Quality Technology*, **32**(4), 418–426.

Mestek, O., Pavlik, J., and Suchanek, M. (1994) Multivariate control charts: control charts for calibration curves. *Fresenius' Journal of Analytical Chemistry*, **350**, 344–351.

Montgomery, D. C. (2009) *Introduction to Statistical Quality Control*. John Wiley and Sons, New York.

Montgomery, D. C., Peck, E. A., and Vining, G. G. (2006) *Introduction to Linear Regression Analysis*, 4th Edition, John Wiley and Sons, New York.

Parker, P. A., Morton, M., Draper, N. R., and Line, W. P. (2001) A single-vector force calibration method featuring the modern design of experiments. Paper presented at the *American Institute of Aeronautics and Astronautics 39th Aerospace Sciences Meeting & Exhibit*, Reno, NV, USA.

Sheriff, D. (1995) Diagnostic procedures facilitate the solving of gas flow problems. *Solid State Technology*, **38**, 63–70.

Staudhammer, C., Maness, T. C., and Kozak, R. A. (2007) New SPC methods for identifying lumber manufacturing defects with real-time laser range sensor data. *Journal of Quality Technology*, **39**, 224–240.

Tsong, Y., Hammerstrom, T., and Chen, J. J. (1997) Multipoint dissolution specification and acceptance sampling rule based on profile modeling and principal component analysis. *Journal of Biopharmaceutical Statistics*, **7**(3), 423–439.

Walker, E. and Wright, S. P. (2002) Comparing curves using additive models. *Journal of Quality Technology*, **34**, 118–129.

Woodall, W. H. (2007) Current research on profile monitoring. *Revista Producão*, **17**(3), 420–425.

Woodall, W. H., Spitzner, D. J., Montgomery, D. C., and Gupta, S. (2004) Using control charts to monitor process and product quality profiles. *Journal of Quality Technology*, **36**, 309–320.

Yeh, A. B., Huwang, L., and Li, Yu-Mei (2009) Profile monitoring for a binary response. *IIE Transactions*, **41**(11), 931–941.

Zhang, X. D., Zhang, C., Wang, B., and Feng, S. C. (2005) Unified functional approach for precision cylindrical components. *International Journal of Production Research*, **43**, 25–47.

Zou, C., Tsung, F., and Wang, Z. (2007) Monitoring general linear profiles using multivariate exponentially weighted moving average schemes. *Technometrics*, **49**(4), 395–408.

Zou, C., Tsung, F., and Wang, Z. (2008) Monitoring profiles based on nonparametric regression methods. *Technometrics*, **50**, 512–526.

CHAPTER 2

Simple Linear Profiles

Mahmoud A. Mahmoud

Department of Statistics, Cairo University, Cairo, Egypt

INTRODUCTION

In Chapter 1 we presented the basic concepts and definitions pertaining to the monitoring of profiles applications. As mentioned in that chapter, the term "profile" refers to the functional relationship between a response variable and one or more explanatory variables. However, the word "profile" is usually used to describe the shape of the response data, regardless of whether there is a set of explanatory variables or not. The focus in this chapter is on profiles that can be adequately represented by a simple linear regression model, in which a single explanatory variable X is used to describe the behavior of the response variable Y.

There has been a growing interest in monitoring processes characterized by simple linear regression profiles. The monitoring of simple linear profiles is a relatively new quality control application, with most of the work done in this application having appeared in the last decade. Most of the studies conducted have been motivated by calibration applications. For example, Mestek et al. (1994) considered the stability of linear calibration curves in the photometric determination of Fe^{3+} with sulfosalicylic acid. Stover and Brill (1998) studied the multilevel ion chromatography linear calibrations to determine instrument response stability and the proper calibration frequency. ISO 5725-6 (1994) recommended a method for monitoring a linear calibration line based on the measurement of three standards at each time period. These standards are at low, medium, and high values, respectively. The three deviations of the measured values from these standards are plotted simultaneously for each sample on a Shewhart-type control chart. They used this charting method to monitor linear calibration profiles of an optical imaging system. Kang and Albin (2000) presented

Statistical Analysis of Profile Monitoring, First Edition. Edited by Rassoul Noorossana, Abbas Saghaei and Amirhossein Amiri.

two examples of process profiles; one of them is a semiconductor manufacturing application in which the process is represented by a linear calibration function.

The literature includes several other simple linear profile monitoring applications. For example, Ajmani (2003) presented a semiconductor manufacturing application from Intel Corporation not involving calibration. Wade and Woodall (1993) considered the simple linear regression model to account for the effect of an input quality variable X on the output quality variable Y when monitoring a particular stage of a manufacturing process. They used one of the cause-selecting control charts considered by Zhang (1992) to monitor the relationship between incoming and outgoing quality measurements of a component part for the braking system for an automobile. In this chapter, we illustrate the use of some simple linear profile techniques using some of these examples.

As discussed in Chapter 1, the monitoring of profiles includes two phases—Phase I and Phase II. The first step in monitoring linear profiles requires the analysis of a preliminary set of data that is assumed to be in a state of statistical control. This analysis, i.e., Phase I analysis is conducted to estimate the profile parameters to be used in designing control charts for Phase II monitoring of profile parameters. The first phase of data collection should utilize a very large sample of data, so that parameters and control limits are well estimated for Phase II. The in-control process parameters are estimated on the basis of a stable historical set of a fixed number of process samples collected over time. In Phase II, the main interest is to quickly detect parameter changes from the in-control parameter values. The distinction between these two phases of data collection is important. Different types of statistical methods are appropriate for the two phases, with each type requiring different measures of statistical performance. The performance of a Phase I control chart method is usually measured in terms of the probability of obtaining at least one charted statistic outside the control limits. This probability is usually referred to as the "probability of signal" in the literature. The performance of control chart methods in Phase II is usually measured in terms of the run-length distribution, where the run length is the number of samples taken until the chart gives an out-of-control signal.

In Section 2.1, we introduce the Phase I simple linear profile model, assumptions, and approaches. We also introduce in this section some illustrative Phase I simple linear profile examples presented in the literature. In Section 2.2, we present the Phase II simple linear profile model, assumptions, approaches, and illustrative applications. Section 2.3 considers some special cases of the monitoring of simple linear profiles and introduces an important simple linear profile application. Diagnostics statistics used to identify the source of profile variation and the location of out-of-control profiles are introduced in detail in Section 2.4. Section 2.5 discusses the consequences of the violation of the model assumptions.

2.1 PHASE I SIMPLE LINEAR PROFILE

In a linear profile data set with a single explanatory variable X and a response Y, the data are m samples in the form $\{(X_{i1}, Y_{i1}), i = 1, 2, \ldots, n_1\}$, $\{(X_{i2}, Y_{i2}),$

$i = 1, 2, \ldots, n_2\}, \ldots, \{(X_{im}, Y_{im}), i = 1, 2, \ldots, n_m\}$. Except in the paper by Mahmoud et al. (2010), it is usually assumed that the profile samples are with $n_j > 2$, $j = 1, 2, \ldots, m$. In most of the work done in the area of simple linear profile monitoring, it was assumed that the X-values are fixed and take the same set of values for each sample. This is usually the case in the calibration applications. If the X-values are random, then under certain conditions one can simply modify the corresponding formulas to account for the varying X-values from sample to sample. For example, Neter et al. (1990, pp. 86–87) stated that the conditional distribution of Y | X must meet the simple linear regression assumptions with the probability distribution of X not dependent on any of the parameters in the regression model.

2.1.1 The Phase I Model and Assumptions

The most efficient procedure used in profile monitoring is to summarize the in-control shape of the profile with a parametric model and monitor for shifts in the parameters of this model. Then the monitoring methods are based on the estimated parameters of the model from successive profile data observed over time. For the simple linear profile case, the model that relates the explanatory variable X to the response Y is in the form

$$Y_{ij} = A_j + B_j X_{ij} + \varepsilon_{ij}, \quad i = 1, 2, \ldots, n_j \text{ and } j = 1, 2, \ldots, m. \tag{2.1}$$

The emphasis in this situation is to detect changes in the regression parameters from sample to sample, assuming that no parameter change has occurred within each sample. Changes in the profile structure introduced by abnormal observations indicate an out-of-control Phase I profile data set. The main objective of Phase I profile monitoring is to identify the out-of-control profile(s) from the baseline data set. Figure 2.1 shows graphically the framework of a Phase I simple linear profile data set. It should be noted that this sampling framework is identical to that of panel data in econometrics.

As far as the monitoring of profile responses is concerned, two important issues need to be addressed. The first issue is related to the process of detecting a change in the profile structure after the process has indeed shifted, locating the position in time of

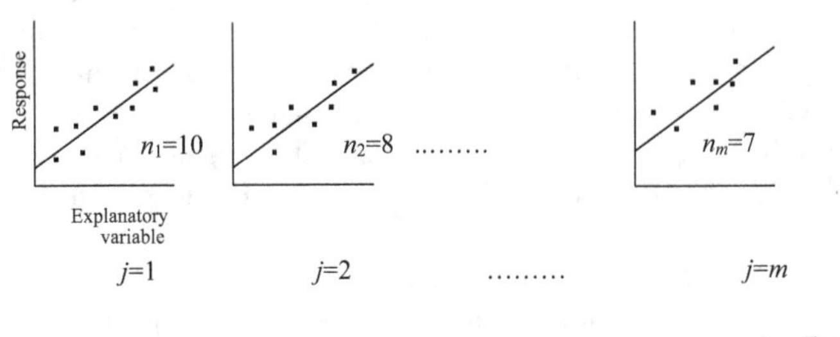

Figure 2.1 The framework of a Phase I linear profile data set.

this change, and identifying clearly the source(s) of the profile variation. In this regard, profile monitoring approaches that are described in this chapter are used effectively to detect changes in the profile structure caused by assignable causes and to accurately identify their positions and sources. The second issue is related to data reduction because a profile response can be of high dimension (i.e., the sample sizes are very large). To address the data-reduction mission, usually one summarizes the in-control shape of the profile with a fitted regression model and monitors for shifts in the parameters of this model. The regression-based approach significantly satisfies the data-reduction mission, since the fitted parameters are usually far fewer than the data points in a profile response. The simple regression model described in Equation (2.1) is particularly appealing because it reduces the data dimension to three: (1) an intercept, (2) slope, and (3) variance of error terms. Profile data sets with more than two observations ($n_j > 2$) per subgroup are consequently needed to estimate the three parameters. In this section, we consider profile approaches suitable for the case of more than two observations per sample. In Section 2.4, we discuss the limitations of these methods when applying to a profile data set with less than two observations per sample, and present appropriate profile monitoring approaches that overcomes these limitations.

2.1.1.1 *The Model Assumptions*

A standard assumption in the monitoring of the simple linear regression profiles model described in Equation (2.1) is that the errors are independent and identically distributed (i.i.d.), usually with an assumed normal distribution with a mean of zero and variance of σ_j^2, i.e., the ε_{ij} are assumed to be i.i.d. N (0, σ_j^2) random variables. The assumptions of the normality and independence of the ε_{ij} are required for determining the statistical performance of any of the Phase I methods presented later in this chapter. Noorossana et al. (2008) studied the effect of ignoring the independence assumption of the errors. Autocorrelated errors frequently occur within profile monitoring data because successive measurements are often made close in space and/or time. Mahmoud (2004) studied the effect of non-normality of the error terms, using double exponential, exponential, and t distributions as alternatives, on the performance of four Phase I simple linear profile methods. He found that the false alarm rate can increase dramatically if the normality assumption is violated. We discuss this issue in detail in Section 2.5.

The in-control values of the parameters A_0, B_0, and σ_0^2 in Equation (2.1) are unknown. If $A_j = A_0$, $B_j = B_0$ and $\sigma_j^2 = \sigma_0^2$, $j = 1, 2, \ldots, m$, then the process is considered to be stable in Phase I. As mentioned before, our main objective in the Phase I analysis is to evaluate the stability of the process and to estimate the in-control parameters. Obtaining a data set reflecting expected in-control performance is usually accomplished by discarding samples associated with assignable causes from the data set, assuming that the associated assignable causes can be identified and removed.

2.1.1.2 *Estimating the Model Parameters*

The most conventional estimation method is that of least squares, which finds the parameter estimates that minimize the sum of the squares of the difference between observed and fitted profiles. It is well known that the least squares estimates (LSE) of the simple linear profile parameters A and B for sample j are

$$a_j = \bar{Y}_j - b_j \bar{X}_j \text{ and } b_j = S_{XY(j)}/S_{XX(j)}, \tag{2.2}$$

where $\bar{Y}_j = \sum_{i=1}^{n_j} Y_{ij}/n_j$, the average response values in sample j, $\bar{X}_j = \sum_{i=1}^{n_j} X_{ij}/n_j$ the average predictor values in sample j, $S_{XY(j)} = \sum_{i=1}^{n_j} \left(X_{ij} - \bar{X}_j\right) Y_{ij}$, and $S_{XX(j)} = \sum_{i=1}^{n_j} (X_{ij} - \bar{X}_j)^2$, see, for example, Myers (1990, Chapter 2). Furthermore, σ_j^2 is estimated by the jth mean square error, MSE_j, where

$$MSE_j = SSE_j/(n_j - 2).$$

Here, $SSE_j = \sum_{i=1}^{n_j} e_{ij}^2$ is the residual sum of squares, where

$$e_{ij} = Y_{ij} - a_j - b_j X_{ij}, \quad i = 1, 2, \ldots, n_j.$$

It is also well known that, under in-control conditions, the least squares estimators a_j and b_j are distributed as a bivariate normal distribution with the mean vector

$$\boldsymbol{\mu} = (A_0 \; B_0)^T$$

and the variance–covariance matrix

$$\sum = \begin{pmatrix} \sigma_a^2 & \sigma_{ab}^2 \\ \sigma_{ab}^2 & \sigma_b^2 \end{pmatrix} = \sigma_0^2 \begin{pmatrix} \dfrac{1}{n_j} + \dfrac{\bar{X}_j^2}{S_{XX(j)}} & \dfrac{-\bar{X}_j}{S_{XX(j)}} \\ \dfrac{-\bar{X}_j}{S_{XX(j)}} & \dfrac{1}{S_{XX(j)}} \end{pmatrix},$$

where σ_a^2, σ_b^2, and σ_{ab}^2 are the variance of a_j, the variance of b_j, and the covariance between them, respectively, when the process is in-control in the Phase I. Also, the quantity

$$\frac{(n_j - 2)MSE_j}{\sigma_j^2}$$

is distributed as a chi-square distribution with $(n_j - 2)$ degrees of freedom independently of a_j and $b_j, j = 1, 2, \ldots, m$.

The first step of a Phase I procedure is to find the LSE of the profile parameters for the m samples. Then a Phase I monitoring approach is applied to assess the stability of the Phase I data set and to locate out-of-control samples. Once a set of data reflecting in-control performance is obtained, one estimates the in-control process parameters, the Y-intercept, the slope, and the variance by the intercept average:

$$\bar{a} = \sum_{j=1}^{m} a_j/m,$$

the slope average:

$$\bar{b} = \sum_{j=1}^{m} b_j/m,$$

and the mean square error average:

$$MSE = \sum_{j=1}^{m} MSE_j/m, \tag{2.3}$$

respectively.

2.1.2 Phase I Simple Linear Profile Approaches

Profile monitoring approaches are used to detect changes in the profile structure caused by assignable causes and to accurately identify their positions and sources. As mentioned before, much of the work on profile monitoring has been for the case in which the profile can be adequately represented by a simple linear regression model. Several approaches for the analysis of Phase I simple linear regression profile data sets were presented in the statistical process control (SPC) literature. In the following subsections, we introduce the most widely known approaches.

2.1.2.1 T^2 Control Chart Approaches

When the true profile parameters are not known, one of the most important approaches used to monitor these parameters is the Hotelling's T^2 control chart. Several versions of the Hotelling's T^2 approach for monitoring Phase I simple linear profiles are available in the literature. The SPC literature includes four T^2 control chart approaches for analyzing these profiles. These are Mestek et al.'s (1994) T^2 chart, Stover and Brill's (1998) T^2 chart, Kang and Albin's (2000) T^2 chart, and the T^2 chart based on successive differences, originally proposed by Hawkins and Merriam (1974) and studied later by Holmes and Mergen (1993) and Williams et al. (2007).

2.1.2.1.1 Mestek et al.'s (1994) T^2 Control Chart

Mestek et al. (1994) proposed a T^2 control chart to check for the stability of the linear calibration curve in the photometric determination of Fe^{3+} with sulfosalicylic acid. This T^2 approach is based on successive vectors containing the absorbances of the calibration curve. This approach is suitable only for the case when the X-values are fixed and constant from sample to sample (i.e., $X_{ij} = X_i, j = 1, 2, \ldots, m$). The T^2 statistics for this method are as follows:

$$T_j^2 = (\mathbf{Y}_j - \bar{\mathbf{Y}})^T \mathbf{S}^{-1} (\mathbf{Y}_j - \bar{\mathbf{Y}}), \quad j = 1, 2, \ldots, m, \tag{2.4}$$

where $\mathbf{Y}_j = (Y_{1j} Y_{2j} \cdots Y_{nj})^T$ is a vector containing the response values of the jth sample, and $\bar{\mathbf{Y}} = (\bar{y}_1 \bar{y}_2 \cdots \bar{y}_n)^T$ is a vector of the response averages, where

$\bar{y}_i = \sum_{j=1}^{m} \mathbf{Y}_{ij}/m$. The pooled sample covariance matrix $\mathbf{S}$ is defined as follows:

$$\mathbf{S}_Y = \begin{bmatrix} s_1^2 & s_{12} & \ldots & s_{1n} \\ s_{21} & s_2^2 & \ldots & s_{2n} \\ . & . & \ldots & . \\ s_{n1} & s_{n2} & \ldots & s_n^2 \end{bmatrix},$$

where

$$s_i^2 = \sum_{j=1}^{m} (Y_{ij} - \bar{y}_i)^2 \Big/ (m-1), \quad i = 1, 2, \ldots, n$$

and

$$s_{il} = \sum_{j=1}^{m} (Y_{ij} - \bar{y}_i)(Y_{lj} - \bar{y}_l) \Big/ (m-1), \quad i, l = 1, 2, \ldots, n (i \neq l).$$

Tracy et al. (1992) and Sullivan and Woodall (1996) stated that if the population covariance matrix $\mathbf{\Sigma}_Y$ is estimated by $\mathbf{S}_Y$, then the T^2 statistic in Equation (2.4) follows a beta distribution. The upper control limit (UCL) for this chart is as follows:

$$UCL = (m-1)^2 B_{n/2, m-n-1/2, \alpha}/m, \tag{2.5}$$

where $B_{n/2,m-n-1/2,\alpha}$ is the $100(1-\alpha)$ percentile of the beta distribution with parameters $n/2$ and $(m-n-1)/2$.

Example 2.1 Photometric determination of Fe^{3+} with sulfosalicylic acid. The purpose of the study by Mestek et al. (1994) was to investigate the stability of the calibration curve in the photometric determination of Fe^{3+} with sulfosalicylic acid. The data set in this calibration application is presented in Table 2.1. This data set includes 22 calibration curves. For each calibration curve, five volumes of 0, 1, 2, 3, and 4 mL of 50 μg/mL Fe^{3+} solution were diluted with water to 25 mL. Then, for each volume, 2.5 mL of a 20% solution of sulfosalicylic acid and 1.5 mL of a concentrated solution of ammonia were added to the diluted solution. Each volume was replicated twice, so that each calibration curve consists of 10 points. For each volume, the absorbance of the solution at 420 nm was measured on a Spekol 11 using 1-cm cells. The average of the X-values in this example is $\bar{X} = 100$ and $S_{XX} = 50{,}000$.

An important step that should be conducted before applying a Phase I method is to evaluate some diagnostics to check for the appropriateness of the normality assumption and the linearity of the relationship between the response and the independent variable for each of the samples. As indicated by Mahmoud and Woodall (2004), these diagnostic statistics suggested the reasonableness of the normality assumption of the error terms and the linearity of the relationship. In Section 2.5, we discuss the consequences of the violations of these assumptions.

Table 2.1 Example Data with the Response Measured According to Each Fe^{3+} Level (Adapted from Mestek et al. 1994)

Sample No.	0 μg Fe^{3+}	50 μg Fe^{3+}	100 μg Fe^{3+}	150 μg Fe^{3+}	200 μg Fe^{3+}	T^2
1	1, 3	104, 104	206, 206	307, 308	409, 412	2.86
2	4, 2	104, 103	206, 204	308, 307	412, 413	11.00
3	3, 2	105, 104	207, 207	311, 309	414, 411	1.42
4	4, 2	104, 104	206, 207	308, 312	411, 413	2.00
5	−9, −8	92, 95	195, 197	296, 299	397, 400	8.50
6	3, 3	107, 105	209, 207	311, 308	412, 410	7.50
7	3, 2	104, 105	207, 208	311, 308	414, 410	1.80
8	2, 2	105, 104	208, 208	310, 309	412, 412	5.00
9	−6, −7	95, 94	196, 197	297, 300	401, 401	5.50
10	2, 4	104, 105	206, 207	311, 310	413, 412	2.90
11	1, 2	103, 104	205, 206	309, 307	412, 411	2.00
12	−7, −7	94, 96	198, 199	298, 301	404, 402	6.80
13	5, 7	105, 107	210, 208	313, 315	415, 415	10.90
14	3, 2	106, 104	208, 207	311, 308	411, 414	1.82
15	−8, −6	94, 95	196, 199	299, 302	400, 404	4.00
16	4, 6	104, 106	207, 210	311, 310	415, 413	8.20
17	2, 4	105, 106	206, 208	308, 310	410, 413	4.00
18	2, 0	104, 103	206, 206	309, 308	414, 409	2.00
19	0, 1	101, 102	203, 206	305, 307	409, 411	3.00
20	1, 4	104, 106	206, 208	311, 309	410, 414	1.90
21	−9, −10	92, 92	194, 194	298, 297	400, 398	6.00
22	−8, −8	95, 95	195, 199	298, 301	401, 403	6.00

Mestek et al. (1994) considered monitoring vectors containing the absorbance averages corresponding to the Fe^{3+} volumes. Thus, in this calibration application, the number of Phase I profiles $m = 22$ with sample size of $n = 5$. Accordingly, the pooled variance–covariance matrix $\mathbf{S}_Y$ is calculated as:

$$\mathbf{S}_Y = \begin{pmatrix} 24.2235 & 23.4080 & 23.4481 & 24.4010 & 25.5276 \\ 23.4080 & 23.3485 & 23.2576 & 23.8712 & 24.8550 \\ 23.4481 & 23.2576 & 23.5455 & 24.0606 & 25.0346 \\ 24.4010 & 23.8712 & 24.0606 & 25.5114 & 26.0839 \\ 25.5276 & 24.8550 & 25.0346 & 26.0839 & 27.6304 \end{pmatrix}.$$

Using this pooled sample covariance matrix, the calculated T^2 statistics in Equation (2.4) are presented in the last column of Table 2.1.

The marginal distribution of these T^2 statistics is the beta distribution with parameters 2.5 and 8. Figure 2.2 shows a control chart for the T^2 statistics with an upper control limit calculated using a nominal overall false alarm probability of $\alpha = 0.05$. Using Equation (2.5), the upper control limit is calculated using

$$UCL = (22 - 1)^2 B_{5/2, 22-5-1/2, 0.00233}/22 = 13.18.$$

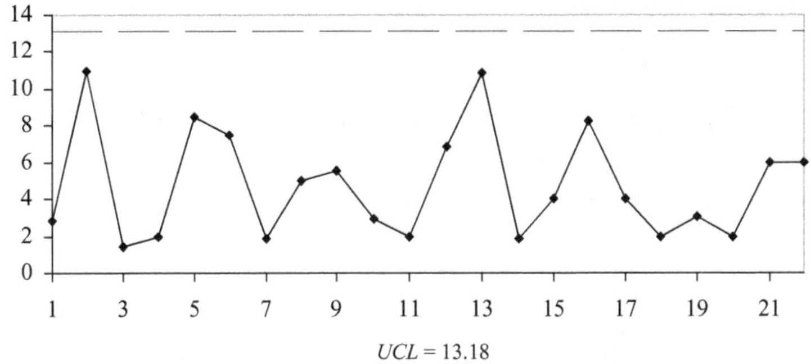

Figure 2.2 Control chart using Mestek et al.s (1994) T^2 approach for the data set in Example 2.1.

The overall false alarm probability produced by this chart is $\alpha = 1 - (1 - 0.00233)^{22} = 0.05$. As shown in Figure 2.2, all the calibration curves appear in-control. Therefore, Mestek et al. (1994) estimated the unknown profile parameters, the intercept, slope, and variance, by the intercept average $\bar{a} = -0.456$, the slope average $\bar{b} = 2.047$, and mean squared error average $MSE = 1.6114$, respectively.

Mahmoud and Woodall (2004) argued that this T^2 control chart is not recommended for monitoring simple linear profile data sets in Phase I for four reasons. First, this chart cannot be applied directly when the X-values are not constant from sample to sample. Second, the T^2 control chart for individual multivariate observations when estimating the population covariance matrix Σ_Y with $\mathbf{S}_Y$ produces a very poor statistical performance in terms of the probability of an out-of-control signal. This was demonstrated by several comparative studies, such as in Sullivan and Woodall (1996), Vargas (2003), and Mahmoud and Woodall (2004). As mentioned in Sullivan and Woodall (1996), the reason is that the population covariance matrix can be poorly estimated by the pooled sample covariance matrix when applying a T^2 control chart with individual vector observations. Third, when $m \leq n$ the sample covariance matrix $\mathbf{S}_Y$ is singular and the beta distribution cannot be used. Finally, with a simple profile relationship, the use of a T^2 chart based on the n Y-values leads to overparameterization. As mentioned before, an important issue in profile monitoring is related to data reduction because a profile response can be of a high dimension. Apparently, this T^2 approach does not tackle the reduction mission.

To analyze the data set in Example 2.1, Mestek et al. (1994) treated each sample as a group of two subsamples, each with five observations. As noted by Mahmoud and Woodall (2004), the sample covariance matrix of vectors containing the group averages would not be the best estimator of the population covariance matrix in this case. If we assume that each sample consists of q subsamples (each with n observations), then a better estimator for the population covariance matrix may be obtained by the pooled covariance matrix

$$\mathbf{S}_p = \sum_{j=1}^{m} S_j \Big/ m,$$

where $\mathbf{S}_j$ is the usual within-group sample covariance matrix for group j. Mason et al. (2001) showed that the following T^2 statistics:

$$T_j^2 = mq(\bar{\mathbf{y}}_j - \bar{\bar{\mathbf{y}}})^T S_p^{-1}(\bar{\mathbf{y}}_j - \bar{\bar{\mathbf{y}}})/(m-1), \quad j = 1, 2, \ldots, m, \tag{2.6}$$

each has a T^2-distribution with n and $m(q-1) - n + 1$ degrees of freedom, where

$$\bar{\mathbf{y}}_j = (\bar{y}_{1j}\bar{y}_{2j}\cdots\bar{y}_{nj})^T$$

is a vector containing the response average values of the jth group,

$$\bar{y}_{ij} = \sum\nolimits_{l=1}^{q} y_{ijl}\Big/q, \quad i = 1, 2, \ldots, n,$$

and y_{ijl} is the ith response value in the subsample l ($l = 1, 2, \ldots, q$) and group j ($j = 1, 2, \ldots, m$). Also,

$$\bar{\bar{y}} = (\bar{y}_1\bar{y}_2\cdots\bar{y}_n)^T$$

is a vector of the overall response averages, where $\bar{y}_i = \sum_{j=1}^{m} \bar{y}_{ij}/m$, ($i = 1, 2, \ldots, n$). Using the relationship between the T^2- and F-distributions, an appropriate UCL for the T^2 values in Equation (2.6) is as follows:

$$UCL = mn(q-1)F_{n,m(q-1)-n+1,\alpha}\Big/[m(q-1) - n + 1].$$

Example 2.2 To illustrate the use of the above T^2 chart, we apply it to the data set in Example 2.1. Figure 2.3 shows a control chart for the T^2 statistics in Equation (2.6) when applied to this data set. In this example, $m = 22$, $n = 5$, and $q = 2$. The UCL of this chart was set on the basis of a nominal overall false alarm probability of

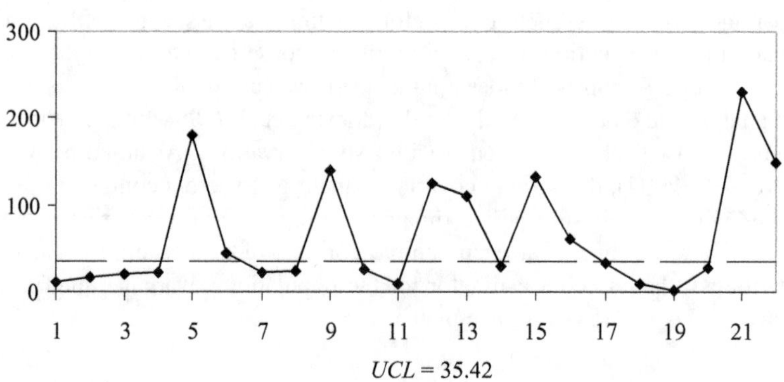

Figure 2.3 Control chart for the T^2 statistics in Equation (2.6) applied to the data set in Example 2.1.

α= 0.05. Thus, the UCL is equal to

$$UCL = 22 \times 5 \times (2-1) F_{5,22\times(2-1)-5+1,0.00233} / [22 \times (2-1) - 5 + 1] = 35.42.$$

As shown in Figure 2.3, the process is not in control. This result shows how the T^2 control chart approach proposed by Mestek et al. (1994) is sensitive to the method used for estimating the population covariance matrix. In general, Mestek et al.'s (1994) T^2 control chart is an ineffective approach because of its use of a poor estimator for the population covariance matrix.

2.1.2.1.2 Stover and Brill's (1998) T^2 Control Chart

Stover and Brill (1998) proposed an alternative T^2 approach. This approach is rather based on vectors containing estimators of the Y-intercept and slope. The T^2 statistics in this method are as following:

$$T_j^2 = (\mathbf{z}_j - \bar{\mathbf{z}})^T \mathbf{S}_1^{-1} (\mathbf{z}_j - \bar{\mathbf{z}}), \quad j = 1, 2, \ldots, m, \tag{2.7}$$

where $\mathbf{z}_j = (a_j b_j)^T$, $\bar{\mathbf{z}} = (\bar{a}\ \bar{b})^T$, and

$$\mathbf{S}_1 = \begin{pmatrix} \dfrac{\sum_{j=1}^{m} (a_j - \bar{a})^2}{m-1} & \dfrac{\sum_{j=1}^{m} (a_j - \bar{a})(b_j - \bar{b})}{m-1} \\ \dfrac{\sum_{j=1}^{m} (a_j - \bar{a})(b_j - \bar{b})}{m-1} & \dfrac{\sum_{j=1}^{m} (b_j - \bar{b})^2}{m-1} \end{pmatrix},$$

where a_j and b_j are as defined in Equation (2.2), and $\bar{a}$ and $\bar{b}$ are as defined in Equation (2.3). The upper control limit for this chart is

$$UCL = (m-1)^2\ B_{1,(m-3)/2,\alpha} / m. \tag{2.8}$$

Example 2.3 Applying this T^2 approach to the calibration application of the photometric determination of Fe^{3+} with sulfosalicylic acid presented in Example 2.1 shows also that the Phase I data set is in-control. Table 2.2 shows the LSE of the intercepts, slopes, and variances of the 22 calibration profiles of size $n = 10$. The estimated variance–covariance matrix of the least squares estimators $\mathbf{S}_1$ is

$$\mathbf{S}_1 = \begin{pmatrix} 22.981645 & 0.006471 \\ 0.006471 & 0.000021 \end{pmatrix}.$$

The last column in Table 2.2 presents the calculated T^2 values using Equation (2.7) and based on this pooled sample covariance matrix. Figure 2.4 shows a T^2 control chart for the photometric determination of Fe^{3+} calibration application. The UCL of this chart was determined on the basis of a false alarm probability of $\alpha_1 = 0.00233$ to produce an overall false alarm probability of $\alpha = 0.05$. Using Equation (2.8), the UCL of this chart is $(22-1)^2 B_{1,22-3/2,0.00233}/22 = 9.456$.

Table 2.2 The Linear Regression Results for the 22 Samples in Example 2.1

Sample	Intercept	Slope	MSE_j	T^2
1	1.900	2.041	1.000	2.219
2	1.700	2.046	2.528	0.266
3	2.200	2.051	1.000	1.040
4	2.300	2.048	1.960	0.357
5	–8.200	2.036	2.190	6.215
6	3.600	2.039	1.563	4.605
7	2.400	2.048	1.796	0.380
8	2.200	2.050	0.325	0.701
9	–7.00	2.038	0.922	4.172
10	2.400	2.050	0.922	0.733
11	1.100	2.049	0.740	0.327
12	–7.100	2.049	1.440	2.917
13	4.800	2.052	2.592	2.043
14	2.500	2.049	1.538	0.525
15	–7.300	2.048	2.657	2.651
16	3.900	2.047	2.250	0.853
17	3.100	2.041	1.440	2.751
18	0.900	2.052	1.960	1.444
19	–0.200	2.047	1.769	0.012
20	2.500	2.048	2.372	0.403
21	–9.900	2.045	0.640	3.952
22	–7.800	2.049	1.850	3.435

This T^2 approach satisfies to a large extent the reduction mission of a profile monitoring approach, because the profile response is monitored using vectors of two observations, representing the intercept and the slope estimates, regardless of the number of observations n_j used to establish each profile. However, as is the case with Mestek et al.'s (1994) T^2 approach, using this approach the covariance matrix of the least square estimators of the regression parameters $\boldsymbol{\Sigma}$ is poorly estimated by the pooled sample covariance matrix $\mathbf{S_1}$.

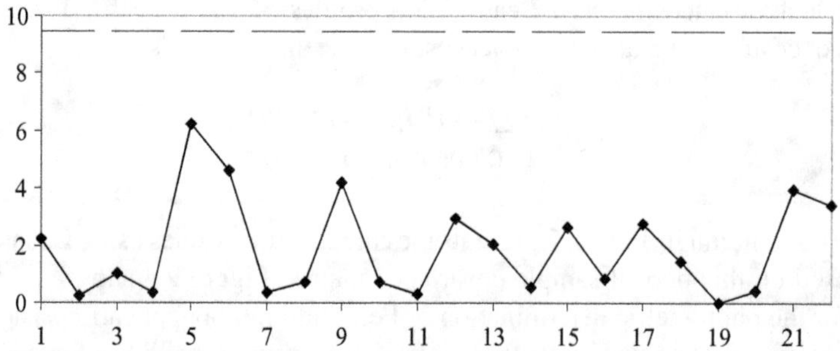

Figure 2.4 Stover and Brill's (1998) T^2 control chart applied to the data set in Example 2.1.

2.1.2.1.3 Kang and Albin's (2000) T^2 Control Chart

Kang and Albin (2000) proposed a T^2 control chart suitable for the case of simple linear profiles with fixed X-values in all samples (i.e., $X_{ij} = X_i$, $j = 1, 2, \ldots, m$). Similar to the T^2 control chart of Stover and Brill (1998), Kang and Albin's (2000) T^2 control chart is based on successive vectors of the least squares estimators of the Y-intercept and slope. The T^2 statistics in this method are as following:

$$T_j^2 = m(\mathbf{z}_j - \bar{\mathbf{z}})^T \mathbf{S}_2^{-1}(\mathbf{z}_j - \bar{z})/(m-1), \quad j = 1, 2, \ldots, m, \tag{2.9}$$

where $\mathbf{z}_j$ and $\bar{\mathbf{z}}$ are as defined in Equation (2.7), and

$$\mathbf{S}_2 = MSE \begin{pmatrix} \dfrac{1}{n} + \dfrac{\bar{X}^2}{S_{XX}} & \dfrac{-\bar{X}}{S_{XX}} \\ \dfrac{-\bar{X}}{S_{XX}} & \dfrac{1}{S_{XX}} \end{pmatrix}, \tag{2.10}$$

where MSE is as defined in Equation (2.3), $\bar{X} = \sum_{i=1}^{n} X_i/n$ and $S_{XX} = \sum_{i=1}^{n} (X_i - \bar{X})^2$. Kang and Albin (2000) proved that the T^2 statistic in Equation (2.9) divided by 2 has an F-distribution with 2 and $m(n-2)$ degrees of freedom. Thus, the appropriate upper control limit of this T^2 chart that produces an overall probability of false alarm of α is obtained by

$$UCL = 2F_{2,m(n-2),\alpha}. \tag{2.11}$$

Obviously, this method is similar to the T^2 method of Stover and Brill (1998), although the marginal distributions of the control statistic used differ because of different estimators for the covariance matrix. As shown in several studies, the T^2 approach of Kang and Albin (2000) outperforms the T^2 approaches of Mestek et al. (1994) and Stover and Brill (1998) in detecting out-of-control profiles; see, for example, Mahmoud and Woodall (2004). The main reason for the superiority of the T^2 approach of Kang and Albin's (2000) over Stover and Brill's (1998) T^2 approach is that the population covariance matrix of the least squares estimators is much better estimated by matrix $\mathbf{S_2}$ than by matrix $\mathbf{S_1}$.

Example 2.4 To illustrate the use of Kang and Albin's (2000) T^2 chart approach, we apply it on the calibration application of the photometric determination of $Fe^{3+}$with sulfosalicylic acid presented in Example 2.1. The estimate of the variance–covariance matrix of the intercept and slope estimators $\mathbf{S_2}$ is calculated as

$$\mathbf{S}_2 = \begin{pmatrix} 0.4834636 & -0.0032231 \\ -0.0032231 & 0.0000322 \end{pmatrix}.$$

Figure 2.5 presents the T^2 values of this chart calculated using Equation (2.11) on a chart with $UCL = 2F_{2,22(10-2),0.00233} = 12.552$. The results obtained by using

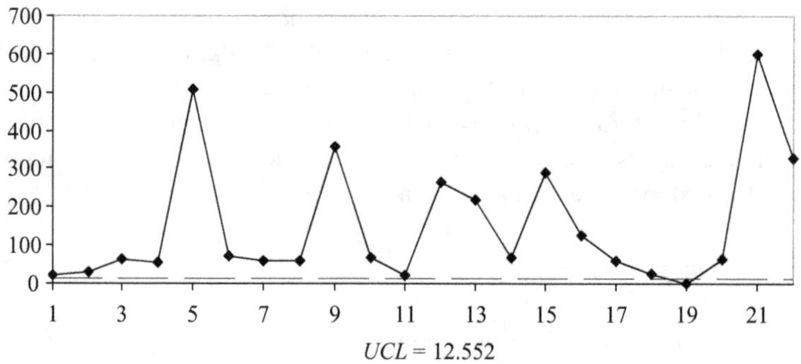

Figure 2.5 Kang and Albin's (2000) T^2 control chart applied to the data set in Example 2.1.

this method were quite contrary to the results obtained by Mestek et al.'s (1994) T^2 and Stover and Brill's (1998) T^2 approaches. Figure 2.5 indicates that the calibration curves are very unstable.

2.1.2.1.4 T^2 Control Chart Based on Successive Differences

An alternative choice of the estimator of the variance–covariance matrix to obtain the T^2 statistics is one based on successive differences. This T^2 chart was originally proposed by Hawkins and Merriam (1974) and studied later by Holmes and Mergen (1993) and Williams et al. (2007). To obtain this estimator, the following matrix of successive differences is defined:

$$\mathbf{V} = \begin{pmatrix} \mathbf{v}_1^T \\ \mathbf{v}_2^T \\ \vdots \\ \mathbf{v}_{m-1}^T \end{pmatrix},$$

where $\mathbf{v}_j = \mathbf{z}_{j+1} - \mathbf{z}_j, j = 1, 2, \ldots, m - 1$, and $\mathbf{z}_j$ is as defined in Equation (2.7). The estimator of the variance–covariance matrix is then obtained by

$$\mathbf{S}_D = \mathbf{V}^T\mathbf{V}/2(m-1),$$

and the T^2 statistics are obtained by

$$T_j^2 = (\mathbf{z}_j - \bar{\mathbf{z}})^T\mathbf{S}_D^{-1}(\mathbf{z}_j - \bar{\mathbf{z}}). \quad j = 1, 2, \ldots, m. \tag{2.12}$$

The marginal distribution of each of these T^2 statistics is unknown. The upper control limit for this chart that gives a specified probability of a false alarm can be obtained by simulation. The upper control limits that give specified overall probabilities of false alarm ($\alpha = 0.01, 0.05$, or 0.10) when different number of profiles m is

Table 2.3 The Upper Control Limits for the T^2 Statistic in Equation (2.12) That Give Approximately an Overall Probability of False Alarm of α Corresponding to Different Values of m

α	0.01	0.05	0.10
m			
5	9.36	8.59	7.67
10	14.40	10.36	8.80
15	14.56	11.00	9.60
20	14.70	11.56	10.00
25	14.90	11.91	10.60
30	15.27	12.30	10.90
35	15.56	12.65	11.20
40	15.81	12.94	11.50
45	16.20	13.18	11.74
50	16.36	13.36	11.92
60	16.70	13.75	12.42
70	17.20	14.05	12.70
80	17.36	14.30	12.90

used in a linear profile data set are presented in Table 2.3. Each of these probabilities was estimated using 20,000 simulations. For each simulation, m profiles were used, where m is as presented in Table 2.3. For each simulation the underlying in-control model $Y_{ij} = X_i + \varepsilon_{ij}$ was considered, $i = 1, 2, \ldots, n$, where the ε_{ij} are i.i.d. $N(0, 1)$ random variables. These upper control limits apply for any values of the explanatory variable X, any sample size n, and any values of the in-control intercept, slope, and variance.

An LSE for the upper control limit of this T^2 chart using the logarithm of m as the independent variable in a simple linear regression model gave the following approximation for UCL according to an overall false alarm probability of $\alpha = 0.01$:

$$UCL_{\alpha=0.01} = 7.38 + 2.33 \log_e m. \tag{2.13}$$

The coefficient of determination R^2 for the regression model in Equation (2.13) is 0.85. For $\alpha = 0.05$, the fitted simple linear regression model is

$$UCL_{\alpha=0.05} = 5.51 + 2.01 \log_e m, \tag{2.14}$$

with a determination coefficient of $R^2 = 0.997$. And finally for $\alpha = 0.10$, the fitted simple linear regression model is

$$UCL_{\alpha=0.10} = 4.39 + 1.93 \log_e m, \tag{2.15}$$

with a determination coefficient of $R^2 = 0.996$.

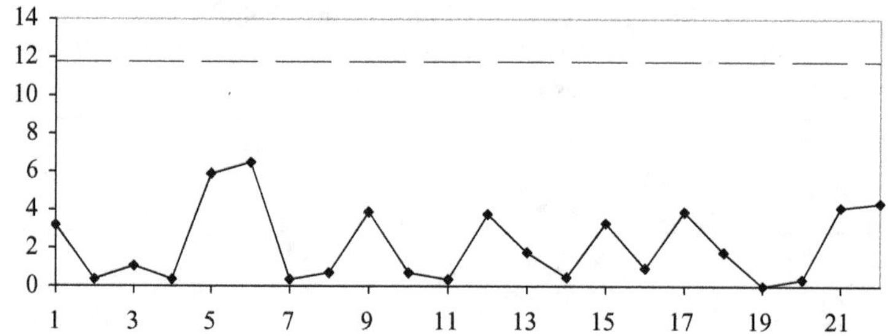

Figure 2.6 The T^2 control chart based on successive differences applied to the data set in Example 2.1.

Example 2.5 To illustrate the use of the T^2 chart based on successive differences we apply it on the calibration application of the photometric determination of Fe^{3+} with sulfosalicylic acid presented in Example 2.1. The estimate of the variance–covariance matrix of the intercept and slope estimators $\mathbf{S}_D$ is calculated as

$$\mathbf{S}_D = \begin{pmatrix} 24.612619 & 0.010750 \\ 0.010750 & 0.000019 \end{pmatrix}.$$

One can observe that the variance–covariance estimator matrix $\mathbf{S}_D$ is very close to the variance–covariance estimator matrix $\mathbf{S}_1$. Figure 2.6 presents the T^2 values of this chart calculated using Equation (2.12) on a chart with $UCL = 11.72$, calculated using Equation (2.14) to give approximately a probability of overall false alarm of 0.05. As was the case with Mestek et al.'s (1994) and Stover and Brill's (1998) T^2 charts, Figure 2.6 indicates that the calibration curves are in-control.

2.1.2.2 Principal Component Approaches

Control chart methods for monitoring Phase I linear profiles based on principal components technique were suggested by Mestek et al. (1994) and Stover and Brill (1998). The principal component method of Mestek et al. (1994) employs a Shewhart-type control chart for the first principal component corresponding to vectors of the Y-values, assuming that the X-values are fixed from sample to sample. The principal components approach of Stover and Brill (1998) is based on control charting the first principal component corresponding to vectors containing the estimates of the regression parameters for each sample. However, these principal component methods are not recommended because they will not be able to detect some out-of-control conditions. The first principal component explains in-control variation in the direction of the major axis corresponding to the first principal component. Therefore, one will not be able to detect combinations of shifts in the Y-intercept and the slope in the direction perpendicular to the major axis corresponding to the first principal component.

On the other hand, the principal components approach of Jones and Rice (1992) is a very useful tool to identify and understand the nature of the variability among the profiles in a Phase I profile data set with equal, equally spaced X-values for each profile. This approach has become a fundamental part of functional data analysis; see, for example, Ramsay and Silverman (2002). The use of this approach as a diagnostic aid to determine the source of profile variations is illustrated in Section 2.5.

2.1.2.3 *Kim et al.'s (2003) Shewhart-Type Control Charts Approach*

Kim, Mahmoud, and Woodall (2003) proposed a more interpretable approach for Phase I analysis of linear profiles than the T^2 approaches. They recommended coding the X-values within each sample by subtracting the sample average $\bar{X}_j$ from the X-values. As a result of this coding approach, the estimators of the Y-intercept and slope are independent. Using their coding recommendation and the fact that the estimator of the variance is independent of the estimators of the Y-intercept and slope, one can monitor each of the three regression parameters using a separate Shewhart-type control chart. If one codes the X-values within each sample so that the average coded value is zero, then the resulting linear regression model is in the form

$$Y_{ij} = A'_j + B'_j X'_{ij} + \varepsilon_{ij}, \quad i = 1, 2, \ldots, n_j, \quad j = 1, 2, \ldots, m, \tag{2.16}$$

where $A'_j = A_j + B_j\bar{X}_j$, $B'_j = B_j$, and $X'_{ij} = (X_{ij} - \bar{X}_j)$, where A_j, B_j, and X_{ij} are as defined in Equation (2.1). In this case, the least squares estimators for the regression parameters for sample j are $a'_j = \bar{Y}_j$, and $b'_j = b_j = S_{XY(j)}/S_{XX(j)}$. It is very well known that for an in-control process, a'_j and b'_j are mutually independent normally distributed random variables with means A'_0 and B'_0 and variances σ^2/n_j and $\sigma^2/S_{XX(j)}$, respectively, where A'_0 and B'_0 are the in-control values of the intercept and slope of the profile model in Equation (2.16).

Mahmoud et al. (2010) recommended that the X-values are always scaled to be zero for three reasons. First, as indicated above, the least squares estimators of the intercept and slope are independent when $\bar{X} = 0$, thus the intercept and slope can be monitored individually and effectively using separate control charts. Second, their simulation results showed that generalization of the properties of the proposed profile monitoring techniques is achieved only if $\bar{X} = 0$. Third, any shifts in the slope parameter become more interpretable since the interest in the linear profile is assumed to be greater within the range of X-values where we usually have data. In practice, if the observed simple linear regression profiles are with $\bar{X} \neq 0$, then one can easily code the X-values so that the average coded value is zero. Kim et al. (2003) showed that changes in the profile parameters (A, B, and σ^2) can be effectively detected by monitoring the simple linear profile model with the coded X-values in Equation (2.16).

Assuming that the X-values are the same in all samples and that the process is in control, then the Shewhart-type control chart for monitoring the intercept of the model in Equation (2.16) is with the following lower and upper control limits:

$$LCL = \bar{a}' - t_{m(n-2), \frac{\alpha}{2}} \sqrt{\frac{(m-1)MSE}{nm}}$$

and

$$UCL = \bar{a}' + t_{m(n-2),\frac{\alpha}{2}} \sqrt{\frac{(m-1)MSE}{nm}}, \tag{2.17}$$

where $t_{m(n-2),\alpha/2}$ is the $100(1-\alpha/2)$ percentile of the t-distribution with $m(n-2)$ degrees of freedom and $\bar{a}' = \sum_{j=1}^{m} a'_j \Big/ m$. The chart statistics in this case are the intercept estimates $a'_j, j = 1, 2, \ldots, m$.

Also assuming that the process is in-control, the upper and lower control limits for a Shewhart control chart for monitoring the slope can be set at

$$LCL = \bar{b}_1 - t_{m(n-2),\frac{\alpha}{2}} \sqrt{\frac{(m-1)\,MSE}{mS_{XX}}}$$

and

$$UCL = \bar{b}_1 + t_{m(n-2),\frac{\alpha}{2}} \sqrt{\frac{(m-1)MSE}{mS_{XX}}}. \tag{2.18}$$

The chart statistics in this case are the slope estimates $b_j, j = 1, 2, \ldots, m$.

Assuming in-control process, it can be shown that the quantity

$$F_j = MSE_j / MSE_{-j} \tag{2.19}$$

has an F-distribution with $n-2$ and $(m-1)(n-2)$ degrees of freedom, where $MSE_{-j} = \sum_{i \neq j}^{m} MSE_i/(m-1)$. Therefore, a Shewhart control chart for monitoring the process variance σ^2 requires plotting the jth mean squared error MSE_j on a chart with the following lower and upper control limits:

$$LCL = \frac{m F_{(n-2),(m-1)(n-2),\frac{\alpha}{2}}}{m - 1 + F_{(n-2),(m-1)(n-2),\frac{\alpha}{2}}} MSE$$

and

$$UCL = \frac{m F_{(n-2),(m-1)(n-2),1-\frac{\alpha}{2}}}{m - 1 + F_{(n-2),(m-1)(n-2),1-\frac{\alpha}{2}}} MSE, \tag{2.20}$$

where $F_{(n-2),(m-1)(n-2),\alpha}$ is the $100(1-\alpha)$ percentile of the F-distribution with $n-2$ and $(m-1)(n-2)$ degrees of freedom.

In Phase I simple linear profile applications, we usually are interested in detecting any parameter changes in the variance, slope, and intercept from their in-control values. For this reason, Kim et al. (2003) recommended the use of three Shewhart control charts to detect changes in the three parameters. This is a generalization of the typical quality control problem in which one tests for changes in the mean and variance of a process characterized by the distribution of a univariate quality characteristic. However, if in a particular application the variance is not expected to change from sample to sample, one can use alternatively two Shewhart control charts to detect changes in the slope and intercept more powerfully. Notice that, when we discussed the use of the different T^2 charts used for monitoring Phase I simple linear profiles, we did not use a specific control chart for monitoring the process variance. However, as recommended by Mahmoud (2008), it is much better to use a T^2 chart in conjunction with a Shewhart control chart for monitoring the process variance.

Since the accuracy of the estimators of the in-control regression coefficients relies heavily on the stability of the process variance, it is recommended that one applies a control chart for the variance before applying the control charts for the Y-intercept and slope. If the variance is shown to be unstable using this control chart, then we do not check whether the regression coefficients are constant. It also can be easily shown that shifts in the slope of the model with original X-values in Equation (2.1) correspond to shifts in both the intercept and slope of the model with coded X-values in Equation (2.16). So, it is also recommended to apply the Shewhart control chart for the slope before applying the control chart for the intercept. Moreover, in practice, if the regression lines are not parallel we may not care whether their intercepts are equal.

Example 2.6 Using the data set in Example 2.1, we illustrate the use of Kim et al. (2003) approach. From Table 2.2, it can be shown that the estimate of the error term variance is

$$MSE = \sum_{j=1}^{22} MSE_j/22 = 1.611545,$$

the estimate for the slope is

$$\bar{b} = \sum_{j=1}^{22} b_j/22 = 2.0465,$$

and the estimate for the intercept

$$\bar{a}' = \sum_{j=1}^{22} \bar{Y}_j/22 = 204.1955.$$

Figures 2.7, 2.8, and 2.9 give control charts for the process variance, slope, and Y-intercept, respectively, using Kim et al. (2003) approach when applied to the data set in Example 2.1. The control limits for the three control charts are calculated using Equations (2.20), (2.18), and (2.17), respectively. Each set of chart limits was determined on the basis of $\alpha_2 = 0.00078$ to produce an overall false alarm probability of 0.05. This false alarm probability was calculated in two steps. The first step is to specify the overall probability of false alarm used by each chart. This probability is calculated by

$$\alpha_1 = 1 - (1 - \alpha)^{1/3},$$

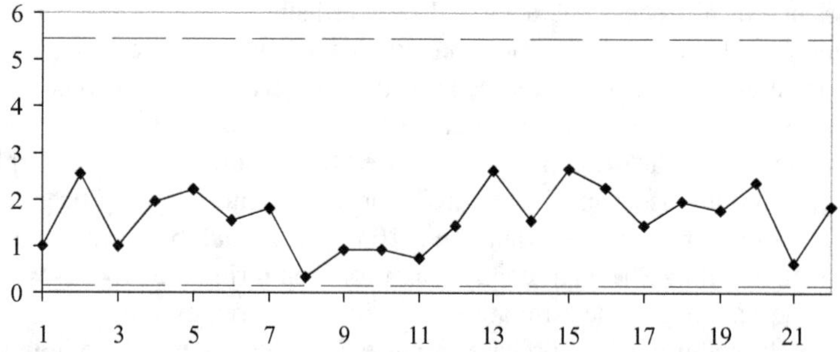

LCL = 0.138 and UCL = 5.45

Figure 2.7 Control chart for the variance applied to the data set in Example 2.1.

where α is the desired overall probability of signal by the three charts. Then, the marginal probability of signal for each control chart statistic is calculated by

$$\alpha_2 = 1 - (1 - \alpha_1)^{1/m},$$

The lower and upper control limits for the variance control chart calculated using Equation (2.20) are as following:

$$LCL = \frac{22 \times F_{(10-2),(22-1)(10-2),\frac{0.00078}{2}}}{22 - 1 + F_{(10-2),(22-1)(10-2),\frac{0.00078}{2}}} \times 1.611545 = 0.138$$

$$UCL = \frac{22 \times F_{(10-2),(22-1)(10-2),1-\frac{0.00078}{2}}}{22 - 1 + F_{(10-2),(22-1)(10-2),1-\frac{0.00078}{2}}} \times 1.611545 = 5.45.$$

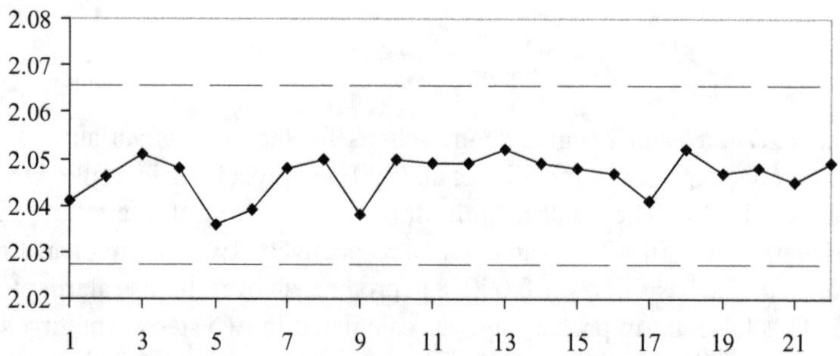

LCL = 2.028 and UCL = 2.065

Figure 2.8 Control chart for the slope applied to the data set in Example 2.1.

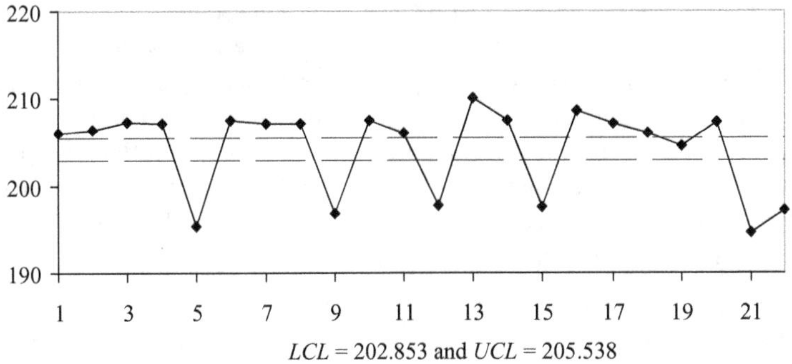

Figure 2.9 Control chart for the intercept applied to the data set in Example 2.1.

Also, the lower and upper control limits for the slope chart calculated using Equation (2.18) are as following:

$$LCL = 2.0465 - t_{22(10-2),\frac{0.00078}{2}}\sqrt{\frac{(22-1)\times 1.611545}{22\times 50000}} = 2.028$$

$$UCL = 2.0465 + t_{22(10-2),\frac{0.00078}{2}}\sqrt{\frac{(22-1)\times 1.611545}{22\times 50000}} = 2.065.$$

Finally, the lower and upper control limits for the intercept chart calculated using Equation (2.17) are as following:

$$LCL = 204.1955 - t_{22(10-2),\frac{0.00078}{2}}\sqrt{\frac{(22-1)\times 1.611545}{10\times 22}} = 202.853$$

$$UCL = 204.1955 + t_{22(10-2),\frac{0.00078}{2}}\sqrt{\frac{(22-1)\times 1.611545}{10\times 22}} = 205.538.$$

According to the chart in Figure 2.7, the error term variance appears stable. Also, the slope, as shown in Figure 2.8, is stable. However, Figure 2.9 shows that the intercept is out-of-control. This result agrees with the result obtained from the Kang and Albin's (2000)T^2 chart. However, using Kim et al. (2003) approach one can more easily explain out-of-control signals. The calibration curves are out-of-control because their intercepts are unstable. The importance of this instability would have to be evaluated on the basis of practical considerations. If this variation were considered to be common cause variation, then this would affect the determination of appropriate control limits for Phase II.

2.1.2.4 Mahmoud and Woodall's (2004) F-Test Approach

Mahmoud and Woodall (2004) proposed an alternative method for data analysis in Phase I. Their method is based on using indicator variables in a multiple regression model combined with a control chart for detecting changes in the process variance. The use of indicator variables in comparing several regression lines is described in many references; see, for example, Myers (1990, p. 135). The first step in the indicator variables technique is to pool all the m samples into one sample of size $N = \sum_{j=1}^{m} n_j$. Then we create $m - 1$ indicator variables such that

$$Z_{ji} = \begin{cases} 1 & \text{if observation } i \text{ is from sample } j \\ 0 & \text{otherwise} \end{cases}, i = 1, 2, \ldots, N, \ j = 1, 2, \ldots, m',$$

where $m' = m - 1$. Finally, we fit to the pooled data the following multiple regression model:

$$y_i = \nu + \theta x_i + \beta_{01} Z_{1i} + \beta_{02} Z_{2i} + \cdots + \beta_{0m'} Z_{m'i} + \beta_{11} Z_{1i} x_i + \beta_{12} Z_{2i} x_i + \cdots$$
$$+ \beta_{1m'} Z_{m'i} x_i + \varepsilon_i, \quad i = 1, 2, \ldots, N, \tag{2.21}$$

where the ε_i are assumed to be i.i.d. $N(0, \sigma^2)$ random variables and (x_i, y_i), $i = 1, 2, \ldots, N$ are N bivariate observations resulting from pooling the m samples into one sample of size N. To test for the equality of the m regression lines we test the following hypotheses:

$$H_0 : \beta_{01} = \beta_{02} = \cdots = \beta_{0m'} = \beta_{11} = \beta_{12} = \cdots = \beta_{1m'} = 0 \text{ and}$$
$$H_1 : H_0 \text{ is not true.}$$

Under the null hypothesis we have the following reduced model:

$$y_i = \nu + \theta x_i + \varepsilon_i, \ i = 1, 2, \ldots, N. \tag{2.22}$$

The usual test statistic for testing H_0 is

$$F = \frac{\{SSE(reduced) - SSE(full)\}/2(m - 1)}{MSE(full)}, \tag{2.23}$$

where $SSE(full)$ and $SSE(reduced)$ are the residual sums of squares resulting from fitting the regression models in Equations (2.21) and (2.22), respectively, and $MSE(full)$ is the mean square error of the full model in Equation (2.21). This test statistic follows an F-distribution with $2(m - 1)$ and $(N - 2m)$ degrees of freedom under the null hypothesis.

Mahmoud and Woodall (2004) recommended applying the above global F-test based on the statistic in Equation (2.23) in conjunction with a univariate control chart to check for the stability of the variation about the regression line. For this purpose, they recommended the third control chart of Kim et al. (2003) that is based on control limits in Equation (2.20). Again, since the accuracy of the estimators of the in-control regression coefficients relies on the stability of the process variance, they recommended that one check for the stability of the error variance before performing the F-test. Note that adding the variance control limits requires decreasing the probability of Type I error used by the F-test in order to maintain the same overall probability of false alarm. Thus, it is recommended to use a probability of false alarm of

$$\alpha_1 = 1 - (1 - \alpha)^{1/3} \tag{2.24}$$

for the variance control chart and a probability of Type I error of

$$\alpha_2 = 1 - (1 - \alpha)^{2/3} \tag{2.25}$$

for the F-test, so that the overall probability of a false alarm produced by this method is equal to α.

Using the data set in Example 2.1 we illustrate the use of the F-test method. Using the control limits for the process variance in Figure 2.7, we find the process variance is stable. Now, to test the coincidence of the 22 calibration regression lines, we pool all the samples into one sample of size $N = 220$ and create $m' = 21$ indicator variables as described in Equation (2.21). Then, the residual sum of squares for the fitted reduced model is $SSE(reduced) = 5448.1$ and the residuals sum of squares for the fitted full mode is $SSE(full) = 285.4$. Thus, the F-value for testing the equality of all the regression lines is

$$F = \frac{\{5448.1 - 285.4\}/2 \times (22 - 1)}{285.4/(220 - 2 \times 22)} = 75.8$$

with a p-value of 0.00. Therefore, we reject the null hypothesis that all of the regression lines are identical, i.e., the process is out-of-control.

For diagnostic purposes, Mahmoud and Woodall (2004) recommended coding the X-values so that the average coded values is zero and then applying 3-sigma control charts for the Y-intercept and slope. Again it is recommended to apply the slope control chart before applying the intercept control chart. All of the slope estimates are within the control limits as was shown in Figure 2.8 while all of the Y-intercept estimates exceed the control limits, except the intercept estimate for sample 19 as shown in Figure 2.9. Hence, using this approach one concludes that the process is out-of-control because the intercept is not stable. In Section 2.4, we discuss in detail other diagnostic statistics.

For the data set in Example 2.1, Mahmoud and Woodall (2004) and Kim et al. (2003) approaches gave the same conclusion. However, an important advantage of the F-test method is that it can be applied directly for the case of simple linear profile data with varying sample size and/or X-values. The formulas for the control limits of the three Shewhart control charts of Kim et al. (2003) method, presented in Equations (2.17), (2.18), and (2.20), should be modified to account for the varying X-values and/or sample size from sample to sample. For example, Mahmoud et al. (2007) stated that if the sample sizes are not the same for all samples, then appropriate control limits for the jth sample variance can be obtained from

$$LCL = \frac{m F_{(n_j-p),\sum_{i\neq j}^{m}(n_i-p),\alpha/2}}{(m-1+F_{(n_j-p),\sum_{i\neq j}^{m}(n_i-p),\alpha/2})} MSE$$

and

$$UCL = \frac{m F_{(n_j-p),\sum_{i\neq j}^{m}(n_i-p),1-\alpha/2}}{(m-1+F_{(n_j-p),\sum_{i\neq j}^{m}(n_i-p),1-\alpha/2})} MSE. \quad (2.26)$$

These limits can vary from sample to sample. However, Mahmoud et al. (2007) showed by simulation means that approximate control limits that produce closely the desired overall false alarm probabilities can be obtained by using $\bar{n} = \sum_{j=1}^{m} n_j/m$ in place of n in Equation (2.20). To illustrate the use of F-test method in conjunction with the variance control chart to monitor Phase I profile data sets with varying sample size, we use the following real application.

Example 2.7 Wind tunnel experiments at NASA Langley Research Center. Another important profile example is a calibration application presented in Parker et al. (2001). The purpose is to investigate replicated calibrations of a force balance used in wind tunnel experiments at NASA Langley Research Center. A force balance is a multiple-axis load cell that provides simultaneous measurement of three orthogonal components of aerodynamic force (normal, axial, and side force) and three orthogonal components of aerodynamic torque (rolling, pitching, and yawing moments) exerted on a scaled aircraft test article. These forces and moments are illustrated in Figure 2.10. Thus, in this example, there are six response variables that are modeled as functions of six explanatory variables. The relative importance of each of these measurements depends on the nature of the aerodynamic investigation.

In most investigations, the axial force component is of primary interest, and it is the only response profile considered in this example. However, an overall assessment of the calibration stability of the force balance would require an analysis of all six responses with six explanatory variables. In Chapter 8, we introduce a full analysis of all forces and moments using a multivariate multiple linear profile approach.

A force balance consists of a structural spring element instrumented with a network of strain gauges and is designed to elastically deform under the application of external forces and moments. This deformation results in differential strain across the structure that is sensed by the strain gauges. Six electrical responses are produced by the strain gauges that are proportional to the magnitude and direction of their respective aerodynamic components. A calibration experiment was performed to model the relationship between the applied forces and moments (explanatory variables) and the electrical responses. There are six prime sensitivities in the model that represent the dominant effect of each electrical response due to the level of one explanatory variable. Mahmoud et al. (2007) used a partial regression approach to isolate the simple linear relationship of the axial force prime sensitivity and account

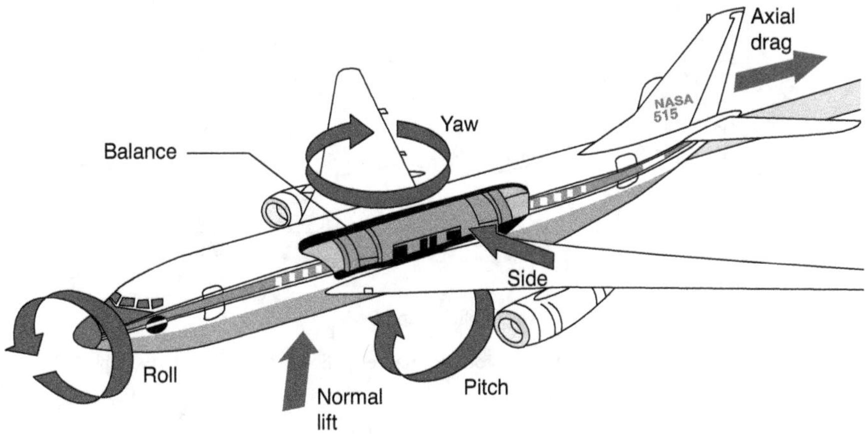

Figure 2.10 Forces and moments. (Adapted from Parker et al. 2001.)

for the influence of the other explanatory variables. The partial regression adjusted axial response and axial force data set is presented in Table A in the Appendix.

The data set used in this example consists of 11 samples of simple linear profiles each with 64, 73, or 74 data points; thus, $m = 11$ and the sample size n is 64, 73, or 74. These samples were collected over 16 months, which has been traditionally considered a reasonable calibration interval. Applying a Phase I monitoring method to historical data gives an indication about the calibration stability within this time interval. In addition, the ability to detect and diagnosis a shift in a profile parameter enables a classification of the nature and severity of the shift. For example, a shift in the slope or variance affects the bias and precision of the predicted forces and moments. Alternatively, small shifts in the Y-intercept are less important due to an offset correction procedure employed during wind tunnel operations. However, a large shift in the intercept may indicate that the structural frame of the balance has been damaged due to an overload condition or that one or more of the strain gauges has sustained physical or electrical damage. Therefore, it is not only necessary to detect a shift, but also to attribute the shift to a specific model parameter.

Since the sample sizes are not the same, the average sample size of 66 was used to derive the control limits of the variance chart. If we use an overall probability of false alarm of 0.05, then the overall probability of false alarm designated for the variance control chart calculated using Equation (2.24) is

$$\alpha_1 = 1 - (1 - 0.05)^{1/3} = 0.016952.$$

Since we have 11 profiles, then the marginal probability of signal used in the variance control chart is equal to $1 - (1 - 0.016952)^{1/11} = 0.001553$. The average of the mean squared error of the 11 profiles is $MSE = 106.64$. The control chart for monitoring the process variance is given in Figure 2.11. Using Equation (2.20) and

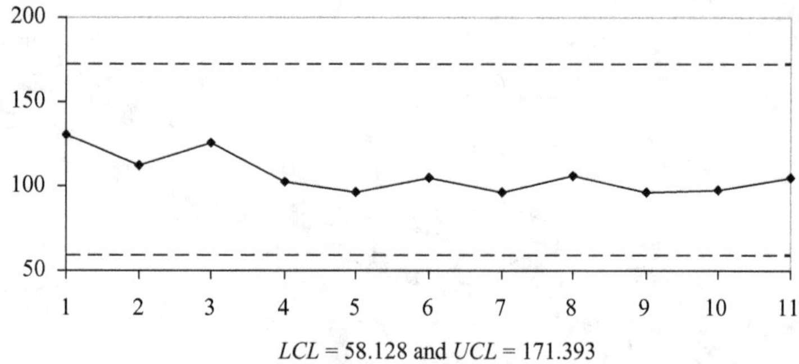

Figure 2.11 Control chart for the process variance applied to the data set in Example 2.7.

a false alarm probability of 0.001553, the control limits for the variance chart are as following:

$$LCL = \frac{11 \times F_{(66-2),(11-1)(66-2),\frac{0.001553}{2}}}{11 - 1 + F_{(66-2),(11-1)(66-2),\frac{0.001553}{2}}} \times 106.64 = 58.128$$

and

$$UCL = \frac{11 \times F_{(66-2),(11-1)(66-2),1-\frac{0.001553}{2}}}{11 - 1 + F_{(66-2),(11-1)(66-2),1-\frac{0.001553}{2}}} \times 106.64 = 171.393.$$

As shown in Figure 2.11, the process variance is stable. The global F-test is applied at a significance level of

$$\alpha_2 = 1 - (1 - 0.05)^{2/3} = 0.033617,$$

so that the overall probability of false alarm is 0.05. The residual sum of squares for the fitted reduced model is $SSE(reduced) = 80860$ and the residuals sum of squares for the fitted full mode is $SSE(full) = 75276$. Thus, the F-value for testing the equality of all the regression lines is

$$F = \frac{\{80860 - 75276\}/2 \times (11 - 1)}{75276/(723 - 2 \times 11)} = 2.600021$$

with a p-value of 0.00017. Therefore, we reject the null hypothesis that all of the regression lines are identical, i.e., the process is out-of-control. Diagnostic aids to determine the source of variability should be followed.

2.1.2.5 *Mahmoud et al.'s (2007) Change Point Approach*

To detect parameter changes in a Phase I simple linear profile data set, Mahmoud et al. (2007) proposed an approach based on the segmented regression model. In a segmented simple linear regression model we assume that the data set consists of a single sample in the form $\{(X_1, Y_1), (X_2, Y_2), \ldots, (X_N, Y_N)\}$. The s-segment regression model with an explanatory variable X and a response Y is assumed to be

$$Y_i = A_j + B_j X_i + \varepsilon_i, \ \theta_{j-1} < i \leq \theta_j, \quad j = 1, \ldots, s, \quad i = 1, 2, \ldots, N, \tag{2.27}$$

where the θ_j are the change points between segments (usually $\theta_0 = 0$ and $\theta_s = N$) and the ε_i are the error terms. Most work on segmented regression has assumed that the ε_i were i.i.d. normal random variables with a mean zero. The segment error term variance, σ_j^2, can be considered to be either constant (homoscedastic model), if changes in the variance are not important to detect, or nonconstant (heteroscedastic model) otherwise. The segmented regression model in Equation (2.27) can be rewritten as

$$\begin{aligned} Y_i &= A_1 + B_1 X_i + \varepsilon_i, \ \theta_0 < i \leq \theta_1 \\ Y_i &= A_2 + B_2 X_i + \varepsilon_i, \ \theta_1 < i \leq \theta_2 \\ &\vdots \\ Y_i &= A_s + B_s X_i + \varepsilon_i, \ \theta_{s-1} < i \leq \theta_s, \end{aligned} \tag{2.28}$$

where $i = 1, 2, \ldots, N$. The segmented regression technique is used to estimate the number of segments s and the locations of the change points θ_j. Generally speaking, the sampling framework of a Phase I simple linear profile data set is identical to that of a segmented regression data set. Thus, Mahmoud et al. (2007) used a well-known change point method based on the segmented regression technique to test the constancy of the Phase I simple linear profile parameters over time. This method can be used to assess the stability of and to detect change points in a Phase I simple linear regression profile data set.

The first step of the change point approach proposed by Mahmoud et al. (2007) is to combine the m samples of profiles into one sample of size N. Then, one applies the segmented simple linear regression model in Equation (2.28) to the combined sample. In a linear profile model, we usually assume that no parameter changes take place within each sample. Thus, the change points θ_j should be restricted to the indices i corresponding to the ends of the profile samples, i.e., $\theta_0 = 0$, $\theta_1 = n_1$, $\theta_2 = n_1 + n_2, \ldots, \theta_m = N$. This is usually called in the literature segmented regression model with restricted change point locations. The formulas for the maximum likelihood estimator (MLE) of the change points of the segmented regression model were given by Hawkins (1976), for both the heteroscedastic and homoscedastic models. In this section, we present only the formulas for the heteroscedastic model. However, if in a particular application the variance is not expected to change from sample to sample, one can use alternative methods on the basis of the homoscedastic segmented regression model to detect changes in the slope and intercept more powerfully.

Mahmoud et al. (2007) introduced a likelihood ratio statistic for testing a single change point ($s = 2$ segments). If a change point is detected, then they suggested using a binary segmentation procedure to detect multiple change points. With a single change in one or more of the regression parameters after sample m_1, consider the following null and alternative hypotheses

$$H_0 : A_1 = A_2 = A,\ B_1 = B_2 = B,\ \text{and}\ \sigma_1^2 = \sigma_2^2 = \sigma^2,\ \ H_1 : H_0 \text{ is not true},$$

then the log of the likelihood function based on the model in Equation (2.28) assuming H_0 to be true is

$$-(\sum_{j=1}^{m} n_j \log(2\pi\sigma^2)/2) - \left(\sum_{j=1}^{m}\sum_{i=1}^{n_j} (Y_{ij} - A - BX_{ij})^2/2\sigma^2\right),$$

or equivalently,

$$-(N\log(2\pi\sigma^2)/2) - \left(\sum_{i=1}^{N} (y_i - A - Bx_i)^2/2\sigma^2\right), \tag{2.29}$$

where $N=\sum_{j=1}^{m} n_j$ and $(x_i,\ y_i)$, $i=1,\ 2,\ldots,\ N$, are N bivariate observations resulting from pooling the m samples into one sample of size N. The maximum likelihood estimators of the parameters A, B, and σ^2 are $\hat{A} = \bar{y} - \hat{B}\bar{x}$, $\hat{B} = S_{xy}\big/S_{xx}$, and $\hat{\sigma}^2 = \sum_{i=1}^{N} (y_i - \hat{A} - \hat{B}x_i)^2\big/N$, respectively, where $\bar{y} = \sum_{i=1}^{N} y_i\big/N$, $\bar{x} = \sum_{i=1}^{N} x_i\big/N$, $S_{xy} = \sum_{i=1}^{N} (x_i - \bar{x})y_i$, and $S_{xx} = \sum_{i=1}^{N} (x_i - \bar{x})^2$. The maximum of the log-likelihood function in Equation (2.29) is

$$l_0 = -(N\log(2\pi) + N\log(\hat{\sigma}^2) + N)/2.$$

On the other hand, the log-likelihood function for all the samples prior to the change point m_1 is

$$-(N_1\log(2\pi\sigma^2)/2) - (N_1\hat{\sigma}_1^2/2\sigma^2) + (\sum_{i=1}^{N_1} [(A - \hat{A}_1) + (B - \hat{B}_1)x_i)]^2/2\sigma^2), \tag{2.30}$$

where

$$N_1 = \sum_{j=1}^{m_1} n_j,\ \hat{A}_1 = \bar{y}_1 - \hat{B}_1\bar{x}_1,\ \hat{B}_1 = S_{xy1}\big/S_{xx1},\ \text{and}$$
$$\hat{\sigma}_1^2 = \sum_{i=1}^{N_1} (y_i - \hat{A}_1 - \hat{B}_1x_i)^2\big/N_1.$$

Also, $\bar{y}_1 = \sum_{i=1}^{N_1} y_i \big/ N_1$, $\bar{x}_1 = \sum_{i=1}^{N_1} x_i \big/ N_1$, $S_{xy1} = \sum_{i=1}^{N_1} (x_i - \bar{x}_1)y_i$, and $S_{xx1} = \sum_{i=1}^{N_1} (x_i - \bar{x}_1)^2$. The maximum of the log-likelihood function in Equation (2.30) is

$$l_1 = -(N_1 \log(2\pi) + N_1 \log(\hat{\sigma}_1^2) + N_1)/2.$$

Similarly, the log-likelihood function for all the samples following the change point m_1 is

$$-(N_2 \log(2\pi\sigma^2)/2) - (N_2\hat{\sigma}_2^2/2\sigma^2) + \left(\sum_{i=N_1+1}^{N} [(A - \hat{A}_2) + (B - \hat{B}_2)x_i)]^2/2\sigma^2\right), \tag{2.31}$$

where

$$N_2 = N - N_1,\ \hat{A}_2 = \bar{y}_2 - \hat{B}_2\bar{x}_2,\ \hat{A}_{12} = S_{xy2}\big/S_{xx2},\ \text{and}$$
$$\hat{\sigma}_2^2 = \sum_{i=N_1+1}^{N} (y_i - \hat{A}_2 - \hat{B}_2 x_i)^2 \big/ N_2.$$

Also, $\bar{y}_2 = \sum_{i=N_1+1}^{N} y_i \big/ N_2, \bar{x}_2 = \sum_{i=N_1+1}^{N} x_i \big/ N_2, S_{xy2} = \sum_{i=N_1+1}^{N} (x_i - \bar{x}_2)y_i$, and $S_{xx2} = \sum_{i=N_1+1}^{N} (x_i - \bar{x}_2)^2$.

Again, the maximum of the log-likelihood function in Equation (2.31) is

$$l_2 = -(N_2 \log(2\pi) + N_2 \log(\hat{\sigma}_2^2) + N_2)/2.$$

Under H_1, the maximum of the log-likelihood function on the basis of the model in Equation (2.28) is

$$l_a = l_1 + l_2.$$

According to the likelihood ratio approach, we reject H_0 if l_a is much larger than l_0. It is well known that the quantity $lrt_{m_1} = -2(l_0 - l_a)$ has asymptotically a chi-square distribution with three degrees of freedom. Thus, it can be shown that the LRT statistic used to test the null hypothesis of a single segment against the alternative of $s = 2$ segments is

$$lrt_{m_1} = N \log \hat{\sigma}^2 - N_1 \log \hat{\sigma}_1^2 - N_2 \log \hat{\sigma}_2^2,\ m_1 = 1, 2, \ldots, m - 1, \tag{2.32}$$

where $\hat{\sigma}^2$ is the MLE of the error term variance for the regression model fitted for all the m samples pooled into one sample of size N, $\hat{\sigma}_1^2$ is the MLE of the error term variance for the regression model fitted for all the samples prior to m_1 pooled into one sample of size N_1, and $\hat{\sigma}_2^2$ is the MLE of the error term variance for the regression model fitted for all the samples following m_1 pooled into one sample of size N_2. The LRT statistic in Equation (2.32) is identical to that of Quandt (1958).

The in-control expected value for the lrt_{m_1} statistic in Equation (2.32) is not the same for all values of m_1 as shown by many researchers; see, for example, Mahmoud et al. (2007). Thus, the Bartlett correction is usually used, i.e., dividing the lrt_{m_1} statistics by a normalizing factor e_{m_1} that makes the expected values the same for all values of m_1. If we let $e_{m_1} = \mathrm{E}(lrt_{m_1})$, then the statistics

$$lrtc_{m_1} = lrt_{m_1}/e_{m_1}, m_1 = 1, 2, \ldots, m-1, \tag{2.33}$$

all have an expected value of unity. The expected value e_{m_1} can be approximated by using the following formula as given in Mahmoud et al. (2007):

$$e_{m_1} \approx 2 - 2\left(\frac{1}{N} - \frac{1}{N_1} - \frac{1}{N_2}\right) - \left(\frac{N}{N-2} - \frac{N_1}{N_1-2} - \frac{N_2}{N_2-2}\right)$$
$$- \frac{1}{3}\left(\frac{N}{(N-2)^2} - \frac{N_1}{(N_1-2)^2} - \frac{N_2}{(N_2-2)^2}\right). \tag{2.34}$$

To estimate the change points' locations and to determine the appropriate number of change points, one can apply the binary segmentation procedure. Using this procedure, one obtains the lrt_{m_1} statistics in Equation (2.32) for all possible values of m_1, $m_1 = 1, 2, \ldots, m-1$ and divides each by the normalizing factor e_{m_1}. The approach signals the presence of a change point if the maximum of these statistics exceeds a threshold. The value of m_1 that maximizes the lrt_{m_1} statistic in Equation (2.32) is the MLE of the change point location. To detect additional change points, one splits the data set into two subsets at m_1 and applies the same binary splitting procedure described above to each subset. This procedure is repeated until no evidence of additional change points is given. The advantage of this change point approach over the competing control chart approaches is that it is not only provides a signal that the process is out-of-control in the SPC context, but also provides an estimate of when it went out-of-control.

The $m-1$ statistics, $lrtc_{m_1}$, $m_1 = 1, 2, \ldots, m-1$, in Equation (2.33) are correlated. Therefore, the distribution of the maximum of the $lrtc_{m_1}$ [$\max(lrtc_{m_1})$] is intractable. Mahmoud et al. (2007) used simulation technique to estimate thresholds that correspond to a specific probability of a Type I error. They stated that the threshold T for the $\max(lrtc_{m_1})$ statistic that corresponds to a probability of a Type I error of α can be approximated using

$$T = \begin{cases} \chi^2_{3,1-\alpha/(m-1)}/3 & \text{if } m \le 6 \\ \chi^2_{3,1-\alpha/r^*}/3 & \text{if } m > 6 \end{cases}, \tag{2.35}$$

where $\chi^2_{3,1-\alpha/(m-1)}$ is the $100(1-\alpha/(m-1))$ percentile of the chi-square distribution with 3 degrees of freedom and

$$r^*(m) = -11.5 + 8.05 \log m.$$

Table 2.4 The Likelihood Ratio Statistics Calculated for the Data Set in Example 2.7

m_1	1	2	3	4	5	6	7	8	9	10
$lrtc_{m_1}$	2.237	0.999	1.670	2.183	5.683	7.393	5.963	7.776	3.397	1.829

Example 2.8 To illustrate the use of the change point approach to monitor simple linear profile we use the data set in Example 2.7. On the basis of a nominal false alarm probability of $\alpha = 0.05$, the value of $r^*(11)$ is

$$r^*(11) = -11.5 + 8.05 \log(11) = 7.8030569.$$

Thus, the threshold for the test statistic is

$$T = \chi^2_{3,1-0.05/7.8030569}/3 = 4.1017119.$$

The values of the $lrtc_{m_1}$ statistics are presented in Table 2.4. As shown in this table, the maximum value of $lrtc_{m_1}$ is 7.776, corresponding to $m_1 = 8$. Thus, the change point method signals the presence of a change point, since this value is greater than $T = 4.1017119$.

Repeating the change point method on the subset of samples following $m_1 = 8$ and prior to it gave no evidence of the presence of an additional change point at $\alpha = 0.025$. Notice that we used here a Bonferroni-corrected significance level of $\alpha =$ 0.025 due to the fact that each segment (subset) represents another opportunity for a Type I erroneous split.

An important advantage of the change point method presented above over the competing approaches is its ability to locate the change point.

2.1.3 Performance Comparisons of Phase I Simple Linear Profile Approaches

Using simulation studies, Mahmoud and Woodall (2004) compared the performance of four Phase I methods for monitoring linear profile processes in terms of the overall probability of a signal. These methods are the T^2 control chart proposed by Stover and Brill (1998), the T^2 control chart proposed by Kang and Albin (2000), the three Shewhart-type control charts proposed by Kim et al. (2003), and the global F-test method proposed by Mahmoud and Woodall (2004). In their study, Mahmoud and Woodall considered the following in-control model:

$$Y_{ij} = X_i + \varepsilon_{ij}, i = 1, 2, \ldots, 10 \text{ and } j = 1, 2, \ldots, 20,$$

with error terms assumed to be i.i.d. normal random variables with a mean zero and variance 1 and fixed X-values of 0, 0.2, 0.4, 0.6, 0.8, 1, 1.2, 1.4, 1.6, and 1.8. However, as mentioned before, generalization of the properties of the profile monitoring techniques is achieved only if $\bar{X} = 0$. Thus, for standard comparisons it is recommended to use X-values with an average of zero.

Note that values of the chart statistics for all of the T^2 methods discussed in this chapter and for the Kim et al.'s (2003) approach are not independent because each of the charted statistics is calculated on the basis of the same sample estimators. To investigate the effect of the dependencies of the control chart statistics on the in-control performance of these methods, Mahmoud and Woodall (2004) calculated using simulation the overall false alarm probabilities produced by each of these methods when the marginal distribution (distributions) of the chart statistic is used to obtain the control chart limits. They found that there was no practical difference between the simulated and nominal overall false alarm probabilities. They concluded that the dependencies of the control chart statistics for the monitoring methods have no practical effect on one's use of the marginal distribution (distributions) to approximate closely the overall false alarm probabilities.

Having shown that, they compared the performance of the competing methods in terms of the overall probabilities of an out-of-control signal indicating instability when there were shifts in model parameters. They considered fixed shifts in k out of the m individual model parameters. Four different types of shifts in the model parameters were considered in their study. These corresponded to shifts in the Y-intercept, in the slope under the model in Equation (2.1), in the slope under the model in Equation (2.16), and in the process standard deviation. Shifts in the intercept were measured in units of $\sigma/\sqrt{n}$, shifts in the slopes were measured in units of $\sigma/\sqrt{S_{xx}}$, and shifts in the process standard deviation were measured as multiples of σ.

Notice that a shift in the Y-intercept in Equation (2.1) from A_j to $A_j + \Delta A$ is equivalent to a shift in the Y-intercept in Equation (2.16) from A'_j to $A'_j + \Delta A$. However, a shift in the slope of the regression model in Equation (2.1) from B_j to $B_j + \Delta B$ leads to shifts in both the Y-intercept and slope in Equation (2.16). In this case, the Y-intercept shifts from A'_j to $A'_j + \bar{X}_j \Delta B$, while the slope shifts from B'_j to $B'_j + \Delta B$. In the special case when $\bar{X}_j = \bar{X}, j = 1, 2, \ldots, m$, if the Y-intercept in Equation (2.1) shifts from A_j to $A_j + \Delta A$ and simultaneously the slope in Equation (2.1) shifts from B_j to $B_j + \Delta B$ so that $\Delta A + \bar{X}_j \Delta B = 0$, then only the slope in Equation (2.16) shifts from B'_j to $B'_j + \Delta B$. However, if $\bar{X}_j$ varies from sample to sample, the shifts in the parameters in Equation (2.16) corresponding to this type of shift is not clearly interpretable. Finally, it is obvious that shifts in the variances of both models in Equations (2.1) and (2.16) would be equivalent.

In general, Mahmoud and Woodall's (2004) simulation study showed that Kim et al. (2003) method and their proposed F-test method have the best overall performance in detecting shifts in a process parameter. For example, Figure 2.12 shows the simulated overall probabilities of an out-of-control signal for shifts in the Y-intercept from A to $A + \lambda\sigma/\sqrt{n}$ for all of the proposed methods when $k = 10$ and $k = 2$. Under a parameter shift affecting much of the Phase I data, the F-test method performs uniformly better than the competing methods. The T^2 chart proposed by Kang and Albin (2000) and Kim et al. (2003) methods have roughly similar performance in detecting shifts in the Y-intercept and in the slope under both models in Equations (2.1) and (2.16). However, Kim et al. (2003) method has greater statistical power in detecting shifts in the process standard deviation. The T^2 chart proposed by Stover and Brill (1998) has very poor performance in detecting shifts in any process

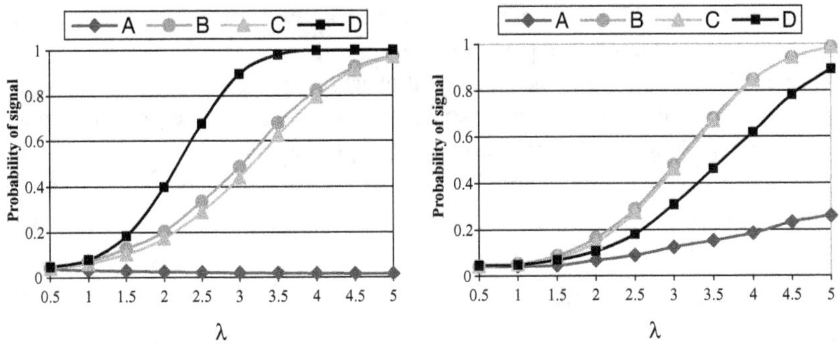

(a) $k = 10$ (b) $k = 2$

Figure 2.12 Probability of out-of-control signal under Y-intercept shifts from A to $A + \lambda\sigma/\sqrt{n}$. (Adapted from Mahmoud 2004.) Method A: Stover and Brill's (1998) T^2 chart, Method B: Kang and Albin's (2000) T^2 chart, Method C: Kim et al.'s (2003) approach, and Method D: the F-test approach.

parameter. Therefore, Mahmoud and Woodall (2004) recommended either their F-test method or the Kim et al. (2003) method for use in Phase I analysis of simple linear profiles.

Mahmoud et al. (2007) compared their proposed change point method to the F-test method proposed by Mahmoud and Woodall (2004) and the Kim et al. (2003) approach using simulation. The authors considered following two types of parameter shifts in this study: sustained step shifts taking place after sample q ($q < m$) and q unsustained shifts scattered randomly among the m samples. Mahmoud et al. (2007) concluded that their proposed change point method performs much better than the two competing control chart methods under sustained parameter shifts and for most of the cases considered in this simulation. On the other hand, both the F-test method and the Kim et al. (2003) approach have much better performance than the change point method in detecting temporary, isolating unsustained parameter shifts in the profile model. To protect against both kind of changes, sustained and randomly occurring unsustained shifts, they recommended employing the change point method in conjunction with either the F-test method or the Kim et al. (2003) approach.

2.2 PHASE II SIMPLE LINEAR PROFILE

The Phase II simple linear profile model is in the form

$$Y_{ij} = A + BX_{ij} + \varepsilon_{ij}, \quad i = 1,\ 2, \ldots, n_j \text{ and } j = 1, 2, \ldots . \tag{2.36}$$

Again, the ε_{ij} are assumed to be i.i.d. $N\,(0, \sigma^2)$ random variables and the X-values in each sample are assumed to be known constants. Here, the in-control values of the parameters A_0, B_0, and σ_0^2 are assumed to be known or estimated from a data set reflecting expected in-control performance.

The performance of a Phase II control charting method is usually measured by some parameter of the run-length distribution. As mentioned before, the run length is the number of samples taken until the chart produces an out-of-control signal. In the literature, often the average run length (*ARL*) is used in performance comparisons studies of Phase II methods.

2.2.1 Phase II Approaches

In general, the approaches proposed in the literature for monitoring Phase II simple linear profiles can be classified into the following categories: (1) omnibus control charts for monitoring simultaneously the intercept and slope of the regression coefficients, and (2) individual control charts for monitoring separately individual regression parameters. In the following subsections, we introduce both types of approaches.

2.2.1.1 Omnibus Control Charts for Monitoring the Regression Coefficients

As mentioned in Section 2.1, the least squares estimators a_j and b_j are distributed as a bivariate normal distribution with the mean vector

$$\mu = (A_0 \; B_0)^T$$

and the variance–covariance matrix

$$\sum = \sigma_0^2 \begin{pmatrix} \dfrac{1}{n_j} + \dfrac{\bar{X}_j^2}{S_{XX(j)}} & \dfrac{-\bar{X}_j}{S_{XX(j)}} \\ \dfrac{-\bar{X}_j}{S_{XX(j)}} & \dfrac{1}{S_{XX(j)}} \end{pmatrix}.$$

Therefore, several authors proposed approaches based on vectors containing estimators of the Y-intercept and slope to monitor simple linear profiles with assumed known values for the intercept, slope, and variance parameters. These approaches are presented in the following sections.

2.2.1.1.1 Kang and Albin's (2000) T^2 Approach

Kang and Albin (2000) proposed two control chart methods for Phase II monitoring of linear profiles. Their first approach is a bivariate T^2 chart based on successive vectors of the least squares estimators of the Y-intercept and slope, assuming known parameter values. Here, the T^2 statistics are

$$\mathbf{T}_j^2 = (z_j - \boldsymbol{\mu})^T \Sigma^{-1} (z_j - \boldsymbol{\mu}), \quad j = 1, 2, \ldots, \tag{2.37}$$

where $z_j = (a_j \; b_j)^T$ is the vector of sample estimators, and $\boldsymbol{\mu}$ and $\boldsymbol{\Sigma}$ are the known in-control mean vector and variance–covariance matrix, respectively. When the process is in-control, the T^2 statistic in Equation (2.37) follows a central chi-square distribution with 2 degrees of freedom. The upper control limit of this chart used by Kang and Albin (2000) is

$$UCL = \chi^2_{2,\alpha}. \tag{2.38}$$

Under out-of-control shifts in the process parameters (assuming that the X-values are the same for all samples), the T^2 statistic in Equation (2.37) follows a noncentral chi-square distribution with 2 degrees of freedom and non-centrality parameter

$$\tau = n(\lambda + \beta\bar{X})^2 + \beta^2 S_{XX},$$

where λ and β are the shifts in the intercept and slope, respectively. It can be shown that the exact ARL of this T^2 control chart is evaluated using the following formula:

$$ARL = \frac{1}{\Pr(T_j^2 > \chi^2_{2,\alpha})}.$$

If the X-values are not the same for all samples, the ARL corresponding to a specified false alarm probability can be estimated using simulation.

Example 2.9 A Phase II example with simulated data set. Kang and Albin (2000) proposed an example with simulated data set to illustrate the use of their Phase II linear profile methods. This example was considered in several articles, for example in Kang and Albin (2000), Kim et al. (2003), Zou et al. (2006), and Mahmoud et al. (2010) to illustrate the use of their proposed Phase II approaches when applied to a simple linear regression profile data. These authors considered the underlying in-control linear profile model $Y_{ij} = 3 + 2X_i + \varepsilon_{ij}$, where the ε_{ij} are i.i.d. $N(0,1)$ random variables. Except for Mahmoud et al. (2010), these authors considered sequence of Phase II profiles with fixed X-values of 2, 4, 6, and 8 (i.e., $n = 4$). However, Mahmoud et al. (2010) considered sequence of Phase II profiles with fixed X-values of -2 and 2. Alternatively, we use in this chapter fixed X-values of -3, -1, 1, and 3 and in-control profile model $Y_{ij} = 13 + 2X_i + \varepsilon_{ij}$. Using simulation, 20 in-control profiles were first generated and then another nine out-of-control future profiles were generated. The first four columns of Table 2.5 give the simulated Phase II data set, while the following three columns give the estimated intercept, slope, and variance of each profile. In this data set, the slope B has shifted from 2.0 to 2.4 after the 20th sample. In this example, the average of the X-values is $\bar{X} = 0$ and the quantity $S_{XX} = 20$. Thus, the known variance–covariance matrix of the regression estimators is

$$\Sigma = 1 \times \begin{pmatrix} \frac{1}{4} + \frac{0^2}{20} & \frac{-0}{20} \\ \frac{-0}{20} & \frac{1}{20} \end{pmatrix} = \begin{pmatrix} 0.25 & 0 \\ 0 & 0.05 \end{pmatrix}.$$

Using $UCL = \chi^2_{2,0.005} = 10.597$ to produce an in-control ARL of

$$ARL = \frac{1}{0.005} = 200,$$

then the T^2 chart of Kang and Albin (2000) applied to this data set is shown in Figure 2.13. Notice that the last column of Table 2.5 gives the T^2 statistics calculated using Equation (2.37). As shown in Figure 2.13, the shift in the process was detected at sample 27. In the previous example, the shift in the process was detected but the source of this shift cannot be identified using this approach.

Table 2.5 The Simulated Phase II Data Set for Example 2.9

Profile Number	X: −3	X: −1	X: 1	X: 3	Intercept Estimate	Slope Estimate	Variance Estimate	T^2
1	6.818	12.644	12.919	19.402	12.946	1.901	3.510	0.206
2	7.723	11.330	16.987	20.322	14.091	2.173	0.487	5.353
3	5.017	10.612	13.934	20.381	12.486	2.471	0.820	5.490
4	7.433	10.399	16.386	17.029	12.812	1.739	2.425	1.507
5	6.834	11.428	16.192	19.110	13.391	2.080	0.453	0.738
6	7.068	12.108	14.919	19.478	13.393	2.002	0.425	0.619
7	8.209	11.732	14.901	19.149	13.498	1.799	0.117	1.796
8	8.515	10.986	15.451	18.639	13.398	1.742	0.331	1.965
9	8.005	9.510	15.926	18.891	13.083	1.954	2.014	0.070
10	6.705	11.311	14.496	17.244	12.439	1.740	0.456	2.610
11	6.327	9.925	14.647	20.649	12.887	2.384	0.723	3.006
12	6.057	9.075	14.573	18.610	12.079	2.158	0.518	3.892
13	9.106	10.037	17.018	18.398	13.640	1.743	3.419	2.958
14	7.616	11.159	15.734	18.687	13.299	1.890	0.219	0.601
15	6.355	10.176	13.908	19.772	12.553	2.199	0.646	1.592
16	6.967	12.525	14.753	17.461	12.927	1.685	1.378	2.000
17	9.686	11.626	15.456	17.735	13.626	1.399	0.310	8.792
18	5.266	12.292	16.461	18.732	13.188	2.228	2.849	1.185
19	7.014	9.930	15.485	18.634	12.766	2.021	0.643	0.228
20	5.912	10.017	13.494	17.924	11.837	1.976	0.076	5.422
21	6.313	10.788	14.682	20.432	13.054	2.313	0.352	1.966
22	7.156	10.523	15.290	21.125	13.523	2.334	0.764	3.322
23	5.498	9.941	15.313	20.957	12.927	2.588	0.191	6.925
24	5.675	10.688	15.865	18.780	12.752	2.225	0.697	1.254
25	5.002	9.833	16.039	20.619	12.873	2.653	0.233	8.587
26	4.009	10.267	17.464	19.556	12.824	2.692	3.082	9.698
27	4.326	10.245	16.378	21.058	13.002	2.817	0.261	13.334
28	5.488	11.946	16.957	20.617	13.752	2.520	0.979	7.666
29	4.361	11.675	16.583	20.282	13.225	2.633	1.670	8.229

2.2.1.1.2 Kang and Albin's (2000) EWMA Approach

It is well known that Shewhart-type control charts, for example the univariate $\bar{X}$-chart and the multivariate Hottling T^2 charts, are not effective in detecting small-to-medium shifts because the decision rule depends only on the current chart statistic. Therefore, several univariate and multivariate control charts that accumulate information from the process history were proposed in the literature, for example, the univariate and multivariate exponentially weighted moving average (EWMA) and cumulative sum (CUSUM) control charts. The second Phase II method of Kang and Albin (2000) is an EWMA chart to monitor the average deviation from the in-control line. The standard deviations from the in-control regression line obtained at sample j are calculated using

$$d_{ij} = (Y_{ij} - A_0 - B_0 X_j)/\sigma_0, \quad i = 1,\ 2, \ldots, n, \tag{2.39}$$

and the average deviation for sample j is

$$\bar{d}_j = \sum\nolimits_{i=1}^{n} d_{ij}/\mathrm{n} = (\bar{Y}_j - A_0 - B_0\bar{X})/\sigma_0. \tag{2.40}$$

The EWMA control chart statistics are given by

$$EWMA(d)_j = \theta\bar{d}_j + (1-\theta)EWMA(d)_{j-1}, \tag{2.41}$$

where $0 < \theta \le 1$ is the smoothing parameter and $EWMA(d)_0 = 0$. The chart signals a change in one or more of the regression parameters if $\left|EWMA(d)_j\right| > h_d$, where $h_d > 0$ is a constant chosen to give a specified in-control ARL. It should be noted that the EWMA chart based just on the average deviations obviously cannot effectively detect a shift in the slope B if one uses X-values with an average of zero.

Kang and Albin (2000) also suggested an R-chart to be used in conjunction with this EWMA chart to monitor the variation about the regression line. For the R-chart,

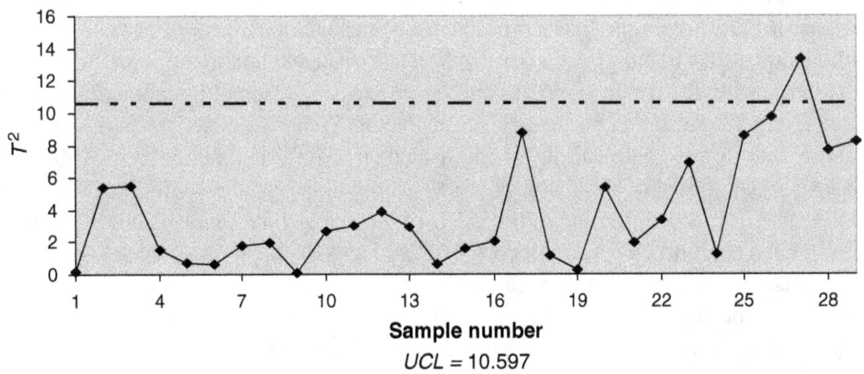

Figure 2.13 The T^2 chart of Kang and Albin (2000) applied to the data set in Example 2.9.

they plotted the sample ranges

$$R_j = \max_i(d_{ij}) - \min_i(d_{ij}), \quad j = 1, 2, \ldots, \tag{2.42}$$

in a chart with the following control limits:

$$LCL = \sigma(d_2 - L_2 d_3) \text{ and } UCL = \sigma(d_2 + L_2 d_3),$$

where $L_2 > 0$ is a constant selected to produce a specified in-control *ARL*, and d_2 and d_3 are constants depending on the sample size n. For example, Ryan (2000) and Montgomery (2001) provide tables for the values of d_2 and d_3 corresponding to different values of the sample size n. A signal is given whenever one of the two charts produces an out-of-control signal. However, a disadvantage of the *R*-chart approach is that, if $n < 7$ there is no lower control limit for the *R*-chart, and consequently one cannot detect decreases in the process variance.

If the sample sizes are not the same for all samples, one can determine the control limits of the EWMA chart and *R*-chart corresponding to a specified overall in-control *ARL* using simulation.

Kang and Albin (2000) recommended using this EWMA approach in Phase I, substituting the values of the unknown parameters by their estimates. However, it is not recommended using this method in Phase I for two reasons. First, EWMA charts are recommended in Phase II because of their power in detecting sustained shifts in parameters, and their quick detection of small-to-moderate process shifts compared to Shewhart-type control charts. However, quick detection is not an issue in Phase I since we have access to a fixed set of historical data. Second, in applying an EWMA chart in Phase I, several samples could be contributing to any out-of-control signal. Therefore, it is not clearly defined how to identify and delete the out-of-control regression lines to achieve stability in the process before estimating the in-control regression parameters.

Example 2.10 To illustrate the use of the EWMA and *R*-charts combination to monitor Phase II simple linear profile data, we use the data set in Example 2.9. For this purpose, a smoothing parameter of $\theta = 0.2$ is used to obtain the charts' statistics. The control limits for the EWMA chart based on the statistic in Equation (2.41) were set at ± 0.48 and the upper control limit for the *R*-chart was set at 4.94. Note that there is no lower control limit for the *R*-chart if $n < 7$, as mentioned in Kim et al. (2003). Using 50,000 simulated Phase II simple linear profiles with fixed *X*-values of -3, -1, 1, and 3, the in-control ARL produced by both charts is approximately 200. Each chart produces an in-control ARL of approximately 380 when used alone. The first two columns of Table 2.6 give the values of the $EWMA(d)_j$ statistics and the ranges of the deviations from the in-control profile R_j, respectively. Figures 2.14 and 2.15 show the EWMA control chart based on the statistics in Equation (2.41) and the *R*-chart, respectively, when applied to the data set in Example 2.9. The EWMA statistics are within the control limits, while the *R*-chart signals a shift in the process

Table 2.6 Some Charts Statistics Calculated Using the Data Set in Example 2.9

Profile Number	$EWMA(d)_j$	R_j	$EWMA_{I(j)}$	$EWMA_{S(j)}$	$EWMA_{E(j)}$	F_j
1	−0.011	3.725	−0.022	−0.088	0.251	0.059
2	0.209	1.656	0.419	0.084	0.057	10.990
3	0.065	3.365	0.129	0.488	0.006	6.699
4	0.014	3.358	0.028	0.157	0.182	0.622
5	0.089	1.358	0.179	0.197	0.000	1.631
6	0.150	1.189	0.300	0.159	0.000	1.458
7	0.220	1.308	0.439	−0.052	0.000	15.340
8	0.255	1.875	0.511	−0.273	0.000	5.925
9	0.221	2.495	0.442	−0.259	0.140	0.035
10	0.064	2.067	0.129	−0.440	0.000	5.727
11	0.029	2.723	0.058	−0.008	0.000	4.158
12	−0.161	1.535	−0.322	0.135	0.000	7.515
13	−0.001	3.069	−0.002	−0.122	0.246	0.866
14	0.059	1.046	0.118	−0.197	0.000	2.742
15	−0.042	1.865	−0.084	0.021	0.000	2.470
16	−0.048	3.064	−0.097	−0.265	0.064	1.451
17	0.086	3.950	0.173	−0.749	0.000	28.333
18	0.107	3.195	0.213	−0.395	0.209	0.416
19	0.039	1.555	0.077	−0.298	0.079	0.355
20	−0.202	0.523	−0.404	−0.260	0.000	71.663
21	−0.151	2.119	−0.301	0.072	0.000	5.588
22	−0.016	2.601	−0.032	0.356	0.000	4.349
23	−0.027	3.459	−0.055	0.810	0.000	36.229
24	−0.071	2.190	−0.143	0.849	0.000	1.800
25	−0.083	3.617	−0.165	1.263	0.000	36.857
26	−0.101	5.455	−0.202	1.629	0.225	3.146
27	−0.081	4.732	−0.161	2.034	0.000	51.008
28	0.086	3.469	0.172	2.092	0.000	7.834
29	0.114	4.222	0.228	2.240	0.103	4.930

at sample 26. This charts' combination approach has detected the assigned shift faster than the T^2 approach of Kang and Albin (2000).

An important conclusion from the previous example is that the EWMA chart based just on the average deviations cannot effectively detect a shift in the slope B if one uses X-values with an average of zero. In this case, shifts in the slope can be detected more effectively by the R-chart.

2.2.1.1.3 Noorossana et al.'s (2004) Multivariate Cumulative Sum (MCUSUM) Approach

To increase the efficiency of the monitoring scheme to small sustained shifts, Noorossana et al. (2004) proposed replacing the Kang and Albin's (2000) T^2 chart by a MCUSUM chart based on vectors of the regression estimators. Also to monitor the

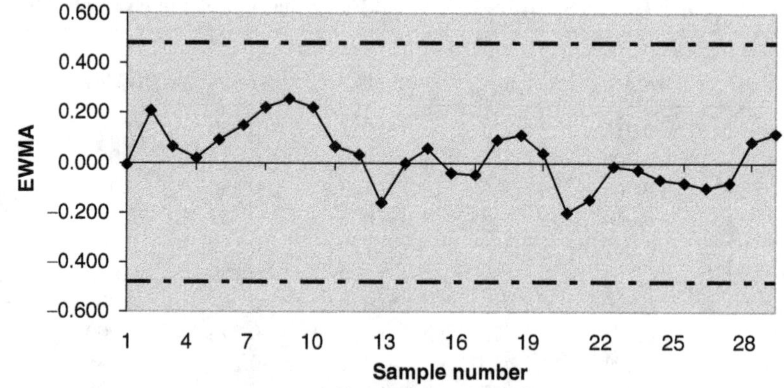

Figure 2.14 The EWMA control chart of Kang and Albin (2000) applied to the data set in Example 2.9.

process standard deviation, they suggested using this MCUSUM chart in conjunction with the R-chart described in Equation (2.42). The MCUSUM chart was proposed first by Healy (1987) for monitoring the mean vector of a multivariate normal process. The statistics of MCUSUM proposed by Noorossana et al. (2004) is

$$S_j = \max(S_{j-1} + a^T (z_j - \boldsymbol{\mu}) - 0.5D, 0), \tag{2.43}$$

where, $S_0 = 0$, $a^T = \frac{\Delta^T \sum^{-1}}{\sqrt{\Delta^T \sum^{-1} \Delta}}$, $D=\sqrt{\Delta^T \sum^{-1} \Delta}$, and Δ is the smallest shift in the process mean vector $\boldsymbol{\mu} = (A\ B)^T$considered important to be detected quickly. The value D is usually referred to as the noncentrality parameter. This MCUSUM chart

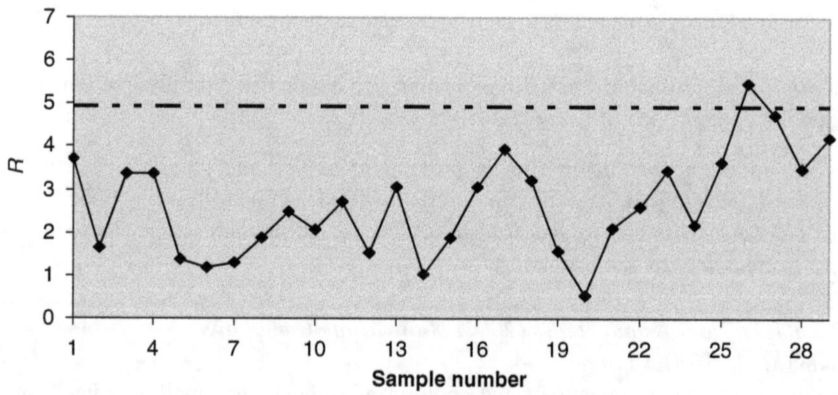

Figure 2.15 The R-chart of Kang and Albin (2000) applied to the data set in Example 2.9.

signals a shift in either the intercept or slope whenever $S_j > H$, where H is chosen to satisfy a specified in-control ARL.

Noorossana et al. (2004) compared the performance of this MCUSUM chart to the T^2 chart of Kang and Albin (2000) using simulation. As expected, they found that the MCUSUM chart outperforms the T^2 chart when small-to-moderate shifts in the process parameters are considered. Otherwise, the T^2 chart performs better.

An alternative approach based on the MEWMA chart scheme to monitor the regression parameters of Phase II linear profiles was proposed by Zou et al. (2007a). This method is described in detail in Chapter 4.

2.2.1.1.4 *Croarkin and Varner's (1982) Scheme*

Another technique for monitoring simple linear profiles was proposed by Croarkin and Varner (1982). This method is based on the principle of inverse calibration and involves plotting the deviations of the measured values from the standards on a Shewhart control chart for lower, middle, and upper values of the standards. This method is now a part of the ISO 5725-6 (1994) "Linear Calibration using Reference Material." This control charting method was initially developed to monitor the calibration curve of optical imaging systems. However, it is not only used in the calibration of line width artifacts, but has also been used in developing control charts in the calibration of phase meters (see Turgel and Vecchia (1987)).

The control chart statistic in Croarkin and Varner's (1982) scheme is obtained by first "correcting" the measured values and then subtracting the standard X-value from it. The quantities plotted on the control chart at the time of the jth sample are

$$z_{ij}^* = \frac{Y_{ij} - A_0}{B_0} - X_i, \quad i = 1, 2, \ldots, n \tag{2.44}$$

assuming that $X_{ij} = X_i, j = 1, 2, \ldots$. The control limits are established at $\pm\frac{\sigma_0}{B_0} Z_\zeta$, where

$$\zeta = \frac{1 - e^{\frac{\log(1-\alpha)}{n}}}{2},$$

and Z_ζ is the value that corresponds to the upper ζ percentage point of the standard normal distribution, n is the number of standards (number of X-values) evaluated at each time period and α is chosen to provide the desired in-control ARL using the relationship $ARL = 1/\alpha$. As mentioned by Woodall (2007), the NIST/SEMATECH e-Handbook of Statistical Methods, which is freely available online at http://www.itl.nist.gov/div898/handbook/, contains a discussion of this scheme. NIST recommends the use of only three standards (one near each end point of the measurement range and one near the center) for checking the calibration relationship.

Example 2.11 Line widths of photo masks reference standards. The data set in this example is presented in the NIST/SEMATECH e-Handbook of

Table 2.7 Line width Measurements for Example 2.11 (Adapted from Gupta et al. 2006.)

	Standards of X			Chart Statistics		
Profile Number	Lower 0.76	Medium 3.29	Upper 8.89	z_{1j}^*	z_{2j}^*	z_{3j}^*
1	1.12	3.49	9.11	0.098	−0.005	0.149
2	0.99	3.53	8.89	−0.035	0.036	−0.076
3	1.05	3.46	9.02	0.027	−0.036	0.057
4	0.76	3.75	9.3	−0.270	0.261	0.343
5	0.96	3.53	9.05	−0.066	0.036	0.087
6	1.03	3.52	9.02	0.006	0.026	0.057

statistical methods at http://www.itl.nist.gov/div898/handbook/mpc/section3/mpc 371.htm.

Gupta et al. (2006) used this data set to illustrate the use of Croarkin and Varner (1982) scheme. The dataset consists of line widths of photo masks reference standards on 10 units (40 measurements) used for monitoring linear calibration profiles of an optical imaging system. The line widths are used to estimate the parameters of the in-control linear calibration profile,

$$Y_{ij} = 0.2817 + 0.9767X_i,$$

with a residual standard deviation of $\sigma = 0.06826\ \mu$m. The monitoring scheme was used to monitor measurements on three units for upper, middle, and lower end of the relevant measurement range of the standards. The data set of six Phase II profiles is presented in the first three columns of Table 2.7. The last three columns present the values of the z_{ij}^* statistics calculated using Equation (2.44). Using $\alpha = 0.05$ to achieve an in-control ARL of 200, then $\zeta = \frac{1-e^{\frac{\log(1-0.05)}{3}}}{2} = 0.008476$. Thus, the upper and lower control limits are set at $\pm\frac{0.06826}{0.9767}Z_{0.008476} = \pm 0.167$. The control chart for this example is presented in Figure 2.16. As shown in this figure, the measurements on the fourth day are out-of-control for all the measurements.

Gupta et al. (2006) compared using simulation the above scheme to the three Shewhart control charts of Kim et al. (2003) using Phase II data sets. They concluded that the Croarkin and Varner (1982) method performed poorly compared to the combined control charting scheme of Kim et al. (2003).

2.2.1.1.5 Niaki et al.'s (2007) Generalized Linear Test

Niaki et al. (2007) proposed a control chart based on the generalized linear test (GLT) to monitor the coefficients of the linear profiles in conjunction with an R-chart to monitor the error variance. The GLT chart statistics are F statistics in the form:

$$F_j = \frac{\sum_{i=1}^{n_j} d_{ij}^2 - SSE_j}{MSE_j},\ j = 1,\ 2,\ \ldots \tag{2.45}$$

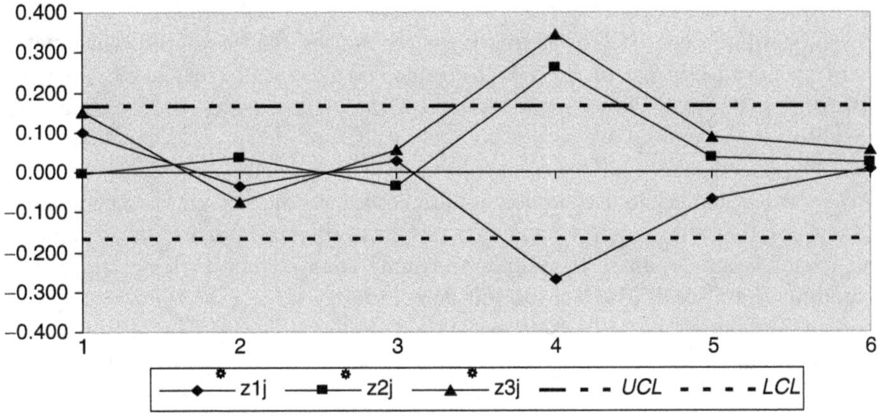

Figure 2.16 NIST chart for monitoring calibration for Example 2.11.

where $SSE_j = \sum_{i=1}^{n_j} (Y_{ij} - a_j - b_j X_{ij})^2$ is the residual sum of squares, $MSE_j = \frac{SSE_j}{n_j - 2}$, and $d_{ij} = \sum_{i=1}^{n_j} (Y_{ij} - A_0 - B_0 X_{ij})$, the nonstandardized deviations from the in-control line. To monitor the intercept and slope of the jth profile, Niaki et al. (2007) proposed plotting the F_j statistic in Equation (2.45) in a chart with an upper control limit of

$$UCL = F_{2,n-2,1-\alpha},$$

assuming that $n_j = n$, j=1, 2, ... The R-chart used for monitoring the process variance is with chart statistic defined in Equation (2.42).

Niaki et al. (2007) proposed also using this GLT statistic in the Phase I application of simple linear profile by replacing the known parameters by initial estimates of the regression coefficients. They compared the performance of the GLT and R-combination approach to the EWMA and R- combination approach of Kang and Albin (2000) using simulation. They concluded that their proposed GLT and R-charts approach is better than the competing approach only when simultaneous shifts in the regression coefficients are considered. Under a sustained shift in a single parameter, the GLT and R approach performs poorer than the competing approach.

Example 2.12 The last column of Table 2.6 gives the F statistics calculated using Equation (2.45) when applied to the data set of Example 2.9. The upper control limit of these F statistics is set at $UCL = F_{2,4-2,1-0.005/2} = 399$. Thus, if we use the R-chart in Figure 2.15 in conjunction with this GLT chart the resulting in-control ARL is approximately 200. It can be shown that all the calculated F statistics are below the control limit, indicating no signal from the GLT chart. Only the R-chart signals a shift at sample 26 as shown in Figure 2.15. Thus, the GLT chart did not detect the indicated shift in this example.

2.2.1.1.6 Zou et al.'s (2007b) Self-Starting Control Chart Approach

A self-starting control chart approach for monitoring the three profile parameters was proposed by Zou et al. (2007b), which avoids, to the extent possible, the distinction between Phase I and Phase II. This control chart based on recursive residuals can be used when the nominal values of the process parameters are unknown. Zou et al. (2007b) assumed that $m-1$ in-control historical samples of size n are available and used to obtain initial estimates of the regression parameters. Then a self-starting control chart that has flexibility in updating the parameters estimates with new Phase II samples and simultaneously checks for out-of-control conditions are used. To describe this approach let $y_{(j-1)n+i} = Y_{ij}$, $i = 1, 2, \ldots, n$, $j = m, m+1, \ldots$. Assume also that $\mathbf{y}_{(j-1)n+i-1} = (y_{(m-1)n+1}\, y_{(m-1)n+2} \ldots y_{(m-1)n+n}\, y_{mn+1}\, y_{mn+2} \ldots y_{mn+n} \ldots y_{(j-1)n+1}\, y_{(j-1)n+2} \ldots y_{(j-1)n+i-1})^T$ and $\mathbf{x}_{(j-1)n+i-1} = (\underbrace{X_1 X_2 \ldots X_n X_1 X_2 \ldots X_n \ldots X_1 X_2 \ldots X_{i-1}}_{(j-1)n+i-1})^T$ are two vectors that contain all the first $(j-1)$ Phase II samples and the first $(i-1)$ observations of the jth Phase II sample pooled into one sample of size $[(j-1)n+i-1]$. Zou et al. (2007b) defined the standardized recursive residuals for the future samples as:

$$e_{ij} = \frac{y_{(j-1)n+i} - (1\ X_i)\hat{\boldsymbol{\beta}}_{(j-1)n+i-1}}{S_{(j-1)n+i-1}\sqrt{1 + (1\ X_i)\mathbf{X}^T_{(j-1)n+i-1}\mathbf{X}_{(j-1)n+i-1}(1\ X_i)^T}}, \quad i = 1, 2, \ldots, n,$$
$$j = m, m+1, \ldots$$

where $\hat{\boldsymbol{\beta}}_{(j-1)n+i-1}$ and $S_{(j-1)n+i}$ are the regression coefficients estimates and square root of the mean squared errors value, respectively, resulting from regressing $\mathbf{y}_{(j-1)n+i-1}$ on $\mathbf{x}_{(j-1)n+i-1}$. Here $\mathbf{X}_{(j-1)n+i-1} = \begin{pmatrix}\mathbf{1} & \mathbf{x}_{(j-1)n+i-1}\end{pmatrix}$, where $\mathbf{1}$ is a unit column vector of size $[(j-1)n+i-1]$. It is well known that under in-control conditions, e_{ij} has a Student-t distribution with $[(j-1)n+i-3]$ degrees of freedom. Zou et al. (2007b) proposed transforming this Student-t statistic by the following statistic:

$$w_{ij} = \Phi^{-1}[T_{(j-1)n+i-3}(e_{ij})], \quad i = 1, 2, \ldots, n, \quad j = m, m+1, \ldots \tag{2.46}$$

where T_v is the CDF of the Student-t distribution with v degrees of freedom and $\Phi^{-1}[.]$ is the inverse of the CDF of the standard normal distribution. For the transformed residuals (w_{ij}, $i = 1, 2, \ldots, n$ and $j = m, m+1, \ldots$) let $\bar{w}_j = \sum_{i=1}^{n} \frac{w_{ij}}{n}$ and $Sw_j = \sum_{i=1}^{n} \frac{(w_{ij}-\bar{w}_j)^2}{n-1}$. The two EWMA statistics used to monitor the regression coefficients and variance used by Zou et al. (2007b) are

$$EWMA_{IS(j)} = \theta\sqrt{n}\bar{w}_j + (1-\theta)EWMA_{IS(j-1)}$$

Table 2.8 The Control Limits of the Self-Starting Control Chart (Adapted from Zou et al. 2007b.)

		In-Control ARL				
Sample Size n	Control Limit	200	300	370	400	500
3	UCL_{IS}	0.9250	0.9741	0.9982	1.0066	1.0320
	UCL_σ	1.3596	1.4577	1.5064	1.5243	1.5769
4	UCL_{IS}	0.9276	0.9746	0.9978	1.0062	1.0302
	UCL_σ	1.2959	1.3794	1.4214	1.4369	1.4812
5	UCL_{IS}	0.9271	0.9748	0.9985	1.0071	1.0303
	UCL_σ	1.2530	1.3318	1.3717	1.3864	1.4280
6	UCL_{IS}	0.9268	0.9745	0.9986	1.0066	1.0311
	UCL_σ	1.2235	1.2983	1.3372	1.3501	1.3895
7	UCL_{IS}	0.9270	0.9746	0.9985	1.0067	1.0310
	UCL_σ	1.2021	1.2739	1.3107	1.3235	1.3615
8	UCL_{IS}	0.9271	0.9746	0.9983	1.0067	1.0310
	UCL_σ	1.1856	1.2549	1.2902	1.3029	1.3397
9	UCL_{IS}	0.9271	0.9745	0.9982	1.0070	1.0310
	UCL_σ	1.1720	1.2396	1.2739	1.2868	1.3233
10	UCL_{IS}	0.9270	0.9745	0.9982	1.0070	1.0311
	UCL_σ	1.1607	1.2269	1.2605	1.2731	1.3078
11	UCL_{IS}	0.9270	0.9748	0.9982	1.0067	1.0310
	UCL_σ	1.1236	1.1860	1.2168	1.2284	1.2601
12	UCL_{IS}	0.9268	0.9747	0.9982	1.0072	1.0313
	UCL_σ	1.1035	1.1621	1.1918	1.2032	1.2340

and

$$EWMA_{\sigma(j)} = \max\left(0, \theta\sqrt{\frac{n-1}{2}}(Sw_j - 1) + (1-\theta)EWMA_{\sigma(j-1)}\right),$$
$$j = m, m+1, \ldots, \tag{2.47}$$

where $EWMA_{IS(m-1)} = 0$, $EWMA_{\sigma(m-1)} = 0$, and $0 < \theta \leq 1$ is a smoothing constant. The method signals a shift in the intercept or slope at the jth future sample if $|EWMA_{IS(j)}| > UCL_{IS}$ and an increase in the process variance if $EWMA_{\sigma(j)} > UCL_\sigma$, where UCL_{IS} and UCL_σ are appropriate control limits. Using a Markov Chain approach, Zou et al. (2007b) calculated the values of UCL_{IS} and UCL_σ according to different values of the in-control ARL and sample size n. Table 2.8 show these values.

2.2.1.1.7 Control Charts Based on Likelihood Ratios

Several control charts based on likelihood ratios to monitor simultaneously the three simple linear profile parameters were proposed in the literature. These charts are presented below.

2.2.1.1.7.1 Zhang et al.'s (2009) Likelihood Ratio Approach Zhang et al. (2009) proposed a control chart based on the likelihood ratio for monitoring simple linear profiles. Their proposed chart integrates the EWMA procedure to detect shifts in the intercept, slope, or standard deviation. They assumed without loss of generality that $\sigma^2 = 1$ and $\bar{X} = 0$. The first step in their approach is to obtain the EWMA statistics of the current process intercept, slope, and variance. These EWMA statistics are used as estimates of the process parameters. Following are the EWMA statistics for the intercept, slope, and variance:

$$\begin{aligned} EI_j &= \theta a_j + (1-\theta)EI_{j-1}, \\ ES_j &= \theta b_j + (1-\theta)ES_{j-1}, \quad \text{and} \\ EE_j &= \theta S_j^* + (1-\theta)EE_{j-1}, \quad j = 1, 2, \ldots \end{aligned}$$

respectively, where $S_j^* = \sum_{i=1}^{n} (Y_{ij} - ES_j X_i - EI_j)^2$, $EI_0 = A_0$, $ES_0 = B_0$, $EE_0 = 1$, and $0 < \theta \leq 1$ is the smoothing parameter. The second step is to calculate the likelihood ratio chart statistic. This statistic is in the form

$$ELR_j = EC_j - n \log EE_j - n, \quad j = 1, 2, \ldots, \tag{2.48}$$

where EC_j is an EWMA statistic in the form

$$EC_j = \theta C_j + (1-\theta)EC_{j-1}, \tag{2.49}$$

where $EC_0 = n$, $0 < \theta \leq 1$, and $C_j = \sum_{i=1}^{n} (Y_{ij} - B_0 X_i - A_0)^2$.

This chart signals a shift in the process intercept, slope, or variance or combination of them if the statistic $ELR_j > h$, where h is an appropriate control limit that gives a specified in-control *ARL*.

Example 2.13 Zhang et al. (2009) used the optical imaging system data set in Table 2.7 to illustrate the use of their proposed approach. They first divided all the Y-values and X-values by $\sigma = 0.06826$. The transformed in-control linear profile model is in the form $Y_{ij}^* = 4.1269 + 14.3085X_i^*$, where $Y_{ij}^* = \frac{Y_{ij}}{0.06826}$ and $X_i^* = \frac{X_i}{0.06826}$. The transformed X- and Y-values are presented in Table 2.9.

The appropriate control limit h that gives a specified in-control *ARL* of 200 is $h = 1.752$ as calculated by Zhang et al. (2009) by simulation. The last five columns of Table 2.9 give the EWMA statistics and the likelihood ratio chart statistic. At sample 4, the $ELR_j = 6.379$ as shown in the last column of Table 2.9 that indicates a shift in the process.

2.2.1.1.7.2 Zou et al.'s (2006) Change Point Approach Zou et al. (2006) considered a change point method for detecting changes in the parameters of a Phase II simple linear regression model when the nominal values of the process parameters

Table 2.9 Transformed Data Set of Example 2.11 (Adapted from Zhang et al. 2009.)

	Standards of X			Chart Statistics				
Profile Number	Lower 11.134	Medium 48.198	Upper 130.237	EI_j	ES_j	EE_j	EC_j	ELR_j
1	16.408	51.128	133.460	66.075	14.309	1.123	3.705	0.357
2	14.503	51.714	130.237	65.957	14.309	1.031	3.304	0.231
3	15.382	50.689	132.142	65.980	14.326	0.881	2.857	0.236
4	11.134	54.937	136.244	66.272	14.510	3.231	12.897	6.379
5	14.064	51.714	132.581	66.241	14.519	2.616	10.859	4.974
6	15.089	51.568	132.142	66.246	14.494	2.115	8.848	3.600

are not known but some historical samples are available. This change point method can be described as follows:

Suppose that there are m in-control historical samples that are used to estimate the unknown regression parameters. Assume that t Phase II (future) profile samples has been collected and let $k = t + m$. Zou et al. (2006) defined the likelihood ratio statistic for testing a change in one or more of the profile parameters after sample k_1, where $m < k_1 < k$ as

$$lr(k_1 n, kn) = kn.\log[\hat{\sigma}^2_{kn}(\hat{\sigma}^2_{k_1 n})^{\frac{-k_1}{k}}(\hat{\sigma}^2_{k_2 n})^{\frac{-k_2}{k}}], \tag{2.50}$$

where $k_2 = k - k_1$, $\hat{\sigma}^2_{kn}$ is the MLE of the error term variance for the regression model fitted for the k samples pooled into one sample of size kn, $\hat{\sigma}^2_{k_1 n}$ is the MLE of the error term variance for the regression model fitted for all the samples prior to k_1 pooled into one sample of size nk_1 and $\hat{\sigma}^2_{k_2 n}$ is the MLE of the error term variance for the regression model fitted for all the samples following k_1 pooled into one sample of size nk_2. Since the expected value and variance of the likelihood ratio statistic in Equation (2.50) varies by k_1, Zou et al. (2006) suggested using the standardized likelihood ratio statistic which is defined as

$$slr(k_1 n, kn) = \frac{lr(k_1 n, kn) - E[lr(k_1 n, kn)]}{\sqrt{Var[lr(k_1 n, kn)]}}, \tag{2.51}$$

where the expectation and variance of the likelihood ratio statistic can be approximated by

$$E[lr(k_1 n, kn)] = k_1 n[\log\frac{k_1 n}{2} - \psi_0(\frac{k_1 n - 2}{2})],$$

$$Var[lr(k_1 n, kn)] = (k_1 n)^2 \psi_1\left(\frac{k_1 n - 2}{2}\right) - 2k_1 n,$$

$$\psi_0(I) = -\gamma + \sum\nolimits_{j=1}^{I-1}\frac{1}{j},\ \psi_0(I + 0.5) = -\gamma - 2\log 2 + 2\sum\nolimits_{j=1}^{I}\frac{1}{2j-1},$$

$\psi_1(I) = \frac{\pi^2}{6} - \sum_{j=1}^{I-1} \frac{1}{j^2}$, $\psi_1(I+0.5) = \frac{\pi^2}{2} - 4\sum_{j=1}^{I} \frac{1}{(2j-1)^2}$, and $\gamma = 0.577215664\ldots$ is the Euler–Mascheroni constant and I is an integer. To increase the efficiency of the change point approach to small shifts, Zou et al. (2006) proposed using an EWMA statistic to cumulate the $slr\ (k_1 n,\ \ kn)$ statistic. The EWMA statistic is in the form $Y_j = \max(0,\ \ \theta slr(k_1 n, kn) + (1-\theta)Y_{j-1}), j = m$, $m+1, \ldots, k-1$, where $Y_0 = 0$ and $0 < \theta \leq 1$ is a smoothing constant. The test statistic for testing a change in the Phase II data set is

$$Y_{\max} = \max_{m \leq j < k} (Y_j). \tag{2.52}$$

This approach signals a change in one or more of the profile parameters if $Y_{\max} > h_t$, where h_t is an appropriate threshold to give a specified in-control ARL. The values of h_t for different values of t and different in-control ARL values when $m = 10$ and 50 and when $n = 4$ were calculated by Zou et al. (2006). These values are given in Table 2.10. Zou et al. (2006) proposed plotting the statistic $\frac{Y_{\max}}{h_t}$ in a chart with $UCL = 1$ to detect changes in the profile parameters. They applied their proposed approach on a simulated data set using the in-control model discussed in Example 2.9.

2.2.1.2 *Individual Control Charting Approaches*

Interpretation of out-of-control signals is a major concern of all types of omnibus control charts described above. If a signal is given when applying an omnibus monitoring scheme, it is not easy to determine the source of variation. Therefore, some authors suggested transforming the X-values in a way that leads to independent estimators of the regression parameters. In this case, each regression parameter can be monitored individually using a univariate control chart.

2.2.1.2.1 *Kim et al.'s (2003) EWMA Charts Approach*

Kim et al. (2003) proposed another method for monitoring a Phase II linear profile process assuming known parameter values. Their idea was to code the X-values within each sample as described in Section 2.1.2. Since coding the X-values this way leads to independent regression estimators, Kim et al. (2003) recommended monitoring the two regression coefficients using separate EWMA charts. They also recommended replacing the R-chart of Kang and Albin (2000) by EWMA charts for monitoring the process standard deviation, including one proposed by Crowder and Hamilton (1992). A signal is produced as soon as any one of the three EWMA charts for the Y-intercept, the slope, and the variation about the regression line produces an out-of-control signal.

The proposed EWMA chart statistic for monitoring the Y-intercept A is

$$EWMA_{I(j)} = \theta\left(\frac{a_j - A_0}{\sigma_0/\sqrt{n}}\right) + (1-\theta)EWMA_{I(j-1)}, \tag{2.53}$$

where $0<\theta\leq 1$ is a smoothing constant and $EWMA_{I(0)} = 0$. An out-of-control signal is given by this chart as soon as $\left|EWMA_{I(j)}\right| > h_I$, where $h_I > 0$ is a constant selected to

Table 2.10 The h_t Values of the EWMA Statistic in Equation (2.52) for $n = 4$ (Adapted from Zou et al. 2006.)

	$m = 10$				$m = 50$			
	In-Control ARL							
t	100	200	370	500	100	200	370	500
1	0.695	0.828	0.938	0.992	0.695	0.828	0.953	0.992
2	0.969	1.125	1.266	1.344	0.969	1.125	1.266	1.344
3	1.219	1.406	1.594	1.660	1.219	1.422	1.594	1.695
4	1.422	1.656	1.875	1.977	1.438	1.688	1.891	1.994
5	1.578	1.844	2.094	2.223	1.609	1.906	2.125	2.258
6	1.719	2.031	2.281	2.398	1.750	2.062	2.344	2.504
7	1.812	2.156	2.438	2.609	1.875	2.219	2.531	2.680
8	1.906	2.250	2.594	2.750	1.969	2.344	2.656	2.820
9	1.969	2.344	2.688	2.855	2.047	2.438	2.812	2.961
10	2.031	2.438	2.781	2.961	2.125	2.562	2.906	3.102
11	2.078	2.500	2.875	3.066	2.188	2.625	3.000	3.172
12	2.125	2.562	2.938	3.137	2.234	2.688	3.094	3.277
13	2.172	2.625	3.000	3.207	2.266	2.750	3.156	3.348
14	2.203	2.656	3.062	3.242	2.297	2.781	3.203	3.383
15	2.250	2.719	3.109	3.312	2.328	2.812	3.250	3.453
16	2.266	2.750	3.156	3.348	2.359	2.844	3.281	3.523
17	2.297	2.781	3.188	3.383	2.391	2.875	3.344	3.559
18	2.312	2.812	3.234	3.418	2.422	2.938	3.375	3.594
19	2.328	2.844	3.281	3.488	2.438	2.969	3.406	3.629
20	2.359	2.875	3.312	3.523	2.469	3.000	3.438	3.664
22	2.375	2.906	3.375	3.594	2.484	3.031	3.500	3.734
24	2.406	2.938	3.406	3.629	2.516	3.062	3.531	3.770
26	2.438	2.969	3.438	3.664	2.531	3.094	3.562	3.805
28	2.453	3.000	3.469	3.699	2.547	3.125	3.594	3.840
30	2.469	3.031	3.500	3.734	2.562	3.156	3.656	3.910
35	2.531	3.094	3.594	3.840	2.594	3.188	3.719	3.980
40	2.562	3.156	3.656	3.910	2.623	3.219	3.750	4.051
50	2.609	3.188	3.719	3.980	2.641	3.250	3.812	4.086
60	2.625	3.219	3.781	4.016	2.656	3.281	3.844	4.121
70	2.641	3.250	3.812	4.051	2.672	3.312	3.875	4.156
80	2.656	3.281	3.844	4.086	2.688	3.328	3.891	4.191
90	2.680	3.312	3.875	4.121	2.703	3.344	3.906	4.221
115	2.703	3.344	3.906	4.191	2.719	3.359	3.938	4.247
140	2.719	3.375	3.938	4.227	2.727	3.375	3.969	4.262
165	2.734	3.391	3.969	4.262	2.734	3.391	4.000	4.279
190	2.750	3.406	3.984	4.297	2.742	3.406	4.016	4.297
240	2.773	3.422	4.000	4.314	2.750	3.422	4.031	4.314
290		3.438	4.031	4.332		3.438	4.039	4.332
390		3.469	4.047	4.350		3.469	4.047	4.350
490			4.062	4.367			4.062	4.367

produce a specified in-control *ARL*. The constant limits $\pm h_I$ are usually referred to as "asymptotic control limits" in the literature. Alternatively, one can use time-varying control limits, which are more effective in detecting shifts in a process parameter that occur in the initial stages; see, for example, Lucas and Saccucci (1990).

The proposed EWMA statistic for monitoring the slope B is given by

$$EWMA_{S(j)} = \theta \left(\frac{b_j - B_0}{\sigma_0/\sqrt{S_{XX}}} \right) + (1-\theta) EWMA_{S(j-1)}, \tag{2.54}$$

where $0<\theta\leq 1$ is again a smoothing constant and $EWMA_{S(0)} = 0$. A signal is produced by this chart as soon as $|EWMA_{S(j)}| > h_S$, where $h_S > 0$ is a constant chosen to produce a specified in-control *ARL*.

It should be noted that, except for the study by Mahmoud et al. (2010), none of the studies of EWMA charts based on regression parameter estimators used standardized values as in Equations (2.53) and (2.54). When using the standardized regression estimates values in the EWMA chart for either regression parameter, the chart statistic follows the standard normal distribution when the process is in-control. As long as the shifts in parameter values are also standardized, one will have the same *ARL* performance regardless of the values of A_0, B_0, σ_0, n, and S_{XX} used in the profile model.

To monitor the process variance, Kim et al. (2003) proposed an EWMA chart based on the mean squared errors of the *jth* profile. The EWMA statistic for monitoring σ^2 is

$$EWMA_{E(j)} = \max\{\theta(\ln MSE_j) + (1-\theta) EWMA_{E(j-1)}, \ln \sigma_0^2\}, \tag{2.55}$$

where $0 < \theta \leq 1$ is again a smoothing constant and $EWMA_{E(0)} = ln\sigma_0^2$. The control limit for detecting increases in the process variance used by Kim et al. (2003) is

$$UCL = \ln \sigma_0^2 + L_E[\theta \operatorname{Var}(\ln MSE_j)/(2-\theta)]^{1/2}, \tag{2.56}$$

where the multiplier $L_E > 0$ is chosen to produce a specified in-control *ARL* and

$$\operatorname{Var}(\ln MSE_j) \approx \frac{2}{n-2} + \frac{2}{(n-2)^2} + \frac{4}{3(n-2)^3} - \frac{16}{15(n-2)^5}. \tag{2.57}$$

Example 2.14 To illustrate the use of the three EWMA charts of Kim et al. (2003) to monitor Phase II simple linear profile data we use the data set in Example 2.9. Again, a smoothing parameter of $\theta = 0.2$ is used to obtain the charts statistics. The control limits for the EWMA charts of both the intercept and slope were set at ± 1.005 while the upper control limit for the variance chart was set at 0.586. Using 50,000 simulations, the in-control ARL produced by this method when these limits are used is 200. The third, fourth, and fifth columns of Table 2.6 show the $EWMA_{I(j)}$, $EWMA_{S(j)}$, and $EWMA_{E(j)}$ statistics calculated for the data set in Example 2.9. Figures 2.17–2.19 show EWMA control charts for the variance, slope, and intercept, respectively. Only

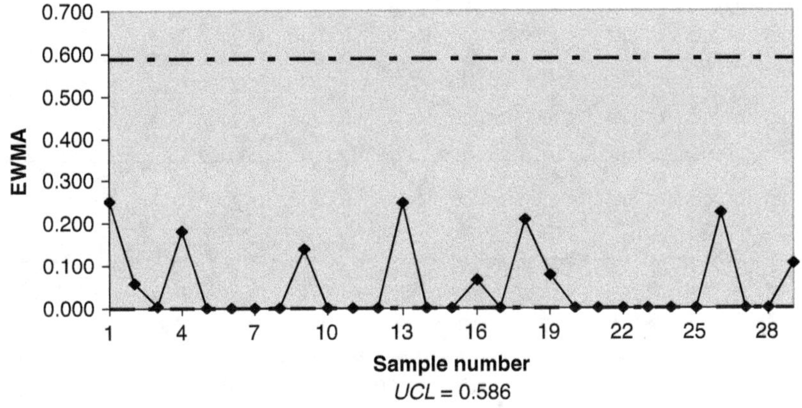

Figure 2.17 EWMA chart for the variance applied to the data set in Example 2.9.

the control chart for the slope signals a shift in the process at sample 25. The other two charts did not signal.

Two important observations are obtained from this example. First, Kim et al. (2003) method is more efficient in detecting the assigned shift in the slope than the previous two approaches, as it detected it faster. Second, Kim et al. (2003) approach is much more interpretable than the omnibus methods described in this chapter. Using Kim et al. (2003) approach we can conclude that the out-of-control condition is attributable to shifts in the process slope, since only the slope chart has signaled.

2.2.1.2.2 Saghaei et al.'s (2009b) CUSUM Charts

An alternative approach to the three EWMA charts scheme of Kim et al. (2003) was proposed by Saghaei et al. (2009b) to monitor Phase II simple linear profiles. They

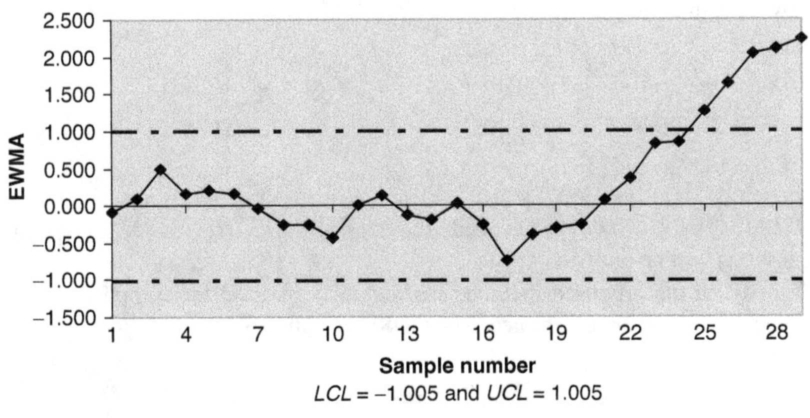

Figure 2.18 EWMA chart for the slope applied to the data set in Example 2.9.

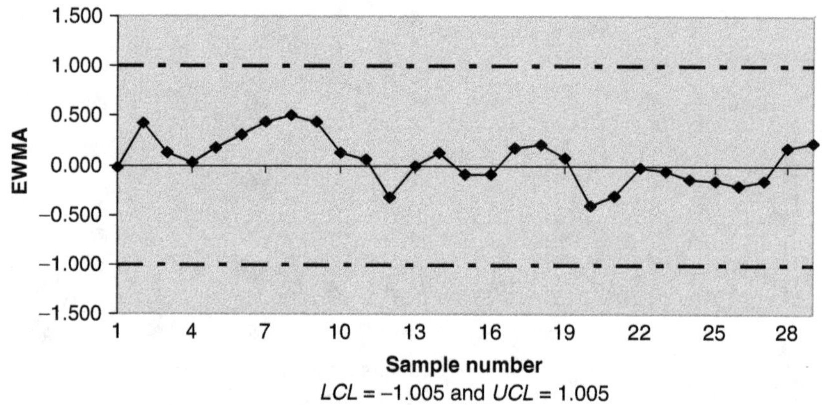

Figure 2.19 EWMA chart for the intercept applied to the data set in Example 2.9.

proposed replacing the three EWMA charts for monitoring the regression parameters by three Cumulative Sums (CUSUM) charts. The CUSUM chart for monitoring the intercept A is as follow:

$$CUSUM_I^+(j) = \max[0, a_j - (A_0 + K_I) + CUSUM_I^+(j-1)]$$
$$CUSUM_I^-(j) = \max[0, (A_0 - K_I) - a_j + CUSUM_I^-(j-1)], \qquad (2.58)$$

where $CUSUM_I^+(0) = CUSUM_I^-(0) = 0$, and $K_I = \Delta_I/2$ is called the reference parameter value, where Δ_I is the size of the intercept shift important to be detected quickly (usually $K_I = 0.5$). The process is out-of-control if either $CUSUM_I^+(j) > H_I^+$ or $CUSUM_I^-(j) < H_I^-$, where H_I^+ and H_I^- are chosen to give a specified in-control *ARL*.

The CUSUM statistics used to detect changes in the slope are computed by the following equations:

$$CUSUM_S^+(j) = \max[0, b_j - (B_0 + K_S) + CUSUM_S^+(j-1)]$$
$$CUSUM_S^-(j) = \max[0, (B_0 - K_S) - b_j + CUSUM_S^-(j-1)], \qquad (2.59)$$

where $CUSUM_S^+(0) = CUSUM_S^-(0) = 0$ and K_S is the slope reference value. If either $CUSUM_S^+(j)$or $CUSUM_S^-(j)$ exceeds the control limits, H_S^+ or H_S^-, respectively, the process will be out-of-control.

Finally, in the proposed method, MSE_j statistic is used for computing *CUSUM* statistic to monitor error variance. The CUSUM statistics are:

$$CUSUM_E^+(j) = \max[0, MSE_j - K_E + CUSUM_E^+(j-1)]$$
$$CUSUM_E^-(j) = \min[0, MSE_j - K_E + CUSUM_E^-(j-1)], \qquad (2.60)$$

where $CUSUM_E^+(0) = CUSUM_E^-(0) = {\sigma_0}^2$, K_E is the error variance reference value and ${H_E}^+$ and ${H_E}^-$ are the decision intervals for error variance control charts. The process is out-of-control if either $CUSUM_E^+(j)$ or $CUSUM_E^-(j)$ exceeds the decision interval ${H_E}^+$ or ${H_E}^-$, respectively.

It is well known that both the CUSUM and EWMA schemes can be designed to have roughly equivalent steady state *ARL* performance by choosing suitable smoothing constants for the EWMA charts and reference values for the CUSUM charts. However, the CUSUM approach has a much better worst-case performance than the EWMA approach, as shown by Woodall and Mahmoud (2005). These authors argued that the EWMA charts could become less effective in reacting to some out-of-control shifts in the process mean. For instance, a trend may occur in the downward direction on the EWMA chart immediately before an upward shift occurs in the process parameter. As a result, the value of the control chart statistic may be close to the lower control limit when the upward shift occurs. Consequently, the chart may take much longer time to produce an out-of-control signal than if the chart statistic were close to the centerline or close to the upper control limit when the shift occurred. Woodall and Adams (1998) referred to this as the "inertia problem" of a control chart. The term inertia in SPC is used to refer to a measure of the resistance that a chart has to signal a particular shift. The CUSUM chart uses resets (to zero) and do not have such a significant inertia problem as the EWMA chart. Thus, in terms of worst-case performance, monitoring schemes based on CUSUM charts are usually recommended over those based on EWMA charts.

2.2.2 Performance Comparisons of Phase II Simple Linear Profile Approaches

Using simulation, Kim et al. (2003) compared the performance of their proposed approach described in Equations (2.55–2.57) to the Phase II methods of Kang and Albin (2000). The *ARL* performance of their three EWMA charts used to monitor separately the Y-intercept, the slope, and the variation about the regression line was shown to be superior to that of the Phase II methods of Kang and Albin (2000) for sustained shifts in the regression parameters. For example, Table 2.11 gives the *ARL* values of the three competing methods when the slope shifts from B_1 to $B_1 + \delta$ using simulated profiles with in-control regression line of $Y_{ij} = 3 + 2X_i + \varepsilon_{ij}$, where the

Table 2.11 ARL Comparisons Under Slope Shifts from B_1 to $B_1+\delta$ (Adapted from Kim et al. 2003.)

	δ									
Chart	0.025	0.050	0.075	0.100	0.125	0.150	0.175	0.200	0.225	0.250
EWMA/R	119.0	43.9	19.8	11.3	7.7	5.8	4.7	3.9	3.4	3.0
T^2	166.0	105.6	60.7	34.5	20.1	12.2	7.8	5.2	3.7	2.7
Kim et al. (2003)	101.6	36.5	17.0	10.3	7.2	5.5	4.5	3.8	3.3	2.9

ε_{ij} are i.i.d. $N(0,1)$ random variables. The *ARL* values in this table were derived by Kim et al. (2003) using simulated Phase II data sets. As shown in Table 2.11, the Kim et al. (2003) approach is superior to the competing approaches for almost all values considered for the shift size δ. Kim et al. (2003) argued also that their method is much more interpretable since each parameter in the model is monitored with a separate control chart.

Also, Gupta et al. (2006) compared the performance of the Kim et al. (2003) method with the Croarkin and Varner's (1982) control chart approach with statistic described in Equation (2.44). Their simulation study showed that the Croarkin and Varner (1982) method performed poorly compared to the combined control charting scheme of Kim et al. (2003).

Most of the performance comparisons studies in the literature of simple linear profile monitoring only investigated the proposed Phase II methods under step shifts. However, other types of shifts such as drift can occur in the process as well. Saghaei et al. (2009a) investigated using simulation the performance of the Kang and Albin's (2000) T^2, the Kang and Albin's (2000) EWMA-R charts, and the Kim et al. (2003) three EWMA charts under drift shift. They concluded that the three EWMA charts scheme of Kim et al. (2003) performs better than the other methods under very small to medium drift shifts. In medium-to-large drift shifts in the intercept and slope the T^2 method roughly performs better than the other methods. However, for standard deviation shift, the EWMA-R approach of Kang and Albin (2000) performs the best. In addition, for very large shifts, the performance of the three methods is approximately the same.

2.3 SPECIAL CASES AND AN IMPORTANT APPLICATION

In this section, we discuss control charting approaches for monitoring simple linear profiles with low sample size per sample. We also present the profile monitoring approach of Wang and Tsung (2005) for monitoring quintile–quintile (Q–Q) plot.

2.3.1 Profile Monitoring Approaches for Data Sets with Low Sample Sizes

As mentioned in Section 2.1, profile data sets with more than two observations ($n_j > 2$) per subgroup are needed to estimate the three regression parameters, the intercept, slope, and variance. All the abovementioned studies of statistical performance of Phase I or Phase II simple linear regression profile methods were reported for the case when $n_j > 2$. However, in practice the available sample size maybe less than three observations per sample. Many reasons may contribute to the low sample size of the observed product or process profile including, for example, high cost of measurements and low production volume. Wade and Woodall (1993) considered a special case of the Phase II simple linear regression profile when the available sample size is $n = 1$ per sample. Mahmoud et al. (2010) evaluated and compared the performance of Phase II simple linear regression profile approaches when only two observations are used to establish each profile. They also proposed an EWMA control chart based on average squared deviations from the in-control line, to be used in conjunction with two EWMA control charts based on the slope and Y-intercept estimators, to monitor changes in the slope, intercept, and variance when $n = 2$. In the following section, we discuss both cases.

2.3.1.1 *The Monitoring of Simple Linear Profiles When n = 1*

An important application of the simple linear regression profile monitoring with a single bivariate observation, i.e., $n = 1$, was studied by Wade and Woodall (1993). These authors considered the simple linear regression model described in Equation (2.1) to account for the effect of an input quality variable X on the output quality variable Y when monitoring a particular stage of a manufacturing process. They used one of the cause-selecting control charts considered by Zhang (1992) to monitor the relationship between incoming and outgoing quality measurements of a component part for the braking system for an automobile. The chart statistics in a cause-selecting control chart are based on the standardized deviations from the in-control profile, d_j, where

$$d_j = \frac{Y_j - A_0 - B_0 X_j}{\sigma_0}, \quad j = 1, 2, \ldots.$$

The cause-selecting control chart is a Shewhart-type control chart with the following control limits:

$$\begin{aligned} UCL &= \bar{d} + 2.66\,\overline{MR} \\ LCL &= \bar{d} - 2.66\,\overline{MR} \\ \overline{MR} &= \frac{\sum_{i=1}^{m-1} |d_{i+1} - d_i|}{m-1}. \end{aligned}$$

The values of $\bar{d}$ and $\overline{MR}$ are calculated from m in-control Phase I bivariate observations of the input and output quality variables, i.e., (X_j, Y_j), $j = 1, 2, \ldots, m$. It can be easily shown that this cause-selecting control chart and the EWMA chart of Kang and Albin (2000) described in Equation (2.41) are based on the same standardized deviations when $n = 1$. Wade and Woodall (1993) proposed using this cause-selecting control chart in conjunction with Shewhart charts for individuals for both the input and output quality variables to explain out-of-control signals.

2.3.1.2 *The Monitoring of Simple Linear Profiles When n = 2*

When we use profile samples with $n = 2$ we can calculate only the estimators of the intercept and slope. No within-sample estimator for the profile variance is available in this case. Mahmoud et al. (2010) investigated the performance and limitations of Phase II profile approaches when $n = 2$. They assumed that the available observations in each sample are (X_1, Y_1) and (X_2, Y_2), such that $X_1 = -X_2 = X$. In this case, the estimated Y-intercept and slope are $\frac{Y_1+Y_2}{2}$ and $\frac{Y_1-Y_2}{2X}$, respectively. Also in this case, the quantity S_{XX} simplifies to $2X^2$. However, with only two observations the mean squared errors MSE_j cannot be calculated, and therefore no sample-specific estimator of the process variance can be given. Thus, the third EWMA control chart proposed by Kim et al. (2003) and described in Equation (2.55) cannot be used to monitor the process variance. It should be also noted that in this case the range of the two deviations is equivalent to the deviation of the slope estimator from the in-control slope. Thus, the *R*-chart proposed by Kang and Albin (2000) is not useful for monitoring the process standard deviation when $n = 2$. In this case, the *R*-chart serves as a Shewhart chart for monitoring the slope of the profile model. If in a particular application the variance is not expected to change from sample to sample, one can use the two EWMA control charts proposed by Kim et al. (2003) to detect changes in the Y-intercept and slope of the model based on the statistics in Equations (2.53) and (2.54). In this case, one can also use the EWMA chart based on average deviations from the in-control line in conjunction with the *R*-chart proposed by Kang and Albin (2000).

2.3.1.3 *An Alternative Phase II Approach for Monitoring Profiles with n = 2*

Mahmoud et al. (2010) argued that the uniformly most powerful test for the process variance in the case of a simple linear profile with two observations can be based on the average squared deviations $\sum_{i=1}^{2} d_{ij}^2 \Big/ 2$, where d_{ij} is as defined in Equation (2.39). In this case, to detect increases in process variance they recommended using an EWMA chart based on these average deviations. The EWMA control statistic for this chart is

$$EWMA(d^2)_j = \theta \sum_{i=1}^{2} d_{ij}^2 \Big/ 2 + (1-\theta) \quad \max\{EWMA(d^2)j - 1,\ 1\}, \qquad (2.61)$$

where $0 < \theta \leq 1$ is again a smoothing constant and $EWMA(d^2)_0 = 1$. A signal is given at sample j if $EWMA(d^2)_j > h_{d^2}$, where h_{d^2} is chosen to give a specified in-control *ARL*. Mahmoud et al. (2010) investigated the performance of this EWMA chart when used in conjunction with either the EWMA chart based on average deviations from in-control line proposed by Kang and Albin (2000) or the two EWMA charts for the intercept or slope proposed by Kim et al. (2003). Their simulation results showed that the combination of two EWMA charts for monitoring the intercept and slope used in conjunction with the EWMA control chart based on average squared deviations statistics is generally more effective than the competing methods for detecting sustained step shifts in the intercept and slope, or increases in the process variance. This combination is also more powerful in detecting simultaneous shifts in the three process parameters.

2.3.2 Control Charting Schemes Based on the Q–Q Plot

Wang and Tsung (2005) used profile monitoring methods to detect changes in a Q–Q plot, which reflected the relationship between the current sample and a baseline sample. The motivation of these authors is that the current practice of monitoring processes with huge sample size is to apply a conventional SPC control chart to the summary statistics of each sample such as sample mean or standard deviation, assuming a univariate quality characteristic. However, some out-of-control situations cannot be detected by a conventional charting method, as the out-of-control data would be averaged out and buried in the huge sample. To monitor univariate quality characteristic using huge sample size, Wang and Tsung (2005) suggested not monitoring the summary statistics of the sample, but characterizing the sample by a Q–Q plot. Then, a linear profile that fits the Q–Q plot of each sample would be monitored via any of the profile charting techniques described in this chapter. The alarms of mean shifts, variance shifts, and distribution shifts are expected to be triggered efficiently and effectively using this control charting scheme based on Q–Q plot.

Suppose that $y_1, y_2, \ldots, y_n$ are n observations from the population distributed as normal with mean μ and standard deviation σ. Let $y_{(1)}, y_{(2)}, \ldots, y_{(n)}$ be the n observations ordered from the smallest to the largest, and let $x_{(1)}, x_{(2)}, \ldots, x_{(n)}$ be the corresponding expected quintiles from the standard normal distribution. To monitor the mean μ and standard deviation σ, Wang and Tsung (2005) proposed monitoring the linear profile model

$$\mathbf{y} = \mu\mathbf{1} + \sigma\mathbf{x} + \boldsymbol{\varepsilon}, \tag{2.62}$$

where $\mathbf{y} = (y_{(1)}y_{(2)} \ldots y_{(n)})^T$, $\mathbf{x} = (x_{(1)}x_{(2)} \ldots x_{(n)})^T$, $E(\boldsymbol{\varepsilon}) = \mathbf{0}$ and $\text{Var}(\varepsilon) = \sigma^2\,\mathbf{V}$, where $\mathbf{V}$ is the $n \times n$ covariance matrix of the quintiles from the standard normal distribution. Wang and Tsung (2005) used the generalized least squares method to fit the model in Equation (2.62), and then proposed three EWMA charts to monitor changes in the intercept, slope, and variance of the profile model in Equation (2.62).

2.4 DIAGNOSTIC STATISTICS

For all of the omnibus control charting approaches described in the previous sections, once an out-of-control signal is given it is difficult to determine the location of the change and which process parameter has indeed shifted. The out-of-control signal indicates that a process change has occurred, but it does not indicate what the special cause is nor does it indicate when the process change actually occurred. Mahmoud et al. (2010) argued that when a chart combination method signals that a special cause is present, it does not provide process engineers with an indication of the source of the unanticipated variation. An important objective of the Phase I and II analyses of linear profiles is to locate the position of the shift after the process has indeed shifted. Also, it is not only important to detect a shift, but also to attribute this shift to a specific model parameter. Locating the position of the change point and classifying its nature would help engineers to identify the underlying cause more quickly. Another important objective of the Phase I analysis is to locate out-of-control samples and remove them in order to obtain a set of data reflecting expected in-control performance. Several

Phase I and II studies in the literature proposed diagnostic statistics to address these objectives. We present here some of these statistics.

2.4.1 Diagnostics Based on Change Point Approach

To determine the source of profile variation, Mahmoud et al. (2007) proposed decomposing their likelihood ratio statistic in Equation (2.33) into three parts representing the relative contributions of the variance, slope, and intercept shifts to this statistic. Assuming that the X-values are the same for all samples and $\bar{X} = 0$, the lrt_{m_1} statistic in Equation (2.32) can be written as:

$$lrt_{m_1} = VAR_{\sigma^2} + VAR_B + VAR_A, \tag{2.63}$$

where $VAR_{\sigma^2} = N\log\{(N_1 r^{2N_2/N} + N_2 r^{-N_1/N})/N\}$,

$$VAR_B = N\log\{1 + (c_2 d_B^2/c_1)\},$$

and

$$VAR_A = N\log\{1 + [c_3 d_A^2/(c_1 + c_2 d_B^2)]\},$$

where $r = \hat{\sigma}_1/\hat{\sigma}_2$, $d_A = \bar{Y}_2 - \bar{Y}_1$, $d_B = S_{XY1}/S_{XX1} - S_{XY2}/S_{XX2}$, $c_1 = N_1\hat{\sigma}_1^2 + N_2\hat{\sigma}_2^2$, $c_2 = S_{xx1}S_{xx2}/S_{xx}$, and $c_3 = N_1N_2/N$. Here, $\bar{Y}_1 = \sum_{i=1}^{N_1} y_i/N_1$, $\bar{Y}_2 = \sum_{i=N_1+1}^{N} y_i/N_2$, $S_{xx} = \sum_{i=1}^{N} x_i^2$, $S_{xx1} = \sum_{i=1}^{N_1} x_i^2$, $S_{xx2} = \sum_{i=N_1+1}^{N} x_i{}^2$, $S_{xy1} = \sum_{i=1}^{N_1} x_i y_i$, and $S_{xy2} = \sum_{i=N_1+1}^{N} x_i y_i$, where (x_i, y_i), $i = 1, 2, \ldots, N$, are $N = nm$ bivariate observations resulting from pooling the m samples into one sample, $N_1 = nm_1$, and $N_2 = n(m - m_1)$. Using this approach, if the change point approach signals a shift in the process, we decompose the $lrt_{\hat{\tau}}$ into the abovementioned three parts, after estimating the location of the change point by the value of m_1 that maximizes the $lrtc_{m1}$ statistic in Equation (2.33). If the value of VAR_{σ^2} is considerably large, we consider this as an evidence that the profile variation is highly attributable to standard deviation shifts. Furthermore, a large value of VAR_B indicates that the regression lines are not parallel. In practice, usually if the regression lines are not parallel, we may not care whether their intercepts are equal. The idea of decomposing the likelihood ratio test statistic into components corresponding to the regression parameters changes has been considered by several authors; see Zou et al. (2006), for example.

Mahmoud et al. (2010) proposed using this change point approach to locate the position of change and the source of profile variation after a Phase II profile monitoring approach signals a change in the relationship between the profile's response and predictor.

2.4.2 Diagnostics Based on F-Statistics

Mahmoud (2008) proposed several F-statistics to determine the locations of out-of-control Phase I samples and the source of profile variation. The first step in this proposed approach is to apply a control chart for monitoring the process variance, such as the one based on the statistic in Equation (2.20). If one or more of the chart statistics are outside the control limits, we consider this as evidence that the variance is not stable. In this case, if assignable causes can be identified, the corresponding samples are to be removed from the data and the method is reapplied. Since the accuracy of the estimators of the in-control regression coefficients relies heavily on the stability of the process variance, we do not test whether the regression coefficients are constant if the variance is shown to be instable.

If the variance is shown to be stable, then the next step is to check whether the regression coefficients are stable. To do that, one may use the global F statistic in Equation (2.23). (Other approaches, such as the Kang and Albin's (2000) T^2 chart, the change point method of Mahmoud et al. (2007), or a combination of these methods may be used as well to determine the stability of the profiles.) If an out-of-control signal is obtained from the global F-test (i.e., the process is out-of-control), then the next step is to test the following two null hypotheses:

H_A: the slopes of the regression lines are all equal (the regression lines are parallel) assuming that the variance is stable, regardless whether the intercept is stable or not.

H_B: the intercepts of the regression lines are all equal assuming that the variance and slopes are stable.

To test H_A, the following F statistic is used:

$$F_S = \{SSE(reduced^*) - SSE(full)\}/[(m-1)MSE(full)], \tag{2.64}$$

where $SSE(full)$ and $MSE(full)$ are the residual sums of squares and the mean square error resulting from fitting the regression model in Equation (2.21), and $SSE(reduced^*)$ is the residual sum of squares of the reduced model:

$$y_i = \nu + \theta x_i + \beta_{01} Z_{1i} + \beta_{02} Z_{2i} + \cdots + \beta_{0m'} Z_{m'i} + \varepsilon_i, \quad i = 1, 2, \ldots, N, \tag{2.65}$$

where Z_{ji} is as defined for Equation (2.21). The test statistic in Equation (2.64) follows an F-distribution with $(m-1)$ and $(N-2m)$ degrees of freedom under the null hypothesis, where $N = n \times m$, assuming equal sample size.

If the value of F_S is significantly large as determined from the F distribution, we consider this as evidence that the regression lines are not parallel and we do not test H_B. However, if F_S is not significantly large, then we use the following F statistic to test the null hypothesis H_B:

$$F_I = \{SSE(reduced) - SSE(full^*)\}/[(m-1)MSE(full^*)], \tag{2.66}$$

where $SSE(full^*)$ and $MSE(full^*)$ are the residual sums of squares and the mean square error resulting from fitting the regression model in Equation (2.65), and $SSE(reduced)$ is the residual sum of squares of the reduced model in Equation (2.22). The test statistic in Equation (2.66) follows an F-distribution with $(m-1)$ and $(N-1-$

m) degrees of freedom under the null hypothesis. If the value of F_I is significantly large as determined from the F-distribution, we attribute the shift to changes in the intercept. This approach is similar in several aspects to that of Gulliksen and Wilks (1950).

Having determined the source of variation, one may proceed to locate the out-of-control samples. The global F statistic can be decomposed into $m - 1$ F statistics, each representing whether or not sample j ($j = 1, 2, \ldots, m - 1$) is out-of-control. To test whether the jth profile is out-of-control, the following F statistic can be used:

$$F_j = \{SSE(r) - SSE(f)\}/[2MSE(f)], \quad j = 1, 2, 3, \ldots, m - 1, \tag{2.67}$$

where $SSE(f)$ and $MSE(f)$ are the residual sums of squares and the mean square error resulting from fitting the regression model

$$y_i = A_0 + A_1 x_i + \beta_0 z_{ji} + \beta_1 z_{ji} x_i + \varepsilon_i, \quad i = 1, 2, \ldots, N^*, \tag{2.68}$$

where (x_i, y_i) are bivariate observations resulting from pooling the jth sample to the last sample into one sample of size $N^* = n_j + n_m$ and the z_{ji} are indicator variables such that

$$z_{ji} = \begin{cases} 1 \text{ if observation } i \text{ is from sample } j \\ 0 \text{ if observation } i \text{ is from sample } m \end{cases}, \quad i = 1, 2, \ldots, N^*.$$

Also, $SSE(r)$ is the residual sum of squares of the reduced model:

$$y_i = A_0 + A_1 x_i + \varepsilon_i, \quad i = 1, 2, \ldots, N^*.$$

The F statistic in Equation (2.67) follows an F-distribution with 2 and (N^*- 4) degrees of freedom under the null hypothesis that sample j is in-control. If the value of F_j ($j = 1, 2, \ldots, m - 1$) is significantly large as determined from the F-distribution, we declare that sample j is out-of-control. If assignable causes can be identified, then out-of-control samples are to be removed. The process is repeated until all out-of-control samples are discarded.

2.4.3 Diagnostics Based on Principal Component Approach

The principal component approach described by Jones and Rice (1992) can be very useful in summarizing and interpreting variation in a set of Phase I profile data with equal, equally spaced values of a dependent variable X for each profile. If the X-values are not naturally equally spaced, one can model or smooth the data and use interpolation to obtain fitted response values for a set of equally spaced values. Using this approach one treats each profile as a multivariate vector of n response Y-values and identifies a few principal components that explain as much variation in the profiles as possible. If these principal components are interpretable, this approach can be very effective in understanding process performance. Jones and

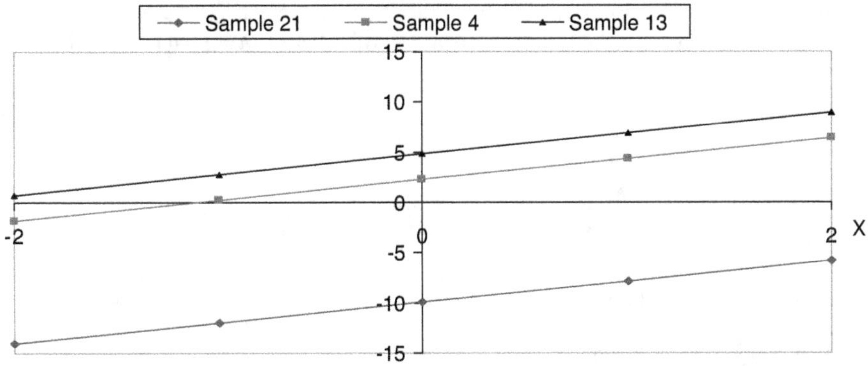

Figure 2.20 Calibration curves corresponding to the minimum, median, and maximum first principal component scores for the data set in Example 2.1.

Rice (1992) proposed plotting the average profile and the profiles with the largest and smallest principal component scores for aiding in the interpretation of the principal components.

Example 2.15 As an illustration, consider the data set of the photometric determination of Fe^{3+} with sulfosalicylic acid presented in Example 2.1. The Phase I analysis of this data set showed that the process is not stable. One may follow by applying the principal component approach of Jones and Rice (1992) to identify the primary patterns of variation among the 22 calibration curves. The first principal component accounts for approximately 95% of the total original variation. Figure 2.20 shows the three fitted calibration curves corresponding to the minimum, median, and maximum first principal component scores. These correspond to samples 21, 4, and 13, respectively. It is clear from Figure 2.20 that these curves differ primarily in the Y-intercept. Therefore, one concludes that 95% of the overall variability is due to the variability in the Y-intercept.

2.5 VIOLATION OF THE MODEL ASSUMPTIONS

The statistical performance of any of the described Phase I and II methods relies heavily in the normality assumption. In their simulation study, Mahmoud and Woodall (2004) reported that, for all of the Phase I approaches considered in their study, departures from normality can lead to dramatic increases in the false alarm rate. Mahmoud (2004) studied the in-control performance of some of the Phase I approaches when the ε_{ij} are i.i.d. double exponential random variables with mean 0 and variance 1, double exponential random variables with mean 0 and variance 2, exponential random variables with mean 1, t random variables with 3 degrees of freedom, and t random variables with 5 degrees of freedom. Mahmoud (2004) concluded that departures from the normality assumption can greatly affect the statistical performance of the Phase I methods as the false alarm rate can increase dramatically if this

Table 2.12 Overall False Alarm Probabilities Produced by Phase I Approaches When the ε_{ij} are i.i.d. Exponential Random Variables with Mean 1 (Adapted from Mahmoud 2004.)

	Method			
Nominal	A	B	C	D
0.01	0.4128	0.3698	0.7801	0.3671
0.02	0.5794	0.499	0.8863	0.4482
0.03	0.7118	0.5918	0.9317	0.5065
0.04	0.7811	0.6642	0.9562	0.553
0.05	0.8562	0.7192	0.9675	0.5839
0.10	0.9797	0.8819	0.994	0.7012

assumption is violated. For example, Table 2.12 shows the overall probability of false alarm produced by four Phase I methods when the ε_{ij} are i.i.d. exponential random variables with mean 1. These methods are as following:

1. Method A: The T^2 control chart proposed by Stover and Brill (1998).
2. Method B: The T^2 control chart proposed by Kang and Albin (2000).
3. Method C: The three Shewhart-type control charts proposed by Kim et al. (2003).
4. Method D: The global F-test proposed by Mahmoud and Woodall (2004).

Mahmoud (2004) calculated the values in Table 2.12 using 50,000 simulated in-control Phase I data sets with $m = 20$ and $n = 10$. As shown in this table, departures from the normality assumption can greatly affect the statistical performance of the four Phase I methods. Hence, it is important to test for the appropriateness of the normality assumption before applying a Phase I approach. There are many statistical approaches to check the normality assumption (see, e.g., Neter et al. (1990, Chapter 4) and Ryan (1997, pp. 52–54)).

Also, departures from linearity affect the performance of the Phase I and II methods. Mahmoud and Woodall (2004) stated that the control chart for the variance may signal instability in the process variance due to a lack-of-fit of the regression model. Therefore, it is imperative to check for the linearity of the m regression lines before applying a Phase I or II method. Neter et al. (1990, p. 131), for example, described a lack-of-fit test appropriate for the case when there are replications at one or more X-levels. Also, see Burn and Ryan (1983) for a lack-of-fit test suitable for the case when no replications are available.

APPENDIX

Table A Adjusted Axial Response and Axial Force for the 11 Calibration Curves of Example 2.7 (Adapted from Mahmoud et al. (2007)).

Profile 1		Profile 2		Profile 3	
Response	Axial Force	Response	Axial Force	Response	Axial Force
8.3700	0.0000	12.0700	0.0000	15.2900	0.0000
8.3700	0.0000	386.4300	18.3501	398.8100	18.9414
420.5800	19.2039	455.4500	21.4144	455.8700	21.4041
428.5600	19.1840	410.3800	18.4088	403.6200	18.1719
−441.5100	−21.7575	−380.1400	−18.6410	−389.8000	−18.9449
−378.9100	−19.3263	−384.1900	−18.5769	−388.3400	−18.8738
−399.6300	−19.3290	−440.7500	−22.0961	−444.5300	−22.1129
−454.2100	−22.1582	−434.8100	−21.3916	−437.4300	−21.4156
445.7000	20.7657	−441.3500	−21.6867	−453.3400	−22.3391
407.3100	18.6251	397.2900	18.0787	410.3400	18.2034
8.4400	0.0000	402.4800	19.2289	404.8600	18.9032
8.3600	0.0000	−379.6400	−19.5331	−373.1200	−18.9983
398.0200	19.3421	12.1000	0.0000	15.0400	0.0000
463.2500	22.4106	469.9300	20.9517	462.1600	20.8403
−471.2700	−22.2840	12.1900	0.0000	15.2600	0.0000
−405.0300	−19.0220	465.7700	21.5740	467.5200	21.3045
−391.2500	−19.0844	−448.8300	−21.4142	−442.5200	−21.5802
−439.2400	−21.5796	455.2700	20.6812	463.4800	20.8460
485.1000	22.4434	358.2400	16.1580	367.9200	16.6628
429.3000	19.4797	−445.7300	−22.2512	−446.6100	−22.3501
8.3600	0.0000	−432.4500	−21.0353	−441.8500	−21.7188
8.4000	0.0000	−382.5700	−18.6441	−389.2500	−19.2402
456.6100	20.6858	−401.3700	−19.4709	−401.9900	−19.8738
432.5300	19.7084	407.0700	18.2059	410.0300	18.1970
−379.0800	−19.2308	456.1900	21.7144	458.5400	21.4201
−440.7800	−21.8201	403.8800	18.4633	373.8400	16.8662
−449.6400	−22.0577	12.1700	0.0000	15.1800	0.0000
−386.7400	−18.6708	−397.1500	−19.6108	−394.5200	−19.7519
419.3000	19.1364	−444.6100	−21.4571	−444.6700	−21.3750
454.7500	21.3032	−363.0000	−17.4398	−389.4700	−19.0233
8.3100	0.0000	454.2000	20.7663	454.9200	20.6254
8.3600	0.0000	399.7000	18.4153	447.9700	20.5442
473.1400	22.7240	−394.2300	−19.9025	−391.6300	−19.9139
403.8100	19.7534	445.9900	20.8320	451.3200	20.6418
−417.9000	−19.7633	468.9400	21.4814	469.4500	21.7133
−474.5800	−22.2624	−452.6100	−22.1165	−446.8600	−22.3848
−457.4200	−22.2825	12.0200	0.0000	15.0500	0.0000
−380.8800	−18.7287	11.9800	0.0000	14.8200	0.0000
432.4500	19.7479	−75.8600	−3.5747	−61.4700	−3.6084

(Cont.)

Table A (*Cont.*)

Profile 1		Profile 2		Profile 3	
Response	Axial Force	Response	Axial Force	Response	Axial Force
483.8100	22.1969	−1.8300	−0.9519	6.4400	−0.2740
8.3200	0.0000	5.5000	−0.9957	2.0300	−0.9199
8.3200	0.0000	−31.6300	−1.6861	−28.9100	−1.5618
8.4700	0.0000	−12.8900	0.5848	−3.3600	0.7392
−24.5300	−2.0712	−32.4600	−3.9910	−36.2000	−3.8591
−36.9900	−1.9946	12.2300	0.0000	15.0600	0.0000
8.4100	0.0000	−35.3500	−2.1831	−38.7700	−1.9109
8.4200	0.0000	−53.7800	−4.1282	−28.7000	−3.5057
−66.3300	−4.4202	−20.0500	−1.8120	−4.9300	−1.2109
−39.8600	−4.3666	−23.0500	−1.7642	−15.3900	−1.5218
−16.2500	−1.2517	12.1800	0.0000	15.0000	0.0000
−13.0400	−1.3194	12.1200	0.0000	15.0000	0.0000
−1065.8900	−51.0501	−6.1300	−0.4496	6.1100	−0.4284
8.0500	0.0000	−32.6400	−2.2736	−38.1200	−2.3660
8.4400	0.0000	12.2500	0.4709	11.1600	0.4500
−39.1500	−2.2527	13.5600	0.2041	5.8800	0.1938
−51.2600	−2.2414	−23.9900	−1.7553	−28.5500	−1.8433
−10.6400	−0.6733	3.5400	0.1805	16.3800	0.1705
−6.0700	−0.7601	1032.7800	48.5288	1036.1100	48.5385
8.3100	0.0000	−1008.7900	−48.5397	−1008.2100	−48.5423
8.0600	0.0000	−29.1100	−1.1703	−36.1200	−1.9337
78.6900	3.2574	−6.6300	−0.5923	−2.5000	−0.9034
26.9200	3.2854	−8.1400	−0.7946	−16.4800	−1.4488
29.3900	1.5813	−0.2100	−0.8658	−5.6000	−1.5485
35.5000	1.5134	12.0900	0.0000	15.0100	0.0000
960.6200	45.5393			14.9300	0.0000
8.4200	0.0000			−34.1900	−2.2768
8.3600	0.0000			−15.7000	0.1442
−23.4700	−1.9755			2.7000	0.0386
−35.2300	−1.9588			14.9600	0.0000
8.4800	0.0000			−42.1900	−4.1393
−30.6900	−1.7664			−34.3100	−1.8152
−27.5600	−1.8356			−27.5900	−3.4495
−40.3200	−1.7364			15.1100	0.0000
−37.0100	−1.8111				

Profile 4		Profile 5		Profile 6	
Response	Axial Force	Response	Axial Force	Response	Axial Force
13.3600	0.0000	15.2400	0.0000	12.9500	0.0000
407.1100	19.3305	396.9900	18.8770	−394.1000	−18.8515
463.1500	21.8822	448.4300	21.1654	−440.5200	−21.1382
402.7400	18.2678	398.5400	17.9243	−379.1800	−19.2991
−381.7900	−18.8445	−380.1300	−18.6369	−384.7100	−18.6357

Table A (*Cont.*)

Profile 4		Profile 5		Profile 6	
Response	Axial Force	Response	Axial Force	Response	Axial Force
−389.3500	−18.7688	−383.8400	−18.5675	405.2000	19.0834
−443.4800	−21.9657	−442.9200	−22.0511	463.2600	20.6791
−424.1400	−20.8852	−430.4500	−21.1205	−435.0500	−21.2202
−439.9200	−21.8450	−440.5600	−21.9645	459.5100	20.6450
405.4600	18.2234	399.3300	17.8017	409.2200	18.2518
405.4000	19.3873	399.9300	18.7959	−389.3100	−18.6310
−379.6400	−19.3634	−378.2400	−19.3537	405.6500	18.3517
13.1300	0.0000	15.1900	0.0000	13.0000	0.0000
464.9800	20.9263	460.4700	20.6882	−439.2600	−21.6161
13.1000	0.0000	15.3300	0.0000	12.9200	0.0000
469.0800	21.5946	466.7300	21.2314	478.2200	21.6996
−439.6500	−21.0198	−434.7200	−21.1350	462.4600	21.7991
458.6200	20.6578	462.3600	20.6723	−448.1500	−22.1241
361.0900	16.2461	359.3200	16.2250	342.5500	15.3388
−440.2800	−21.8849	−445.4800	−22.1607	−446.3200	−21.9615
−428.4500	−21.0417	−428.1700	−21.2034	−432.1900	−21.1394
−381.3300	−18.6741	−378.6600	−18.7771	−379.9900	−18.6850
−397.2000	−19.5849	−393.8500	−19.4667	−396.1700	−19.3132
408.5600	18.3179	399.8100	17.8553	402.1100	18.0165
455.8100	21.5865	457.7800	21.4217	−446.9700	−21.2553
361.6100	16.3146	375.7700	16.8345	333.0200	14.8597
13.2400	0.0000	15.2700	0.0000	12.9400	0.0000
−398.6800	−19.5830	−389.3100	−19.2932	−393.3600	−19.3322
−440.8300	−21.2355	−440.5600	−21.1688	456.3900	21.4631
−380.8500	−18.3425	−388.2600	−18.9421	395.4100	18.8657
451.5500	20.7934	454.5800	20.8245	455.6000	20.7956
401.5600	18.1367	395.8300	17.8179	−386.8700	−19.4981
−384.6800	−19.5869	−375.3100	−19.2784	398.1100	17.8853
454.2400	20.8710	454.3100	20.7883	450.8400	20.6667
475.4100	21.7407	462.4100	21.2732	466.9400	21.4325
−446.7800	−22.1326	−448.8200	−22.3964	−444.1200	−21.9820
13.1300	0.0000	15.1200	0.0000	12.9100	0.0000
12.9800	0.0000	15.0800	0.0000	12.8200	0.0000
−51.6900	−2.9089	−52.4200	−3.2159	−24.5600	−1.7718
11.7600	−0.0378	1.1100	−0.4690	5.7500	−0.1912
2.6200	−0.6497	−4.4800	−1.0801	−1.0400	−0.8689
−21.5200	−1.0870	−35.8800	−1.9685	−23.8500	−1.1986
−2.6400	1.1864	−15.5900	0.3085	1130.7500	52.7636
−23.8300	−3.4509	−38.7800	−4.2829	−31.1700	−3.7895
12.9900	0.0000	15.2800	0.0000	13.0000	0.0000
−73.9900	−3.6050	−46.0000	−2.2603	−47.9500	−2.2333
−54.9800	−4.3242	−49.8300	−4.2081	−50.2400	−4.1192
−19.3400	−2.0001	−18.2500	−1.9113	−19.4800	−1.8313

(*Cont.*)

Table A (*Cont.*)

Profile 4		Profile 5		Profile 6	
Response	Axial Force	Response	Axial Force	Response	Axial Force
−31.3100	−1.9493	−22.8400	−1.8451	−24.7800	−1.7986
12.9300	0.0000	15.3400	0.0000	12.9100	0.0000
12.9300	0.0000	15.3300	0.0000	−30.9400	−1.5969
−0.9500	−0.5431	7.0500	−0.4854	8.6000	−0.2393
−41.1300	−2.5156	−41.2800	−2.4009	−22.6100	−1.5289
11.9100	0.2920	14.6600	0.3935	29.7900	1.1196
10.1600	0.1265	6.0700	0.1681	11.6700	0.5450
−29.0800	−1.9651	−24.4300	−1.8419	−15.9600	−1.1622
4.4900	0.0789	11.0600	−0.0992	18.2600	0.4642
1025.6200	48.1228	−989.0800	−47.7648	−3.6200	1.1206
−999.6300	−48.1226	1019.4000	47.7512	−1105.1000	−52.7628
−47.7200	−2.2431	−41.3400	−2.3347	12.6900	0.0000
−4.4000	−1.0222	−2.2200	−1.0560	−20.6200	−1.7882
−25.4400	−1.7634	−25.7500	−1.8420	−48.4200	−2.8521
−18.9800	−1.8333	−16.7800	−1.9023	−11.7500	−1.6793
12.9900	0.0000	15.1100	0.0000	12.8800	0.0000

Profile 7		Profile 8		Profile 9	
Response	Axial Force	Response	Axial Force	Response	Axial Force
10.0300	0.0000	11.2300	0.0000	6.9000	0.0000
−389.0000	−18.5708	−438.4500	−21.1129	−418.1800	−20.1242
−444.4900	−21.1540	−513.3900	−24.6448	−489.8300	−23.4565
−382.8900	−19.3767	−453.2000	−22.9985	−426.7100	−21.4099
−388.4200	−18.7376	−443.4100	−21.3106	−425.6000	−20.1840
397.7700	18.8748	448.0200	21.3330	420.6300	20.0581
461.2800	20.7086	531.5700	23.9347	502.4300	22.8498
−434.5200	−21.0488	−515.3400	−24.8993	−477.1700	−22.7351
466.0800	20.9151	533.5600	24.2133	485.3200	22.2965
400.6300	18.1146	456.5300	20.4839	422.8700	19.0558
−389.8400	−18.6662	−456.8000	−22.0290	−413.1600	−19.8308
406.5900	18.3955	453.4400	20.6285	420.0700	19.2726
10.1000	0.0000	11.4600	0.0000	7.0900	0.0000
−449.0500	−22.0561	−507.5000	−25.0173	−480.7500	−23.5612
10.0100	0.0000	11.3400	0.0000	6.9600	0.0000
468.1200	21.5983	530.6800	24.6427	495.1600	23.2177
456.3300	21.5168	520.8400	24.6004	492.6100	23.4133
−447.8200	−22.0701	−527.5100	−25.9149	−486.8900	−23.5887
333.8300	15.0925	448.4600	20.5504	424.5200	19.8365
−446.7500	−21.8126	−518.4700	−25.4916	−487.9200	−23.7927
−436.5400	−21.0730	−505.1400	−24.3787	−481.1400	−23.0226
−388.1400	−18.7603	−438.0600	−21.1189	−417.9900	−19.8807
−394.7700	−19.0326	−456.2100	−22.3040	−428.0000	−20.8476

Table A (*Cont.*)

Profile 7		Profile 8		Profile 9	
Response	Axial Force	Response	Axial Force	Response	Axial Force
399.0600	18.0481	458.6300	20.7052	423.9400	19.2916
−447.8400	−21.2983	−510.0400	−24.4061	−484.3000	−23.0848
386.2200	17.6572	431.1400	19.8556	425.5100	19.8422
10.0300	0.0000	11.3400	0.0000	6.9300	0.0000
−400.6400	−19.4026	−454.2700	−22.1667	−432.7100	−20.8983
454.4100	21.5092	520.2200	24.6105	486.9600	23.0526
395.5500	18.8125	451.1600	21.5136	426.4700	20.3856
447.4100	20.7028	516.2300	23.8623	482.7200	22.5582
−384.2300	−19.3909	−437.5700	−22.0955	−411.3300	−20.4674
398.8100	17.9528	448.7900	20.4313	418.5700	19.4045
443.7300	20.5823	516.5500	23.8550	496.0300	22.9277
455.9900	21.0046	531.8500	24.7543	488.5500	23.0317
−456.1300	−22.2422	−513.5500	−25.3961	−483.5800	−23.8014
9.9900	0.0000	11.3300	0.0000	6.7700	0.0000
10.1000	0.0000	11.2200	0.0000	7.1300	0.0000
−36.3100	−1.8096	−39.2900	−2.1787	−10.5400	−0.6299
4.8900	−0.2121	0.5300	−0.3765	8.1300	0.3069
−3.7600	−0.8844	−11.6300	−1.2092	−3.3900	−0.4627
−21.5200	−1.2292	−22.1300	−1.2619	−17.3100	−0.8814
1128.2000	52.7638	647.3800	30.1895	1131.0800	53.3461
−32.6300	−4.0126	−26.9100	−3.6379	1.1200	−1.9924
10.2900	0.0000	11.3900	0.0000	7.4300	0.0000
−35.3700	−1.8796	−10.3400	−0.5891	−25.6400	−1.0637
−39.3000	−3.2053	−26.6700	−3.0174	−30.8000	−3.1562
−6.8900	−0.9061	2.7800	−0.5822	−4.6000	−0.7368
−4.0900	−0.8706	−11.2400	−1.2592	−4.2800	−0.7326
10.1500	0.0000	11.2200	0.0000	7.1900	0.0000
−44.1600	−2.2386	−21.7700	−1.2824	−49.4600	−2.5624
5.5700	−0.1255	9.5900	−0.0693	7.3200	−0.1048
−24.2000	−1.3040	−21.8300	−1.2606	−30.9800	−1.2807
26.6200	1.3275	0.7000	0.3197	17.0300	1.5150
16.1700	0.6518	3.6400	0.0898	11.0000	0.7335
−7.6400	−0.9510	−21.7900	−1.6634	−8.0800	−0.9219
13.3400	0.5833	8.3700	0.3125	14.1000	0.6992
3.2300	1.3372	−11.7900	0.4596	10.9300	1.5179
−1107.8200	−52.7618	−624.6700	−30.1894	−1116.6100	−53.3463
9.9800	0.0000	11.1700	0.0000	6.9000	0.0000
−2.9700	−0.7830	−5.3400	−1.1449	−8.6800	−1.2967
−16.2800	−0.9527	−33.5900	−1.8627	−17.8400	−0.8796
−6.6600	−1.0143	−5.7000	−1.0845	−9.9700	−0.9491
10.1300	0.0000	11.2400	0.0000	7.0200	0.0000

(*Cont.*)

Table A (*Cont.*)

Profile 10		Profile 11	
Response	Axial Force	Response	Axial Force
8.6800	0.0000	7.9600	0.0000
−431.7900	−20.6753	−427.5700	−20.3568
−492.6000	−23.4537	−486.5700	−23.2636
−431.6200	−21.6238	−428.7500	−21.6624
−427.0800	−20.5354	−417.9700	−19.9804
422.8100	20.2473	423.7100	20.2423
500.8400	22.8232	503.7800	22.7530
−481.9000	−23.2444	−478.3100	−23.0542
497.0500	22.5903	497.0800	22.8287
428.3000	19.3494	437.8400	19.6712
−410.4700	−19.7220	−413.7100	−19.8863
437.3600	20.0747	418.5300	19.1761
9.0700	0.0000	8.2000	0.0000
−483.5700	−23.6669	−482.2800	−23.6283
9.0300	0.0000	8.1000	0.0000
493.9000	23.0858	495.6700	23.0515
487.5600	23.0952	487.1200	23.1036
−483.9100	−23.6779	−488.6100	−23.8166
439.5600	20.2598	423.1900	19.4618
−490.4700	−24.1091	−485.9600	−23.6469
−481.8200	−23.1491	−483.2700	−23.1946
−418.7300	−20.0257	−427.9800	−20.5810
−422.0600	−20.5774	−432.5400	−21.0118
436.0300	19.6670	443.0600	20.2835
−487.3700	−23.3096	−485.7200	−23.0529
413.7300	19.1249	436.8700	20.3094
8.7900	0.0000	7.9900	0.0000
−437.5900	−21.2958	−424.8800	−20.6290
494.5800	23.5109	483.9800	23.0903
424.8200	20.2733	419.6500	20.1897
484.3600	22.4442	489.2000	22.7688
−406.9700	−20.4619	−416.7700	−20.7187
425.0000	19.5169	436.9400	20.1026
491.8700	22.8250	486.7800	22.5408
500.3900	23.3020	495.0300	23.1959
−485.0300	−23.8225	−481.1400	−23.7856
8.7000	0.0000	8.0600	0.0000
8.5100	0.0000	8.0500	0.0000
−27.2700	−1.3973	−23.9900	−1.4330
7.1000	−0.0705	1.8300	−0.0838
−3.1100	−0.8305	−11.0400	−0.8596
−19.2500	−0.9305	−20.2000	−0.9380
1132.4600	53.3462	1132.0700	53.3461

Table A (*Cont.*)

Profile 10		Profile 11	
Response	Axial Force	Response	Axial Force
−30.9700	−3.5995	−19.8300	−3.0774
8.7900	0.0000	8.3900	0.0000
−26.7900	−1.3046	−19.6800	−0.7477
−34.5900	−3.3717	−26.3200	−2.9165
−4.5200	−0.9427	0.7900	−0.4931
−11.2600	−0.9493	3.3500	−0.4733
8.6300	0.0000	8.2600	0.0000
−10.3100	−0.3821	−21.7500	−1.1054
−9.2400	−0.8308	9.3600	−0.0244
−27.1900	−1.5353	−22.7400	−1.0632
19.2700	1.2676	15.1700	1.0476
13.9900	0.6005	8.7400	0.4850
−13.4200	−1.1587	−22.7400	−1.3683
8.2700	0.5532	7.9400	0.4260
1.4400	1.2656	−4.2300	1.0602
−1115.1500	−53.3460	−1115.4400	−53.3460
8.4700	0.0000	8.0200	0.0000
−17.7600	−1.4905	−9.7100	−1.4058
−20.7100	−1.1779	−35.9700	−1.8837
−3.4500	−0.8277	−23.2400	−1.9442
8.6200	0.0000	8.1900	0.0000

REFERENCES

Ajmani, V. (2003) Using EWMA control charts to monitor linear relationships in semiconductor manufacturing. Paper presented at the *Joint Statistical Meetings,* San Francisco, USA.

Burn, D. A. and Ryan, T. A., Jr. (1983) A Diagnostic Test for Lack of Fit in Regression Models. *ASA Proceedings of the Statistical Computing Section*, 286–290.

Croarkin, C. and Varner, R. (1982) Measurement Assurance for Dimensional Measurements on Integrated-Circuit Photomasks. NBS Technical Note 1164, U.S. Department of Commerce, Washington D.C., USA.

Crowder, S. V. and Hamilton, M. D. (1992) An EWMA for Monitoring a Process Standard Deviation. *Journal of Quality Technology*, **24**(1), 12–21.

Gulliksen, H. and Wilks, S. S. (1950) Regression tests for several samples. *Psychometrika*, **15**, 91–114.

Gupta, S., Montgomery, D. C., and Woodall, W. H. (2006) Performance evaluation of two methods for online monitoring of linear calibration profiles. *International Journal of Production Research*, **44**(10), 1927–1942.

Hawkins, D. M. (1976) Point estimation of the parameters of piecewise regression models. *Applied Statistics*, **25**(1), 51–57.

Hawkins, D. M. and Merriam, D. F. (1974) Zonation of multivariate sequences of digitized geologic data. *Mathematical geology*, **6**(3), 263–269.

Healy, J. D. (1987) A note on multivariate cusum procedures. *Technometrics*, **29**, 409–412.

Holmes, D. S. and Mergen, A. E. (1993) Improving the performance of the T^2 control chart. *Quality Engineering*, **5**(4), 619–625.

ISO 5725-6. Accuracy (trueness and precision) of measurement methods and results – part 6. *International Organization for Standardization*: Geneva, Switzerland, 1994.

Jones, M. C. and Rice, J. A. (1992) Displaying the important features of large collections of similar curves. *The American Statistician*, **46**, 140–145.

Kang, L. and Albin, S. L. (2000) On-line monitoring when the process yields a linear profile. *Journal of Quality Technology*, **32**(4), 418–426.

Kim, K., Mahmoud, M. A., and Woodall, W. H. (2003) On the monitoring of linear profiles. *Journal of Quality Technology*, **35**(3), 317–328.

Lucas, J. M. and Saccucci. M. S. (1990) Exponentially weighted moving average control schemes: properties and enhancements. *Technometrics*, **32**(1), 1–29.

Mahmoud, M. A. (2004) *The Monitoring of Linear Profiles and the Inertial Properties of Control Charts*, Ph.D. Dissertation, submitted to the Faculty of Virginia Polytechnic Institute and State University, Department of Statistics.

Mahmoud, M. A. (2008) Phase I Analysis of multiple linear regression profiles, communications in statistics. *Simulation and Computation*, **37**(10), 2106–2130.

Mahmoud, M. A., Morgan, J. P., and Woodall, W. H. (2010) The monitoring of simple linear regression profiles with two observations per sample. *Journal of Applied Statistics*, **37**(8), 1249–1263.

Mahmoud, M. A., Parker, P. A., Woodall, W. H., and Hawkins, D. M. (2007) A change point method for linear profile data. *Quality and Reliability Engineering International*, **23**(2), 247–268.

Mahmoud, M. A. and Woodall, W. H. (2004) Phase I analysis of linear profiles with calibration applications. *Technometrics*, **46**(4), 380–391.

Mason, R. L., Chou, Y. M., and Young, J. C. (2001) Applying Hotelling's T^2 statistic to batch processes. *Journal of Quality Technology*, **33**(4), 466-479.

Mestek, O., Pavlik, J., and Suchánek, M. (1994) Multivariate control charts: control charts for calibration curves. *Fresenius' Journal of Analytical Chemistry*, **350**(6), 344–351.

Montgomery, D. C. (2001) *Introduction to Statistical Quality Control*, 4th Edition. John Wiley & Sons, Inc., New York.

Myers, R. H. (1990) *Classical and Modern Regression with Applications*, 2nd Edition. PWS-Kent Publishing Company, Boston, MA.

Neter, J., Wasserman, W., and Kutner, M. H. (1990) *Applied Linear Statistical Models*, 3rd edition. Richard D. Irwin, Inc., Boston, MA.

Niaki, S. T. A., Abbasi, B., and Arkat, J. (2007) A generalized linear statistical model approach to monitor profiles. *International Journal of Engineering, Transactions A: Basics*, **20**(3), 233–242.

Noorossana, R., Amiri, A., and Soleimani, P. (2008) On the monitoring of autocorrelated linear profiles. *Communications in Statistics-Theory and Methods*, **37**(3), 425–442.

Noorossana, R., Amiri, A., Vaghefi, S. A., and Roghanian, E. (2004) Monitoring quality characteristics using linear profile. *Proceedings of the 3rd International Industrial Engineering Conference,* Tehran, Iran.

Parker, P. A., Morton, M., Draper, N. R., and Line, W. P. (2001) A Single-vector force calibration method featuring the modern design of experiments. Paper presented at the *American Institute of Aeronautics and Astronautics 39th Aerospace Sciences Meeting & Exhibit,* Reno, NV, USA.

Quandt, R. E. (1958) The estimation of the parameter of a linear regression system obeying two separate regimes. *Journal of the American Statistical Association*, **53**, 873–880

Ramsay, J. O. and Silverman, B. W. (2002) *Applied Functional Data Analysis: Methods and Case Studies*, Springer: New York, NY.

Ryan, T. P. (1997). *Modern Regression Methods*, John Wiley & Sons, Inc., New York.

Ryan, T. P. (2000) *Statistical Methods for Quality Improvement*, 2nd Edition, John Wiley & Sons, Inc., New York.

Saghaei, A., Amiri, A., and Mehrjoo, M. (2009a) Performance evaluation of control schemes under drift in simple linear profiles. Paper presented at the *International Conference of Manufacturing Engineering and Engineering Management,* London, England.

Saghaei, A., Mehrjoo, M., and Amiri, A. (2009b). A CUSUM-based method for monitoring simple linear profiles. *The International Journal of Advanced Manufacturing Technology*, **45**(11), 1252–1260.

Stover, F. S., and Brill, R. V. (1998) Statistical quality control applied to ion chromatography calibrations. *Journal of Chromatography A*, **804**(1–2), 37–43.

Sullivan, J. H. and Woodall, W. H. (1996) A comparison of multivariate control charts for individual observations. *Journal of Quality Technology*, **28**, 398–408.

Tracy, N. D., Young, J. C., and Mason, R. L. (1992) Multivariate control charts for individual observations. *Journal of Quality Technology*, **24**(2), 88–95.

Turgel, R. S. and Vecchia, D. F. (1987) Precision Calibration of Phase Meters. *IEEE Transactions on Instrumentation and Measurement*, **36**(4), 918–922.

Vargas, N. J. A. (2003). Robust estimation in multivariate control charts for individual observations. *Journal of Quality Technology*, **35**(4), 367–376.

Wade, M. R. and Woodall, W. H. (1993) A review and analysis of cause-selecting control charts. *Journal of Quality Technology*, **25**(3), 161–169.

Wang, K. and Tsung, F. (2005) Using profile monitoring techniques for a data-rich environment with huge sample size. *Quality and Reliability Engineering International*, **21**(7), 677–688.

Williams, J. D., Woodall, W. H., and Birch, J. B. (2007) Statistical monitoring of nonlinear product and process quality profiles. Quality and Reliability Engineering International, **23**(7), 925–941.

Woodall, W. H. (2007) Current research in profile monitoring. *Revista Producão,* **17**, 420–425.

Woodall, W. H. and Adams, B. M. (1998) *Statistical Process Control*, In: *Handbook of Statistical Methods for Engineers and Scientists,* edited by H. M. Wadsworth, Jr., McGraw-Hill Companies, Inc., New York.

Woodall, W. H. and Mahmoud, M. A. (2005) The inertial properties of quality control charts. *Technometrics*, **47**, 425-436.

Zhang, G. X. (1992) *Cause-Selecting Control Chart and Diagnosis, Theory and Practice.* Aarhus School of Business, Department of Total Quality Management, Aarhus, Denmark.

Zhang, J., Li, Z., and Wang, Z. (2009) Control chart based on likelihood ratio for monitoring linear profiles. *Computational Statistics and Data Analysis*, **53**, 1440–1448.

Zou, C., Zhang, Y., and Wang, Z. (2006) Control chart based on change-point model for monitoring linear profiles. *IIE Transactions*, **38**(12), 1093–1103.

Zou, C., Tsung, F., and Wang, Z. (2007a) Monitoring general linear profiles using multivariate exponentially weighted moving average schemes. *Technometrics*, **49**(4), 395-408.

Zou, C., Zhou, C., Wang, Z., and Tsung, F. (2007b) A self-starting control chart for linear profiles. *Journal of Quality Technology*, **39**(4), 364–375.

CHAPTER 3

Multiple Linear and Polynomial Profiles

Amirhossein Amiri
Industrial Engineering Department, Faculty of Engineering, Shahed University, Tehran, Iran

Reza Baradaran Kazemzadeh
Industrial Engineering Department, Faculty of Engineering, Tarbiat Modares University, Tehran, Iran

INTRODUCTION

As discussed in Chapters 1 and 2, sometimes quality of a process or product is better characterized by function or profile rather than by a univariate or a multivariate quality distribution. The different models of profiles are introduced in Chapter 1, and simple linear profile monitoring is covered completely in Chapter 2. However, as discussed in Chapter 1, sometimes more complicated models such as multiple linear regression or polynomial regression, rather than the simple linear regression, should be used to model some real cases. There are some examples in the literature that show the application of multiple linear regression and polynomial regression in real cases. For example, Amiri et al. (2010) discuss a case where the relationship between the torque produced by an automobile engine and the engine speed in revolutions per minute (RPM) can be modeled by a second-order polynomial regression. Another example is the calibration application at National Aeronautics and Space Administration (NASA) Langley Research Center, first introduced by Parker et al. (2001). As discussed in Chapter 2, in this case, the replicated calibrations of a force balance used in wind tunnel experiments are considered and the relationships between response variables and explanatory variables are modeled by multiple linear

Statistical Analysis of Profile Monitoring, First Edition. Edited by Rassoul Noorossana, Abbas Saghaei and Amirhossein Amiri.

regression models. Parker and Finley (2007) discuss an example of calibration model with multiple linear regression. Zou et al. (2007) investigate an example from semiconductor manufacturing that is modeled by a polynomial regression profile.

This chapter contains two sections: Section 3.1 and Section 3.2 in which monitoring multiple linear profiles and monitoring polynomial profiles are discussed, respectively. In each section, both Phases I and II are covered. As discussed in the previous chapters, one checks for the stability of the process and estimates the parameters in Phase I. However, the main purpose of Phase II analysis is to detect shifts in the parameters as quickly as possible. In Section 3.1.1, Phase I monitoring schemes are introduced and an illustrative example from the literature is presented. In Section 3.1.2, some diagnostic aids based on F statistics are introduced that are useful in identifying the out-of-control parameters as well as the locations of out-of-control samples. Phase II methods for monitoring multiple linear profiles as well as a real case example are discussed in Section 3.1.3. In Section 3.1.4, some diagnostic methods in Phase II monitoring of multiple linear profiles are introduced. In Section 3.2.1, three methods for Phase I monitoring of polynomial profiles are explained and the performance comparison of the methods is discussed. In addition, a change point estimator in this phase is introduced. Methods for Phase II monitoring of polynomial profiles are discussed in Section 3.2.2 along with a numerical example.

3.1 MONITORING MULTIPLE LINEAR PROFILES

In this section, we assume that there is a relationship between a response variable, y, and p explanatory variables, which is modeled by a multiple linear regression. First, we discuss Phase I monitoring schemes and then explain Phase II monitoring methods.

3.1.1 Phase I Schemes for Monitoring Multiple Linear Profiles

Assume that there are m samples in the form of $(x_{1ij}, x_{2ij}, \ldots, x_{pij}, y_{ij}), i = 1, 2, \ldots, n_j, j = 1, 2, \ldots, m$ and $n_j > p$ with one response variable y and p explanatory variables $X_1, X_2, \ldots, X_p$. The model that relates the response variable and explanatory variables in each sample is as follows:

$$Y_{ij} = \beta_{0j} + \beta_{1j}x_{1ij} + \beta_{2j}x_{2ij} + \ldots + \beta_{pj}x_{pij} + \varepsilon_{ij}$$
$$i = 1, 2, \ldots, n_j \quad j = 1, 2, \ldots, m, \tag{3.1}$$

or equivalently,

$$\mathbf{Y}_j = \mathbf{X}_j\boldsymbol{\beta}_j + \boldsymbol{\varepsilon}_j \qquad j = 1, 2, \ldots, m, \tag{3.2}$$

where $\mathbf{Y}_j = (y_{1j}, y_{2j}, \ldots, y_{n_j j})^T$, $\boldsymbol{\varepsilon}_j = (\varepsilon_{1j}, \varepsilon_{2j}, \ldots, \varepsilon_{n_j j})^T$, $\boldsymbol{\beta}_j = (\beta_{1j}, \beta_{2j}, \ldots, \beta_{pj})^T$, and matrix $\mathbf{X}_j$ is as follows:

$$\mathbf{X}_j = \begin{pmatrix} 1 & x_{11j} & x_{21j} & \cdots & x_{p1j} \\ 1 & x_{12j} & x_{22j} & \cdots & x_{p2j} \\ \vdots & \vdots & \vdots & \vdots & \vdots \\ 1 & x_{1n_j j} & x_{2n_j j} & \cdots & x_{pn_j j} \end{pmatrix} \tag{3.3}$$

It is assumed that the x values are known constants in each sample and ε_{ij} are independent, identically normal distribution with mean zero and variance σ_j^2. If the process is in statistical control, then the regression parameters are constant in each sample.

The ordinary least squares (OLS) estimators of the regression parameters for sample j are given as follows:

$$\hat{\boldsymbol{\beta}}_j = (\mathbf{X}^T\mathbf{X})^{-1}\mathbf{X}^T\mathbf{Y}_j \tag{3.4}$$

Under in-control situation, the estimators of the regression parameters follow a $(p+1)$-variate normal distribution with mean vector $\boldsymbol{\mu} = (\beta_0, \beta_1, \ldots, \beta_p)$ and variance–covariance matrix $\boldsymbol{\Sigma} = \sigma^2(\mathbf{X}^T\mathbf{X})^{-1}$.

Kang and Albin (2000) proposed a T^2 control chart for Phase I monitoring of simple linear profiles. The generalized formula for this T^2 statistic given in Mahmoud (2008) is as follows:

$$T_j^2 = m(\hat{\boldsymbol{\beta}}_j - \bar{\hat{\boldsymbol{\beta}}})^T \mathbf{S}_1^{-1}(\hat{\boldsymbol{\beta}}_j - \bar{\hat{\boldsymbol{\beta}}})/(m-1), \quad j = 1, 2, \ldots, m, \tag{3.5}$$

where $\hat{\boldsymbol{\beta}}_j = (\hat{\beta}_{0j}, \hat{\beta}_{1j}, \ldots, \hat{\beta}_{pj})$, $\bar{\hat{\boldsymbol{\beta}}} = (\bar{\beta}_0, \bar{\beta}_1, \ldots, \bar{\beta}_p)$, and $\mathbf{S}_1 = MSE(\mathbf{X}^T\mathbf{X})^{-1}$. It can easily be shown that $T_j^2/(p+1)$ follows an F-distribution with $(p+1)$ and $m(n-p-1)$ degrees of freedom.

Hence, the upper control limit for the T^2 statistic in Equation (3.5) is as follows:

$$UCL = (p+1)F_{p+1, m(n-p-1), \alpha}. \tag{3.6}$$

Mahmoud and Woodall (2004) showed that this method was poor in detecting shifts in the standard deviation. Mahmoud (2008) recommended the use of a univariate control chart along with the T^2 control chart to improve the performance of the method for detecting standard deviation shift. He modified the control chart proposed by Kim et al. (2003) for monitoring the process variance and proposed monitoring mean square error of jth sample by using the following control limits:

$$LCL = mF_{(n-p-1),(m-1)(n-p-1),\alpha/2}\, MSE/(m - 1 + F_{(n-p-1),(m-1)(n-p-1),\alpha/2}) \text{ and}$$
$$UCL = mF_{(n-p-1),(m-1)(n-p-1),(1-\alpha/2)}\, MSE/(m - 1 + F_{(n-p-1),(m-1)(n-p-1),(1-\alpha/2)}) \tag{3.7}$$

Mahmoud (2008) also presents the control limits for MSE_j when the sample size varies from sample to sample as follows:

$$LCL = mF_{(n_j-p-1),\sum_{i\neq j}^{m}(n_i-p),\alpha/2}\,MSE/(m-1+F_{(n_j-p-1),\sum_{i\neq j}^{m}(n_i-p),\alpha/2})$$
$$UCL = mF_{(n_j-p-1),\sum_{i\neq j}^{m}(n_i-p),(1-\alpha/2)}\,MSE/(m-1+F_{(n_j-p-1),\sum_{i\neq j}^{m}(n_i-p),(1-\alpha/2)}) \quad (3.8)$$

The control limits in Equation (3.8) vary in each sample. Mahmoud (2008) notes that one can replace n in Equation (3.7) by $\bar{n} = \sum_{j=1}^{m} n_j$ when the discrepancy in sample sizes is small. He recommends the use of a probability of Type I error equal to $1-(1-\alpha)^{1/(p+2)}$ and $1-(1-\alpha)^{(p+1)/(p+2)}$ for variance control chart and the T^2 control chart, respectively, to obtain the overall probability of Type I error equal to α.

Using a mean square successive difference estimator is another choice to estimate the sample covariance matrix of the T^2 statistics. This T^2 control chart was originally proposed by Hawkins and Merriam (1974) and was used first by Williams et al. (2007) in the area of profile monitoring. Williams et al. (2007) defined the following matrix of successive differences:

$$\hat{\mathbf{V}} = \begin{bmatrix} \hat{\mathbf{v}}_1^T \\ \hat{\mathbf{v}}_2^T \\ \vdots \\ \hat{\mathbf{v}}_{m-1}^T \end{bmatrix}, \quad (3.9)$$

where $\hat{\mathbf{v}}_j = \hat{\boldsymbol{\beta}}_{j+1} - \hat{\boldsymbol{\beta}}_j$, $j = 1, 2, \ldots, m-1$ and $\hat{\boldsymbol{\beta}}_j$ is the vector of least squares estimators for the parameters of jth regression line. Then, the covariance matrix is estimated by

$$\mathbf{S}_2 = \frac{\hat{\mathbf{V}}^{\mathbf{T}} \times \hat{\mathbf{V}}}{2(m-1)} \quad (3.10)$$

and the T^2 statistics are obtained by the following equation:

$$T_j^2 = \left(\hat{\boldsymbol{\beta}}_j - \overline{\hat{\boldsymbol{\beta}}}\right)^{\mathbf{T}} \mathbf{S}_2^{-1} \left(\hat{\boldsymbol{\beta}}_j - \overline{\hat{\boldsymbol{\beta}}}\right), \quad j = 1, 2, \ldots, m. \quad (3.11)$$

The upper control limit for the T^2 statistics in Equation (3.11) is obtained by simulation to give a specified probability of Type I error.

The third method is the extension of F method by Mahmoud and Woodall (2004) to the multiple linear regression case. In this method, one pools all the m samples into one sample of size N and creates $m-1$ indicator variables such that

$$\begin{cases} Z_{ji} = 1 & \text{if observation } i \text{ is from sample } j \\ Z_{ji} = 0 & \text{otherwise, } i = 1, 2, \cdots, N \quad j = 1, 2, \ldots, m-1 \end{cases}$$

Finally, one fits the following multiple linear regression to the pooled data:

$$
\begin{aligned}
y_i &= \beta_0 + \beta_1 X_{1i} + \cdots + \beta_p X_{pi} + \beta_{01} Z_{1i} + \beta_{11} Z_{1i} X_{1i} + \cdots + \beta_{p1} Z_{1i} X_{pi} \\
&\quad + \cdots + \beta_{0m'} Z_{m'i} + \beta_{1m'} Z_{m'i} X_{1i} + \cdots + \beta_{pm'} Z_{m'i} X_{pi} + \varepsilon_i \\
i &= 1, 2, \ldots, N \quad m' = m - 1 \qquad (3.12)
\end{aligned}
$$

To estimate the regression parameters in Equation (3.12), one should use the following multiple linear regression model and fit it m times to each m sample separately:

$$
\begin{aligned}
y_{ij} &= \beta_{0j} + \beta_{1j} x_{1ij} + \beta_{2j} x_{2ij} + \cdots + \beta_{pj} x_{pij} + \varepsilon_{ij} \\
i &= 1, 2, \ldots, n_j \quad j = 1, 2, \ldots, m \qquad (3.13)
\end{aligned}
$$

To test for equality of the m regression lines, the following hypotheses are tested:

$H_0 : \beta_{01} = \beta_{11} = \ldots = \beta_{p1} = \ldots = \beta_{0m'} = \beta_{1m'} = \ldots = \beta_{pm'} = 0$
$H_1 : H_0$ is not true.

Under the null hypothesis, the reduced model is as follows:

$$
y_i = \beta_0 + \beta_1 X_{1i} + \cdots + \beta_p X_{pi} + \varepsilon_i \qquad i = 1, 2, \ldots, N \qquad (3.14)
$$

The standard test statistic for testing H_0 is

$$
F = \frac{[SSE(R) - SSE(F)]/(p+1)(m-1)}{SSE(F)/[N - m(p+1)]}, \qquad (3.15)
$$

where $SSE(F)$ and $SSE(R)$ are the residual sums of squares obtained by fitting the regression models in Equations (3.12) and (3.14), respectively. The statistic in Equation (3.15) has an F-distribution with $(p + 1)(m - 1)$ and $[N - m(p + 1)]$ degrees of freedom under the null hypothesis.

Mahmoud and Woodall (2004) proposed the use of a univariate control chart in conjunction with global F-test in Equation (3.15) to monitor standard deviation of simple linear profiles. They assumed that the sample sizes were constant from sample to sample. Kazemzadeh et al. (2008) extended their method to the polynomial case where the sample size was variable. The statistics that are plotted on the univariate control chart in the case of multiple linear regression are:

$$
F_j = \frac{(n_j - p - 1)MSE_j/\sigma^2(n_j - p - 1)}{\sum_{i \neq j}^{m} (n_i - p - 1)MSE_i/\sigma^2 \sum_{i \neq j}^{m} (n_i - p - 1)} \qquad (3.16)
$$

The statistics in Equation (3.16) follow an F-distribution with $(n_j - p - 1)$ and $\sum_{i \neq j}^{m} (n_i - p - 1)$ degrees of freedom under the null hypothesis. The lower and upper control limits for the F statistics are as follows:

$$
\begin{aligned}
LCL &= F_{(n_j-p-1),\sum_{i\neq j}^{m}(n_i-p-1),1-(\alpha/2)} \\
UCL &= F_{(n_j-p-1),\sum_{i\neq j}^{m}(n_i-p-1),\alpha/2}
\end{aligned} \tag{3.17}
$$

Another method is based on the likelihood ratio test, first proposed by Mahmoud et al. (2007) for monitoring simple linear profiles and then extended to monitoring multiple linear profiles and polynomial profiles by Mahmoud (2008) and Kazemzadeh et al. (2008), respectively. With a step shift in one or more regression parameters after sample m_1 the following models are assumed:

$$
\begin{aligned}
Y_{ij} &= \beta_{01} + \beta_{11}x_{1ij} + \beta_{21}x_{2ij} + \cdots + \beta_{p1}x_{pij} + \varepsilon_{ij1} \\
i &= 1, 2, \ldots, n_j \quad j = 1, 2, \ldots, m_1 \\
Y_{ij} &= \beta_{02} + \beta_{12}x_{1ij} + \beta_{22}x_{2ij} + \cdots + \beta_{p2}x_{pij} + \varepsilon_{ij2} \\
i &= 1, 2, \ldots, n_j \quad j = m_1 + 1, \ldots, m,
\end{aligned} \tag{3.18}
$$

where $\varepsilon_{ij1} \sim N(0, \sigma_1^2)$ and $\varepsilon_{ij2} \sim N(0, \sigma_2^2)$.

To test the following hypotheses,

$H_0 : \beta_{01} = \beta_{02} = \beta_0, \beta_{11} = \beta_{12} = \beta_1, \ldots, \ \beta_{p1} = \beta_{p2} = \beta_p,$ and $\sigma_1^2 = \sigma_2^2 = \sigma^2$
and $H_1 : H_0$ is not true,

one first obtains the following likelihood ratio statistic, lrt_{m1}, for all possible values of m_1, i.e., $m_1 = 1, 2, \ldots, m - 1$:

$$
lrt_{m1} = N \log \hat{\sigma}^2 - N_1 \log \hat{\sigma}_1^2 - N_2 \log \hat{\sigma}_2^2, \tag{3.19}
$$

where $\hat{\sigma}^2$ is the maximum likelihood estimator of the error variance for the multiple linear regression model fitted to all the m samples pooled into one sample of size N and computed as follows:

$$
\hat{\sigma}^2 = \sum_{i=1}^{N} (y_i - \hat{\beta}_0 - \hat{\beta}_1 x_i - \hat{\beta}_2 x_{2i} - \cdots - \hat{\beta}_p x_{pi})^2 / N \tag{3.20}
$$

$\hat{\sigma}_1^2$ and $\hat{\sigma}_2^2$ are the maximum likelihood estimators of the error variances for the multiple linear regression fitted to all the samples prior to and following m_1 pooled into one sample of size $N_1 = \sum_{j=1}^{m_1} n_j$ and $N_2 = N - N_1$, respectively, given by

$$\hat{\sigma}_1^2 = \sum_{i=1}^{N_1} (y_i - \hat{\beta}_{01} - \hat{\beta}_{11}x_{1i} - \hat{\beta}_{21}x_{2i} - \cdots - \hat{\beta}_{p1}x_{pi})^2 / N_1$$

$$\hat{\sigma}_2^2 = \sum_{i=N_1+1}^{N} (y_i - \hat{\beta}_{02} - \hat{\beta}_{12}x_{1i} - \hat{\beta}_{22}x_{2i} - \cdots - \hat{\beta}_{p2}x_{pi})^2 / N - N_1 \quad (3.21)$$

Then one divides each likelihood ratio statistic by its expected value under the null hypothesis, which can be approximated using the following equation:

$$E(lrt_{m1}) \approx p' - \frac{p'^2}{2}\left(\frac{1}{N} - \frac{1}{N_1} - \frac{1}{N_2}\right) - \left(\frac{N}{N-p'} - \frac{N_1}{N_1-p'} - \frac{N_2}{N_2-p'}\right)$$
$$-\frac{1}{3}\left(\frac{N}{(N-p')^2} - \frac{N_1}{(N_1-p')^2} - \frac{N_2}{(N_2-p')^2}\right), \quad (3.22)$$

where $p' = p + 1$ and obtains the following statistics:

$$slrt_{m1} = lrt_{m1} / E(lrt_{m1}) \quad (3.23)$$

The method signals the presence of one or more assignable causes if the maximum of the statistics in Equation (3.23) exceeds a threshold. The threshold can be determined by simulation or approximated as in Mahmoud et al. (2007) to give a specified probability of Type I error.

The last method is the method proposed by Mahmoud (2008). In this method, one monitors the parameters of the following simple linear profile including the intercept, slope, and error variance instead of p parameters of the multiple linear regression model:

$$y_{ij} = \theta_0 + \theta_1 \bar{\hat{y}}_i + \varepsilon_{ij} \quad i = 1, 2, \ldots, n \quad j = 1, 2, \ldots m, \quad (3.24)$$

where y_{ij} is ith observation in jth sample, $\bar{\hat{y}}_i = \sum_{j=1}^{m} \hat{y}_{ij}$ is the average of fitted responses for ith observation, where $\hat{y}_{ij} = \hat{\beta}_{0j} + \hat{\beta}_{1j}x_{1ij} + \hat{\beta}_{2j}x_{2ij} + \cdots + \hat{\beta}_{pj}x_{pij} + \varepsilon_{ij}$, and θ_0 and θ_1 are the intercept and the slope of the simple linear regression model, respectively.

The method is based on the idea that if the parameters of multiple linear regression model are in-control, then the intercept and the slope of simple linear regression, θ_0 and θ_1, are also in-control. On the other hand, any changes in the intercept and the slope mean that at least one of the parameters of the multiple linear regression model is out-of-control. One of the assumptions of the method is that there is the same set of x values in all samples. Since the model in Equation (3.24) is a simple linear regression, it can be monitored by any Phase I methods discussed thoroughly in Chapter 2. Mahmoud (2008) used the F method proposed by Mahmoud and Woodall

(2004) to monitor the intercept and the slope of the model in Equation (3.24). To monitor the error variance of the model, in conjunction with the method, Mahmoud (2008) used a univariate control chart based on the F-distribution discussed earlier in this chapter as a method along with the T^2 method.

Example 3.1 In this section, an example by Mahmoud (2008) is explained to clarify the Phase I methods introduced earlier in this chapter. The example is a real calibration case at NASA. In this calibration case at NASA, presented first by Parker et al. (2001), the replicated calibrations of a force balance used in wind tunnel experiments are investigated. A force balance provides six measurements of forces and torques including normal, axial, and side forces, and pitching, rolling, and yawing moments. Six electrical responses are produced that correspond to the six mentioned forces and moments. The relationships between six response variables and six explanatory variables are modeled by multiple linear regression models. An overall assessment of the calibration case needs an analysis of all models using a multivariate multiple linear regression structure, which will be discussed in detail in Chapter 8. Mahmoud (2008) only investigated the relationship between the axial electrical response and six forces and moments, and modeled this relationship by a multiple linear regression. The data set used in this example included 11 samples, $m = 11$, each with n equal to 64, 73, or 74 observations. After fitting the multiple linear regression model to each sample, model adequacy checking was done and Mahmoud (2008) reported that there was no significant departure from normality assumption for each sample and minor departures from linearity in few samples, which would not affect the Phase I methods as discussed by Mahmoud and Woodall (2004). The least squares estimators of the regression parameters as well as the corresponding averages for all samples are reported in Table 3.1.

Mahmoud (2008) applied first method (T^2 control chart along with a univariate control chart based on an F-distribution) as well as the third method (global

Table 3.1 OLS Estimators of the Regression Parameters for 11 Calibration Samples

Sample	Intercept	Normal	Axial	Filth	Roll	Yaw	Side	MSE_j
1	0.4800	0.2370	21.0087	−0.0853	0.0255	−0.1213	0.0106	75.3100
2	0.4624	0.2368	21.0204	−0.0846	0.0271	−0.1201	0.0096	82.7193
3	0.4114	0.2363	21.0230	−0.0849	0.0262	−0.1215	0.0100	116.5910
4	0.3474	0.2358	21.0253	−0.0853	0.0270	−0.1203	0.0101	82.0526
5	0.4159	0.2349	21.0191	−0.0852	0.0253	−0.1187	0.0109	79.9825
6	0.4495	0.2368	21.1466	−0.0858	0.0261	−0.1238	0.0122	82.7544
7	0.4956	0.2370	21.1438	−0.0865	0.0221	−0.1196	0.0116	82.6667
8	0.8199	0.2357	21.0475	−0.0848	0.0257	−0.1187	0.0109	104.8947
9	0.5972	0.2353	21.0487	−0.0855	0.0252	−0.1190	0.0119	98.3684
10	0.7336	0.2360	21.0491	−0.0851	0.0248	−0.1195	0.0081	98.2281
11	0.5689	0.2354	21.0518	−0.0852	0.0228	−0.1191	0.0116	97.2632
Average	0.5256	0.2361	21.0531	−0.0853	0.0255	−0.1201	0.0107	90.9846

Source: Adopted from Mahmoud 2008.

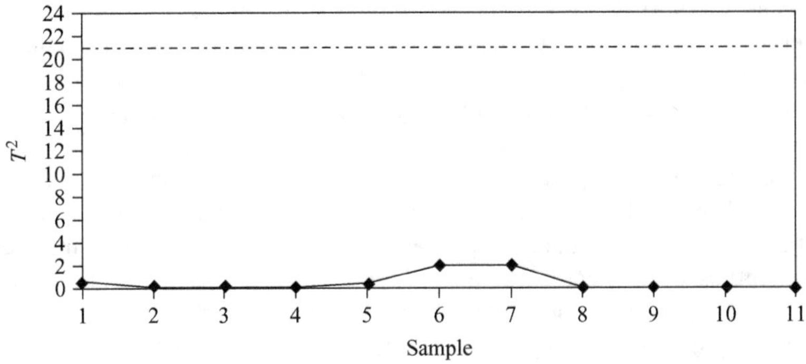

Figure 3.1 T^2 control chart for 11 samples. (Adopted from Mahmoud 2008.)

F-test) in combination with a univariate control chart discussed in the first method for monitoring the error variance. Mahmoud (2008) used the average sample size of 66 in the example and reported the following T^2 control chart to monitor the regression parameters as well as a univariate control chart to monitor the error variance in Phase I analysis.

The upper control limit for the T^2 control chart is equal to 21 and is obtained on the basis of the probability of Type I error equal to 0.0040718 using Equation (3.6). The control limits of the variance control chart are $LCL = 46.78$ and $UCL = 151.80$ on the basis of the probability of Type I error equal to 0.0005827 by using Equation (3.7). An overall probability of Type I error equal to 0.05 is obtained for joint T^2 and variance control charts. As shown in Figures 3.1 and 3.2, all samples are in-control and show that the calibration process is stable.

3.1.2 Diagnostic Aids for Phase I Analysis

The methods discussed in the previous section can only determine whether a process is in-control or not. When each of the Phase I methods shows that the process is

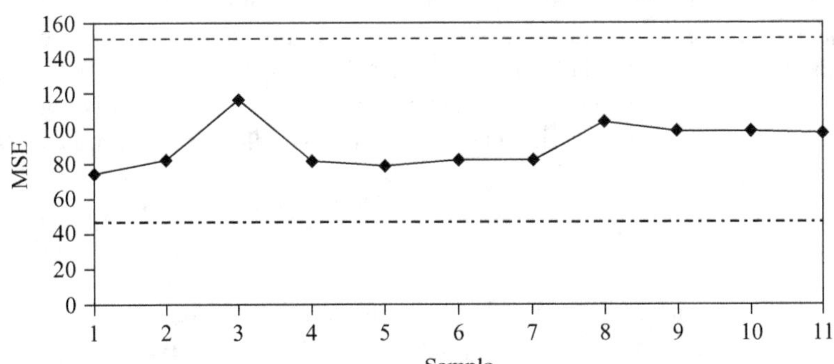

Figure 3.2 Univariate control chart for monitoring error variance. (Adopted from Mahmoud 2008.)

not stable, the next step is to find the out-of-control samples because of assignable cause(s) and remove them to obtain an in-control data set. Moreover, finding the out-of-control parameter can be useful and help in quicker detection of assignable causes. In this section, we explain the procedure proposed by Mahmoud (2008) for these purposes.

Mahmoud (2008) proposed to check first the stability of variance with the statistics and control limits given in Equations (3.16) and (3.17), respectively. If the control chart shows that one or more samples are out-of-control, one should find their corresponding assignable causes, remove the out-of-control samples, and reapply the method. If the variance control chart shows that there is no shift in variance, then one should apply the following hypothesis to find the source of profile variation:

H_{0A}: The slopes of the regression lines are all equal assuming that the variance is stable, regardless of whether the intercept is stable or not.

H_{1A}: Otherwise.

H_{0B}: The intercepts of the regression lines are all equal assuming that the variance and slopes are stable.

H_{1B}: Otherwise.

Mahmoud (2008) proposed to use the following statistics to test H_A:

$$F_S = \frac{[SSE(R^*) - SSE(F)]/p(m-1)}{SSE(F)/[N - m(p+1)]}, \tag{3.25}$$

where $SSE(F)$ and $SSE(R^*)$ are the residual sums of squares obtained by fitting the regression models in Equations (3.12) and (3.26), respectively:

$$y_i = \beta_0 + \beta_1 X_{1i} + \cdots + \beta_p X_{pi} + \beta_{01} Z_{1i} + \beta_{02} Z_{2i} + \cdots + \beta_{0m'} Z_{m'i} + \varepsilon_i$$
$$i = 1, 2, \ldots, N \tag{3.26}$$

The statistic in Equation (3.25) has an F-distribution with $p(m - 1)$ and $[N - m(p + 1)]$ degrees of freedom under the null hypothesis. If the H_{0A} hypothesis is rejected, we can conclude that one or more slopes of the regression lines are different and the lines are not parallel. Mahmoud (2008) stops the procedure if this happen and does not check the equality of the intercept. However, if the H_{0A} is not rejected, he uses the following F statistic to test the hypothesis H_{0B} versus hypothesis H_{1B}:

$$F_I = \frac{[SSE(R) - SSE(F^*)]/(m-1)}{SSE(F^*)/(N - p - m)}, \tag{3.27}$$

where $SSE(F^*)$ and $SSE(R)$ are the residual sums of squares obtained by fitting the regression models in Equations (3.26) and (3.14), respectively. The statistic in Equation (3.27) has an F-distribution with $(m - 1)$ and $(N - p - m)$ degrees of freedom under the null hypothesis.

To find the out-of-control samples, Mahmoud (2008) decomposes the global F statistics into $(m - 1)$ F statistics, each representing whether or not sample j is out-of-control ($j = 1, 2, \ldots, m - 1$). To do this, the statistics in Equation (3.28) can be used:

$$F_j = \frac{[SSE(r) - SSE(f)]/(p+1)}{SSE(f)/[N^* - 2(p+1)]}, \tag{3.28}$$

where $SSE(f)$ is the residual sum of squares obtained by fitting the regression models in Equation (3.29):

$$y_i = \beta_0 + \beta_1 X_{1i} + \cdots + \beta_p X_{pi} + \beta_{0j} Z_{ji} + \beta_{1j} Z_{ji} X_{1i} + \cdots + \beta_{pj} Z_{ji} X_{pi} + \varepsilon_i$$
$$i = 1,\ 2, \ldots, N^* = n_m + n_j \tag{3.29}$$

where Z_{ji} is defined as follows:

$$\begin{cases} Z_{ji} = 1 & \text{if observation } i \text{ is from sample } j \\ Z_{ji} = 0 & \text{if observation } i \text{ is from sample } m \end{cases}, \quad i = 1, 2, \ldots, N^*$$

and $SSE(r)$ is the residual sum of squares obtained by fitting the regression model in Equation (3.30):

$$y_i = \beta_0 + \beta_1 X_{1i} + \cdots + \beta_p X_{pi} + \varepsilon_i \quad i = 1, 2, \ldots, N^* = n_m + n_j \tag{3.30}$$

The statistic in Equation (3.28) has an F-distribution with $(p + 1)$ and $(N^* - 2(p + 1))$ degrees of freedom under the null hypothesis that sample j is in-control. If the null hypothesis is rejected, we can conclude that the jth sample is out-of-control. If we can find any assignable causes for out-of-control samples, we should omit them and repeat the procedure up to the time all out-of-control samples are removed.

3.1.3 Phase II Methods for Monitoring Multiple Linear Profiles

In this section, some methods for monitoring multiple linear profiles in Phase II are introduced and one of them is explained in detail. Zou et al. (2007) use a multivariate exponentially weighted moving average (MEWMA) control chart to monitor multiple linear profiles in Phase II. It should be noted that the other methods in the literature such as MCUSUM/χ^2 by Noorossana and Amiri (2007) and T^2 control chart by Kang and Albin (2000) can also be used for this purpose. The latter has already been discussed in detail in Chapter 2 and the former will be reviewed in Section 3.2.2.

Zou et al. (2007) proposed a MEWMA control chart for monitoring general linear profiles. They first defined

$$\mathbf{Z}_j(\boldsymbol{\beta}) = (\hat{\boldsymbol{\beta}}_j - \boldsymbol{\beta})/\sigma \tag{3.31}$$

$$Z_j(\sigma) = \phi^{-1}\{F((n-p)\hat{\sigma}_j/\sigma^2; n-p)\} \tag{3.32}$$

where $\hat{\boldsymbol{\beta}}_j = (\mathbf{X}^{\mathrm{T}}\mathbf{X})^{-1}\mathbf{X}^{\mathrm{T}}\mathbf{Y}_j$, $\hat{\sigma}_j^2 = \frac{1}{n-p-1}\sum_{i=1}^{n}(y_{ij} - \hat{\beta}_{0j} - \hat{\beta}_{1j}x_i - \hat{\beta}_{2j}x_{2i} - \cdots - \hat{\beta}_{pj}x_{pi})^2$, $\phi^{-1}(.)$ is the inverse of the standard normal cumulative distribution function and $F(.;\nu)$ is the chi-square cumulative distribution function with ν degrees of freedom. Then they defined $\mathbf{Z}_j$ as $(\mathbf{Z}_j^T(\boldsymbol{\beta}), Z_j(\sigma))^T$, which is known as a $(p+2)$-variate random vector. The vector is assumed to be multivariate normally distributed with a mean zero and a covariance matrix $\boldsymbol{\Sigma} = \begin{pmatrix} (\mathbf{X}^T\mathbf{X})^{-1} & 0 \\ 0 & 1 \end{pmatrix}$ when the process is in-control.

Finally, they defined the following MEWMA statistic:

$$\mathbf{W}_j = \theta \mathbf{Z}_j + (1-\theta)\mathbf{W}_{j-1} \quad j = 1, 2, \ldots, \tag{3.33}$$

where $\mathbf{W}_0$ is a $(p+2)$-dimensional starting vector and θ is a smoothing parameter $(0<\theta \leq 1)$. The chart signals if

$$U_j = \mathbf{W}_j^T \boldsymbol{\Sigma}^{-1} \mathbf{W}_j > L\frac{\theta}{2-\theta}, \tag{3.34}$$

where $L(>0)$ is chosen to obtain a specified in-control ARL.

Example 3.2 In this section, a real example by Zou et al. (2007) is explained. This example, which is on deep reactive ion etching (DRIE) process, comes from semiconductor manufacturing. In the DRIE process, one of the most important quality characteristics is the profile of a trench and the desired profile is the one with smooth and vertical sidewall as shown in the middle of Figure 3.3. The sidewalls of the trench should be perpendicular to its bottom with certain smoothness around the corners. Other shapes of the profile including positive and negative ones, which are due to underetching and overetching, are unacceptable.

The whole DRIE profile cannot be modeled by a general linear profile. Engineers believe that the left and right corners of each profile contain complete information to distinguish the out-of-control conditions. Because of symmetric shapes in the right and left corners, they choose one side, the left corner, to monitor the process (see Figure 3.4).

Zou et al. (2007) found a reference point A according to the initial coordinate and then rotated the left corner of the profile by 45° counterclockwise as shown in Figures 3.4b, c. Then they modeled the profile in Figure 3.4c by a quadratic polynomial model.

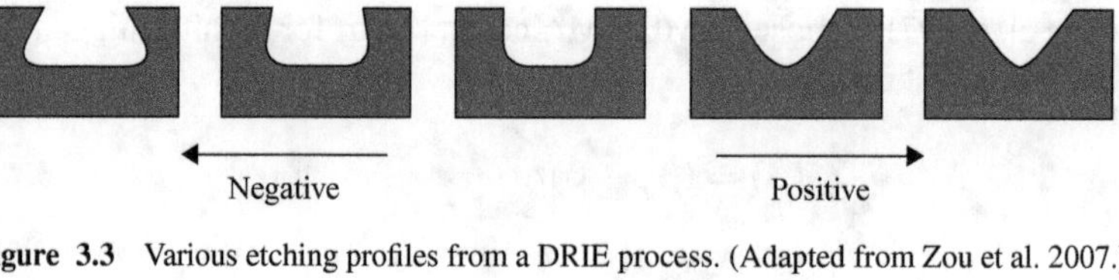

Figure 3.3 Various etching profiles from a DRIE process. (Adapted from Zou et al. 2007.)

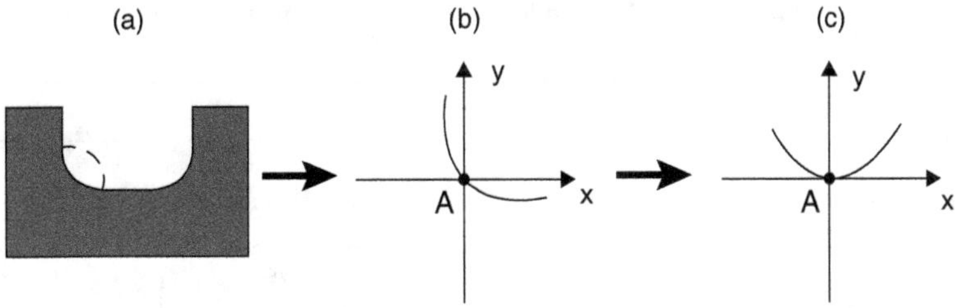

Figure 3.4 Modeling the DRIE profile. (Adapted from Zou et al. 2007.)

They fixed 11 equally spaced x_i values, –2.5, –2, –1.5, –1, –0.5, 0, 0.5, 1, 1.5, 2, 2.5, and obtained the readings of the corner profiles after rotation. The y values are given in Table 3.2.

On the basis of the data set in Table 3.2, they obtained the estimated in-control model: $y_{ij} = 0.62x_i^2 + \varepsilon_{ij}$ $i = 1, \ldots, n$ and $\sigma = 0.4$. Assuming $\theta = 0.2$, they set L equal to 15.41 to obtain in-control ARL of 370. Hence, $L\frac{\theta}{2-\theta}$ will be 1.71. They used this control chart to monitor the process. Based on Zou et al. (2007), since it is rather difficult and expensive to generate the out-of-control samples in the laboratory, they artificially made a shift in the curvature coefficient from 0.62 to 0.67 after fifth sample and generated out-of-control profiles via simulations. The simulated y_{ij} and corresponding U_j are given in Table 3.3.

Table 3.2 The y Values of DRIE Profile

j	y_{ij} (unit: μm)										
1	3.47	1.95	1.43	0.38	−0.20	−0.32	0.20	0.57	1.50	2.64	4.02
2	3.92	2.16	1.53	0.03	0.15	−0.22	0.70	0.62	1.16	1.68	3.98
3	4.04	2.68	1.10	0.51	0.26	0.17	−0.56	0.92	1.15	1.69	3.71
4	3.76	2.54	1.91	1.21	−0.07	−0.09	0.06	0.63	1.17	2.26	4.05
5	2.75	2.82	1.26	0.92	0.11	−0.71	−0.02	1.39	1.14	1.80	3.75
6	3.60	3.00	1.02	0.55	−0.09	1.24	−0.10	0.53	1.51	2.20	4.28
7	3.70	1.55	1.51	0.48	−0.95	−0.37	−0.18	0.47	1.65	2.56	3.91
8	3.69	2.78	0.30	0.53	0.93	−0.10	0.62	1.00	1.68	2.62	4.05
9	3.94	3.31	2.21	0.11	0.50	−0.84	0.13	0.90	1.30	2.42	3.55
10	3.72	3.01	0.96	0.40	−0.27	0.75	0.07	0.27	1.36	2.46	3.72
11	3.97	2.89	1.61	0.68	0.52	0.36	−0.45	0.46	0.80	1.99	3.35
12	3.89	2.12	1.15	0.94	0.13	0.09	−0.01	0.68	1.77	2.83	4.26
13	4.23	2.34	1.32	0.72	0.13	0.68	1.02	0.03	1.35	3.20	4.34
14	3.95	2.20	1.02	0.78	0.64	0.52	0.01	0.83	1.23	2.39	3.72
15	4.28	2.97	1.41	0.48	0.49	−0.84	−0.07	0.26	1.09	3.06	4.08
16	4.18	3.23	1.74	0.70	0.31	0.53	−0.36	0.58	1.38	2.73	4.32
17	4.35	2.44	1.61	0.84	1.21	0.30	0.49	1.30	1.31	2.14	4.01
18	3.19	2.71	1.20	0.57	0.31	−0.51	0.49	−0.75	1.78	1.53	4.30

Source: Adapted from Zou et al. 2007.

Table 3.3 The y Values of DRIE Profile with a Shift in Curvature Coefficient after the Fifth Sample

J						y_{ij}						U_j
0												
1	2.64	2.70	2.10	0.35	0.32	0.35	0.94	0.31	1.15	2.70	4.09	0.29
2	4.60	2.51	1.25	0.94	−0.09	−0.29	0.96	0.82	1.38	1.66	3.57	0.33
3	4.21	2.09	1.36	1.18	0.34	−0.67	0.54	−0.03	1.51	2.82	3.70	0.33
4	3.11	2.46	1.59	0.46	−0.45	0.44	−0.05	0.49	1.71	2.79	4.06	0.19
5	4.14	2.19	1.03	0.47	0.29	−0.08	0.00	0.80	1.05	3.02	4.00	0.08
6	3.64	2.87	0.25	0.51	0.24	0.37	−0.53	0.61	1.40	2.35	4.31	0.27
7	4.40	2.58	1.25	0.41	0.13	0.19	−0.31	−0.09	1.15	3.42	4.07	0.46
8	3.92	2.78	1.74	0.10	0.61	−0.99	−0.02	0.30	2.05	2.56	3.61	0.62
9	4.66	2.97	1.50	1.06	0.12	−0.48	−0.54	0.48	1.21	3.01	3.73	0.93
10	4.34	2.27	1.31	0.52	0.19	−0.10	0.14	1.07	1.04	3.02	3.86	0.76
11	3.89	1.82	1.31	0.08	−0.11	0.49	−0.18	−0.28	1.68	2.62	4.11	0.30
12	4.52	2.43	1.50	0.46	−0.38	−0.61	0.01	1.32	1.20	2.71	4.42	1.38
13	3.53	2.59	2.57	0.26	0.01	0.13	0.12	0.43	2.02	2.39	4.20	1.07
14	4.42	3.02	1.93	−0.37	0.30	−0.95	−0.65	1.14	0.76	1.55	3.93	2.00

Source: Adapted from Zou et al. 2007.

The corresponding control chart is drawn in Figure 3.5 based on the statistics U_j and the upper control limit is equal to 1.71. As shown in the figure, the MEWMA control chart signals the presence of an assignable cause at sample 14 showing an out-of-control state.

3.1.4 Diagnostic Aids for Phase II Analysis

In this section, we first introduce a change point estimator to find the location of shift, which helps quality practitioners to limit the time scope while searching for assignable cause when an out-of-control signal is given by the control chart. Then, we

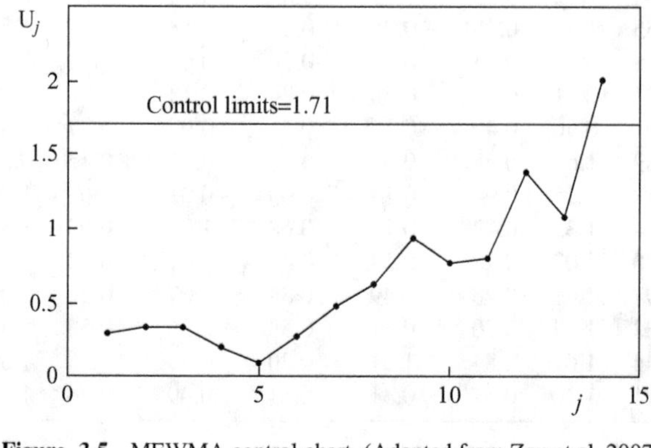

Figure 3.5 MEWMA control chart. (Adapted from Zou et al. 2007.)

explain about a parametric test method to find which parameters in the profile have been shifted after getting the signal from the control chart. Zou et al. (2007) proposed the use of maximum likelihood estimator of the change point to find the location of shift. If an out-of-control signal is given by the MEWMA control chart at subgroup k, the change estimator, τ, of a step shift proposed by Zou et al. (2007) is as follows:

$$\begin{aligned} \hat{\tau} &= \arg\max\{lr(tn, kn)\} \\ 0 &\le t < k, \end{aligned} \tag{3.35}$$

where $lr(tn, kn)$ is the generalized likelihood ratio statistics and is computed as follows (the proof is given in Zou et al. 2007):

$$lr(tn, kn) = \sum_{j=t}^{k} \frac{1}{\sigma^2}(\mathbf{Y}_j - \mathbf{X}\boldsymbol{\beta})^T(\mathbf{Y}_j - \mathbf{X}\boldsymbol{\beta}) - (k-t)n\left[\ln\left(\frac{\tilde{\sigma}^2_{(t,k)}}{\sigma^2}\right) + 1\right], \tag{3.36}$$

where

$$\tilde{\sigma}^2_{(t,k)} = \frac{(k-t)n - p}{(k-t)n}\tilde{\sigma}^2_{t,k}$$

and

$$\tilde{\sigma}^2_{t,k} = \frac{1}{(k-t)n - p}\sum_{j=t+1}^{k}(\mathbf{Y}_j - \mathbf{X}\tilde{\boldsymbol{\beta}}_{t,k})^T(Y_j - \mathbf{X}\tilde{\boldsymbol{\beta}}_{t,k})$$

in which

$$\tilde{\boldsymbol{\beta}}_{t,k} = \frac{1}{k-t}(\mathbf{X}^T\mathbf{X})^{-1}\mathbf{X}^T\sum_{j=t+1}^{k}\mathbf{Y}_j$$

Zou et al. (2007) proposed the following parametric tests to determine the out-of-control profile parameters. The T statistic in Equation (3.37) which has a t distribution with $(k-\hat{\tau})n - p$ degrees of freedom is used to find the Y-intercept change.

$$T_{test} = \frac{\sqrt{(k-\hat{\tau})n}\left(\tilde{\beta}^{(1)}_{\hat{\tau},k} - \beta^{(1)}\right)}{\tilde{\sigma}_{\hat{\tau},k}}, \tag{3.37}$$

where $\tilde{\beta}^{(1)}_{\hat{\tau},k}$ denotes the first components of the vector $\tilde{\boldsymbol{\beta}}_{\hat{\tau},k}$.

The change in the standard deviation can be found by the following chi-square statistics with $(k-\hat{\tau})n - p$ degrees of freedom:

$$\chi^2_{Test} = \frac{[(k-\hat{\tau})n - p](\tilde{\sigma}^2_{\hat{\tau},k})}{\sigma^2} \tag{3.38}$$

Finally, Zou et al. (2007) followed the work by Jensen et al. (1984) to find the change in the other regression parameters. The proposed test statistic and the rejection region are as follows:

$$F_{Test}^{(i)} = \frac{(k - \hat{\tau})(\tilde{\beta}_{\hat{\tau},k}^{(i)} - \beta^{(i)})^2}{m_{ii}\tilde{\sigma}_{\hat{\tau},k}^2} > F_\alpha\left[p - 1, (k - \hat{\tau})n - p, \mathbf{R}\right] \quad i = 2, \ldots, p, \tag{3.39}$$

where the m_{ii} are diagonal elements of $\mathbf{M} = (\mathbf{X}^T\mathbf{X})^{-1}$, $\mathbf{R} = \text{diag}\{m_{11}^{-0.5}, \ldots, m_{pp}^{-0.5}\}\, M\, \text{diag}\{m_{11}^{-0.5}, \ldots, m_{pp}^{-0.5}\}$ is the correlation matrix for $\hat{\boldsymbol{\beta}}$, and $F_\alpha(p - 1, (k - \hat{\tau})n - p, \mathbf{R})$ is the upper α percentile of the multivariate F-distribution with parameters $(p - 1, (k - \hat{\tau})n - p, \mathbf{R})$.

3.2 MONITORING POLYNOMIAL PROFILES

In this section, we assume that there is a kth-order polynomial relationship between a response variable, y, and one explanatory variable. Similar to Section 3.1, we first discuss Phase I monitoring schemes and then review monitoring methods in Phase II.

3.2.1 Phase I Methods for Monitoring Polynomial Profiles

Kazemzadeh et al. (2008) proposed an approach based on the generalized likelihood ratio test for monitoring polynomial profiles in Phase I, which is an extension of the approach by Mahmoud et al. (2007). In this method, it is assumed that there are m samples of size n_j in the form of $[(x_{ij}, y_{ij}), i = 1, 2, \ldots, n_j, j = 1, 2, \ldots, m]$. With a step shift in one or more regression parameters after sample m_1, the following models are assumed:

$$\begin{aligned} Y_{ij} &= \beta_{01} + \beta_{11}x_{ij} + \beta_{21}x_{ij}^2 + \cdots + \beta_{k1}x_{ij}^k + \varepsilon_{ij1} \\ i &= 1, 2, \ldots, n_j \quad j = 1, 2, \ldots, m_1 \\ Y_{ij} &= \beta_{02} + \beta_{12}x_{ij} + \beta_{22}x_{ij}^2 + \cdots + \beta_{k2}x_{ij}^k + \varepsilon_{ij2} \\ i &= 1, 2, \ldots, n_j \quad j = m_1 + 1, \ldots, m, \end{aligned} \tag{3.40}$$

where $\varepsilon_{ij1} \sim N(0, \sigma_1^2)$ and $\varepsilon_{ij2} \sim N(0, \sigma_2^2)$.

Considering the following null and alternative hypotheses,

$$\begin{aligned} &H_0 : \beta_{01} = \beta_{02} = \beta_0, \beta_{11} = \beta_{12} = \beta_1, \ldots, \beta_{k1} = \beta_{k2} = \beta_k, \text{and } \sigma_1^2 = \sigma_2^2 = \sigma^2, \\ &\text{and } H_1 : H_0 \text{ is not true,} \end{aligned}$$

they proposed the standardized likelihood ratio statistic as follows:

$$slrt_{m1} = \frac{lrt_{m1} - E(lrt_{m1})}{\sqrt{Var(lrt_{m1})}}, \tag{3.41}$$

where lrt_{m1} is calculated by Equation (3.42) as follows:

$$lrt_{m1} = N \log \hat{\sigma}^2 - N_1 \log \hat{\sigma}_1^2 - N_2 \log \hat{\sigma}_2^2 \tag{3.42}$$

In Equation (3.42), $\hat{\sigma}^2$, $\hat{\sigma}_1^2$, and $\hat{\sigma}_2^2$ are computed by Equations (3.43), (3.44), and (3.45), respectively, as follows:

$$\hat{\sigma}^2 = \sum_{i=1}^{N} \left(y_i - \hat{\beta}_0 - \hat{\beta}_1 x_i - \hat{\beta}_2 x_i^2 - \cdots - \hat{\beta}_k x_i^k\right)^2 / N \tag{3.43}$$

$$\hat{\sigma}_1^2 = \frac{\sum_{i=1}^{N_1} \left(y_i - \hat{\beta}_{01} - \hat{\beta}_{11} x_i - \hat{\beta}_{21} x_i^2 - \cdots - \hat{\beta}_{k1} x_i^k\right)^2}{N_1} \tag{3.44}$$

$$\hat{\sigma}_2^2 = \frac{\sum_{i=N_1+1}^{N} \left(y_i - \hat{\beta}_{02} - \hat{\beta}_{12} x_i - \hat{\beta}_{22} x_i^2 - \cdots - \hat{\beta}_{k2} x_i^k\right)^2}{N - N_1} \tag{3.45}$$

and $N = \sum_{j=1}^{m} n_j$, $N_1 = \sum_{j=1}^{m_1} n_j$, $N_2 = N - N_1$.

In $\hat{\sigma}^2$, the regression parameters are estimated by the maximum likelihood estimator and are computed using all N observations. In $\hat{\sigma}_1^2$ and $\hat{\sigma}_2^2$, the regression parameters are estimated on the basis of MLE using the observations prior to and after the change point, respectively.

$E(lrt_{m1})$ and $Var(lrt_{m1})$ in Equation (3.41) are the expected value and the variance of lrt statistic under null hypothesis, respectively. The $slrt_{m1}$ statistic in Equation (3.41) is calculated for all possible values of m_1, $m_1 = 1, 2, \ldots, m-1$ and the method signals if the maximum of these statistics exceeds a threshold. Kazemzadeh et al. (2008) showed that the values of $E(lrt_{m1})$ can be determined by simulation or be approximated by using the following equation:

$$E(lrt_{m1}) \approx k' - \frac{k'^2}{2}\left(\frac{1}{N} - \frac{1}{N_1} - \frac{1}{N_2}\right) - \left(\frac{N}{N-k-1} - \frac{N_1}{N_1-k-1} - \frac{N_2}{N_2-k-1}\right) - \frac{1}{3}$$
$$\left(\frac{N}{(N-k-1)^2} - \frac{N_1}{(N_1-k-1)^2} - \frac{N_2}{(N_2-k-1)^2}\right) \tag{3.46}$$

By using simulation, the values of $Var(lrt_{m1})$ are determined. A threshold can also be chosen by simulation to give a specified probability of Type I error.

Kazemzadeh et al. (2008) proposed an approximate thresholds for the max($slrt_{m1}$) statistic assuming that there is a second-order polynomial regression between response variable and explanatory variable. To obtain a probability of Type I error of α, one can use the following thresholds (the proof is given in Kazemzadeh et al. 2008):

$$T = \frac{F^{-1}[(1-\alpha)^{\frac{1}{r^*}}] - 4}{\sqrt{8}}, \tag{3.47}$$

where $F(.)$ is the cumulative distribution function of chi-distribution with 4 degrees of freedom and r^* is estimated by Equation (3.48).

$$r^* = 12.4 + 8.05\,[(\log m) - 3.069] + 1.41\,[(\log m) - 3.069]^2\, m > 2, \tag{3.48}$$

where m is the number of profiles in Phase I analysis. If $m = 2$, then there is only one statistic; hence, $r^* = 1$.

Kazemzadeh et al. (2008) used the following estimator to find the location of shift when the likelihood ratio approach signals:

$$\begin{gathered} \hat{m}_1 = \arg\max(slrt_{m_1}) \\ 1 \le m_1 < m, \end{gathered} \tag{3.49}$$

where $slrt_{m_1}$ is computed by using Equation (3.41).

The likelihood ratio approach can also be used to detect multiple change points. After finding the first change point at m_1, one can split the data set into two subsets at this point and apply the same change point procedure to each subset. This procedure is repeated until no evidence of additional change points is given. Kazemzadeh et al. (2008) evaluated the performance of the change point estimator by simulation studies. Simulation studies showed the suitable performance of the change point estimator, particularly in medium and large shifts. Moreover, the performance of the change point estimator depends on the location as well as the magnitude of the shift.

The second method for monitoring polynomial profiles in Phase I is an F-approach proposed by Kazemzadeh et al. (2008). This method is an extension of the F-approach by Mahmoud and Woodall (2004), and similar to the method explained for monitoring multiple linear profiles in Section 3.1.1.

The third method used by Kazemzadeh et al. (2008) is based on the T^2 statistic, first discussed by Williams et al. (2007) in the area of profile monitoring. The T^2 statistic for observation j is given by the following equation:

$$T_j^2 = \left(\hat{\boldsymbol{\beta}}_j - \bar{\hat{\boldsymbol{\beta}}}\right)^{\mathbf{T}} \mathbf{S}^{-1} \left(\hat{\boldsymbol{\beta}}_j - \bar{\hat{\boldsymbol{\beta}}}\right), \tag{3.50}$$

where $\hat{\boldsymbol{\beta}}_j$ is the vector of parameters estimators and $\bar{\hat{\boldsymbol{\beta}}}$ is the average vector and computed by using $\bar{\hat{\boldsymbol{\beta}}} = \frac{1}{m}\sum_{j=1}^{m} \hat{\boldsymbol{\beta}}_j$. The $\mathbf{S}$ matrix is the estimator of covariance matrix based on successive differences of the regression parameters' estimators and is equal to

$$\mathbf{S} = \frac{\hat{\mathbf{V}}^{\mathrm{T}} \times \hat{\mathbf{V}}}{2(m-1)}, \tag{3.51}$$

where $\hat{\mathbf{V}}^T = \left[\hat{\mathbf{v}}_1 \; \hat{\mathbf{v}}_2 \; \cdots \; \hat{\mathbf{v}}_{m-1}\right]$ and $\hat{\mathbf{v}}_i = \hat{\boldsymbol{\beta}}_{i+1} - \hat{\boldsymbol{\beta}}_i, \quad i = 1, 2, \ldots, m-1.$

The upper control limit is chosen by simulation to give a specified probability of Type I error.

Kazemzadeh et al. (2008) proposed using the standardized likelihood ratio method in conjunction with the F method to improve the performance of monitoring procedure in detecting both step shift and drift. Moreover, the F method is better than the T^2 method except in very small shifts and at the beginning and end of samples. The T^2 method is ineffective for detecting shifts in standard deviation. Hence, to detect shift in the standard deviation, another control chart along with it should be used. Under randomly occurring outliers, the F method performs better that the standardized likelihood ratio method. Hence, Kazemzadeh et al. (2008) propose using the standardized likelihood ratio method in conjunction with the F method to improve our abilities to detect both step shift and drift.

3.2.2 Phase II Methods for Monitoring Polynomial Profiles

There are some methods in the literature of profile monitoring that can be used for monitoring polynomial profiles in Phase II. These methods are discussed by Kazemzadeh et al. (2009). One of the methods, which can be used for monitoring polynomial profiles in Phase II, is T^2 control chart developed by Kang and Albin (2000). This method is discussed in detail in Chapter 2. Another method for monitoring kth-order polynomial profiles is the MEWMA control chart proposed by Zou et al. (2007) for monitoring general linear profiles and discussed in detail in Section 3.1.3. Kazemzadeh et al. (2009) extended the MCUSUM/χ^2, proposed first by Noorossana and Amiri (2007), to monitor polynomial profiles in Phase II. This method has also been discussed in detail in Chapter 2. The last method for monitoring polynomial profiles is the orthogonal polynomial method proposed by Kazemzadeh et al. (2009), which is discussed in detail in this section.

3.2.2.1 Orthogonal Polynomial Method

In this method, the polynomial regression is transformed to orthogonal polynomial regression model. As a result, the regression parameters are independent and can be monitored separately. Kazemzadeh et al. (2009) finds this method suitable when the order of polynomial regression (k) is not very large. In the case of large order, using this method may be a bit difficult since we deal with a lot of control charts. However, the quality of a process or product is usually characterized by a low-order polynomial regression and this method can easily be used in most applications.

If the process is in-control, the main and transformed models in Phase II are, respectively, as follows:

$$\begin{aligned} y_{ij} &= \beta_0 + \beta_1 x_i + \beta_2 x_i^2 + \cdots + \beta_k x_i^k + \varepsilon_{ij} \\ y_{ij} &= B_0 P_0(x_i) + B_1 P_1(x_i) + B_2 P_2(x_i) + \cdots + B_k P_k(x_i) + \varepsilon_{ij} \\ i &= 1, 2, \ldots n \quad j = 1, 2, \ldots \end{aligned} \tag{3.52}$$

$P_u(x_i)$ in Equation (3.52) is a uth-order orthogonal polynomial defined such that

$$\begin{aligned} &\sum_{i=1}^{n} P_r(x_i) P_s(x_i) = 0 \quad r \neq s, r, s = 0, \ldots, k \\ &P_0(x_i) = 1 \quad i = 1, 2, \ldots, n \end{aligned} \tag{3.53}$$

The least squares estimators of B_{lj} are calculated by

$$\hat{B}_{lj} = \frac{\sum_{i=1}^{n} P_l(x_i) y_{ij}}{\sum_{i=1}^{n} P_l^2(x_i)} \quad j = 1, 2, \ldots \quad l = 0, 1, \ldots, k \tag{3.54}$$

The estimators $\hat{B}_{lj}$ are normally distributed as follows (the proof is given in Kazemzadeh et al. 2009):

$$\hat{B}_{lj} \sim N\left(B_l, \frac{\sigma^2}{\sum_{i=1}^{n} P_l^2(x_i)} \right) \quad j = 1, 2, \ldots \quad l = 0, 1, \cdots, k \tag{3.55}$$

After transforming the polynomial regression to orthogonal polynomial regression model and making the regression parameter's estimators independent, Kazemzadeh et al. (2009) proposed using univariate EWMA control charts to monitor the regression parameters separately. The *EWMA* statistic is calculated as follows:

$$EWMA_l(j) = \theta \hat{B}_{lj} + (1 - \theta) EWMA_l(j - 1), \quad j = 1, 2, \ldots \tag{3.56}$$

where $\theta (0 < \theta \leq 1)$ is a smoothing parameter and $EWMA_l(0) = B_l$.

The control limits for the statistic in Equation (3.56) are

$$LCL = B_l - K_l \sqrt{\frac{\theta}{(2 - \theta)} \cdot \frac{\sigma^2}{\sum_{i=1}^{n} P_l^2(x_i)}}$$

$$UCL = B_l + K_l \sqrt{\frac{\theta}{(2-\theta)} \cdot \frac{\sigma^2}{\sum_{i=1}^{n} P_l^2(x_i)}}, \tag{3.57}$$

where K_l (>0) is chosen to obtain a specified in-control ARL.

Kazemzadeh et al. (2009) also used the *EWMA* statistic in Equation (3.58), based on the approach proposed by Crowder and Hamilton (1992), to monitor the error variance:

$$EWMA_E(j) = \max\{\theta\,(MSE_j - 1) + (1-\theta)EWMA_E(j-1),\ 0\} \quad j = 1, 2, \ldots \tag{3.58}$$

where $\theta (0 < \theta \leq 1)$ is a smoothing parameter and $EWMA_E(0) = 0$.

The upper control limit for the *EWMA* statistic in Equation (3.58) is

$$UCL = L_E \sqrt{\frac{\theta Var(MSE_j)}{(2-\theta)}}, \tag{3.59}$$

where L_E (>0) is chosen to achieve a specified in-control ARL, and MSE_j and $Var(MSE_j)$ are given as follows:

$$MSE_j = \sum_{i=1}^{n} e_{ij}^2 \quad \text{and} \quad Var(MSE_j) = \frac{2\sigma^4}{n} \tag{3.60}$$

In Equation (3.60), e_{ij} is calculated as follows:

$$\begin{aligned} e_{ij} &= y_{ij} - \hat{y}_{ij} = y_{ij} - B_0 - B_1 P_1(x_i) - B_2 P_2(x_i) - \cdots - B_k P_k(x_i) \\ i &= 1, 2, \ldots, n \quad j = 1, 2, \ldots \end{aligned} \tag{3.61}$$

Kazemzadeh et al. (2009) compared the performance of the orthogonal method with the other methods introduced in the first paragraph of this section through simulation studies. They used a second-order polynomial regression profile as an example for this purpose. The results showed that under the intercept shift, except in very small shifts in which MCUSUM/χ^2 performs better than the other methods, the performance of orthogonal polynomial, MCUSUM/χ^2, and MEWMA methods is roughly the same. Under shift in the second and third parameters, orthogonal polynomial method performs uniformly better than the other methods. This method also performs better than the other methods in small to medium shifts in the standard deviation. However, as the magnitude of shift increases, the performance of all methods is roughly the same.

Example 3.3 We use the following example presented first by Kazemzadeh et al. (2009) to show how one can use the orthogonal method in practice. Assume the following in-control model:

$$y_{ij} = 3 + 2x_i + x_i^2 + \varepsilon_{ij} \quad i = 1, 2, \ldots, 10 \quad j = 1, 2, \ldots,$$

where ε_{ij} are normally distributed with mean zero and variance 1, and the fixed x values are 1, 2, ..., 10.

As a first step, we should transform the main model to the following orthogonal polynomial model:

$$y_{ij} = 52.5 + 6.5P_1(x_i) + 2P_2(x_i) + \varepsilon_{ij} \quad i = 1, 2, \ldots, 10 \quad j = 1, 2, \cdots,$$

where ε_{ij} are normally distributed with mean zero and variance 1. The fixed $P_1(x_i)$ values are –9, –7, –5, –3, –1, 1, 3, 5, 7, 9 and the fixed $P_2(x_i)$ values are 6, 2, –1, –3, –4, –4, –3, –1, 2, 6. After that we should choose the desired in-control ARL value and smoothing parameter, θ, and determine parameters of the control limits, i.e., K_1 and L_E, based on in-control ARL and θ values. The values ARL $= 200$ and $\theta = 0.2$ lead to $K_1 = 3.1$ and $L_E = 3.59$. In this example, the control limits for the parameters of the transformed model are calculated by using Equation (3.57) and are summarized in Table 3.4.

Finally, we should start monitoring the process. To do that, we collect one sample from the process and obtain observations y_{ij} at fixed values of x_i for $i = 1, 2, \ldots, 10$. Then, we compute the statistics and compare them with the corresponding control limits. We follow this procedure until an out-of-control signal is detected by the control charts. At this time, a corrective action is required. To show the last step of the method clearly, it is assumed that the intercept of the main model is shifted from 3 to 3.5. As a result, the intercept of the transformed model will change form 52.5 to 53. Hence, after the shift, the transformed model is:

$$y_{ij} = 53 + 6.5P_1(x_i) + 2P_2(x_i) + \varepsilon_{ij}$$
$$i = 1, 2, \ldots, 10 \quad j = 1, 2, \ldots$$

Some observations are generated using the above model and the statistics are computed. The intercept control chart signals at the sixth profile indicating an out-of-control condition. The calculated statistics are summarized in Table 3.5.

Table 3.4 Control Limits for the Parameters of the Transformed Model

Parameter	*LCL*	*UCL*
Intercept	52.173	52.827
Second parameter	6.4431	6.5569
Third parameter	1.9101	2.0899
Standard deviation	–	0.53517

Source: Adopted from Kazemzdeh et al. 2009.

Table 3.5 Calculated Statistics for Six Profiles

J	1	2	3	4	5	6
$\hat{B}_{0j}$	53.01	52.98	52.79	52.92	52.73	53.20
$EWMA_0(j)$	52.60	52.68	52.70	52.74	52.74	52.83
$\hat{B}_{1j}$	6.51	6.41	6.55	6.45	6.49	6.36
$EWMA_1(j)$	6.50	6.48	6.50	6.49	6.49	6.46
$\hat{B}_{2j}$	2.01	2.08	2.02	2.08	1.88	2.10
$EWMA_2(j)$	2.00	2.02	2.02	2.03	2.00	2.02
MSE_j	0.95	1.23	0.51	1.15	0.94	1.71
$EWMA_E(j)$	0.00	0.05	0.00	0.03	0.01	0.15

Source: Adopted from Kazemzdeh et al. 2009.

The observations and $P_i(x_i)$ values of the sixth profile are summarized in Table 3.6.

The calculations for the last column of Table 3.5 on the basis of the observed values of y_{ij} and the corresponding $P_i(x_i)$ values from Table 3.6 are as follows:

$$\hat{B}_{06} = \sum_{i=1}^{10} P_0(x_i) y_{i6} \Big/ \sum_{i=1}^{10} P_0^2(x_i) = 53.2$$

$$EWMA_0(6) = \theta \hat{B}_{06} + (1-\theta) EWMA_0(5) = (0.2)(53.2) + (0.8)(52.74) = 52.83$$

$$\hat{B}_{16} = \sum_{i=1}^{10} P_1(x_i) y_{i6} \Big/ \sum_{i=1}^{10} P_1^2(x_i) = 6.36$$

$$EWMA_1(6) = \theta \hat{B}_{16} + (1-\theta) EWMA_1(5) = (0.2)(6.36) + (0.8)(6.49) = 6.46$$

$$\hat{B}_{26} = \sum_{i=1}^{10} P_2(x_i) y_{i6} \Big/ \sum_{i=1}^{10} P_2^2(x_i) = 2.1$$

$$EWMA_2(6) = \theta \hat{B}_{26} + (1-\theta) EWMA_2(5) = (0.2)(2.1) + (0.8)(2) = 2.02$$

Table 3.6 Dependent and Independent Observations for the Sixth Profile

i	y_{i6}	$P_1(x_i)$	$P_2(x_i)$
1	8.24	−9	6
2	13.44	−7	2
3	20.14	−5	−1
4	26.24	−3	−3
5	38.29	−1	−4
6	51.30	1	−4
7	66.81	3	−3
8	82.93	5	−1
9	101.52	7	2
10	123.05	9	6

Source: Adopted from Kazemzdeh et al. 2009.

REFERENCES

Amiri, A., Jensen, W. A., and Kazemzadeh, R. B. (2010). A case study on monitoring polynomial profiles in the automotive industry. *Quality and Reliability Engineering International*, **26**(5), 509–520.

Crowder, S. V. and Hamilton, M. D. (1992). An EWMA for monitoring a process standard deviation. *Journal of Quality Technology*, **24**(1), 12–21.

Hawkins, D. M. and Merriam, D. F. (1974). Zonation of multivariate sequences of digitized geologic data. *Mathematical Geology*, **6**(3), 263–269.

Jensen, D. R., Hui, Y. V., and Ghare, P. M. (1984). Monitoring an input-output model for production. I: the control charts. *Management Science*, **30**(10), 1197–1206.

Kang, L. and Albin, S.L. (2000). On-Line Monitoring When the process yields a linear profile. *Journal of Quality Technology*, **32**(4), 418–426.

Kazemzadeh, R. B., Noorossana, R., and Amiri, A. (2008). Phase I monitoring of polynomial profiles. *Communications in Statistics-Theory and Methods*, **37**(10), 1671–1686.

Kazemzadeh, R. B., Noorossana, R., and Amiri, A. (2009). Monitoring polynomial profiles in quality control applications. *The International Journal of Advanced Manufacturing Technology*, **42**(7), 703–712.

Kim, K., Mahmoud, M. A., and Woodall, W. H. (2003). On the monitoring of linear profiles, *Journal of Quality Technology*, **35**(3), 317–328.

Mahmoud, M. A. (2008). Phase I analysis of multiple linear regression profiles, *Communications in Statistics, Simulation and Computation*, **37**(10), 2106–2130.

Mahmoud, M. A., Parker, P. A., Woodall, W. H., and Hawkins, D. M. (2007). A change point method for linear profile data. *Quality and Reliability Engineering International*, **23**(2), 247–268.

Mahmoud, M. A. and Woodall, W. H. (2004). Phase I analysis of linear profiles with calibration applications. *Technometrics*, **46**(4), 380–391.

Noorossana, R. and Amiri, A. (2007). Enhancement of linear profiles monitoring in Phase II. *AmirKabir Journal of Science and Technology*, **18**(66-B), 19–27. (In Persian)

Parker P. A. and Finley, T. D. (2007). Advancements in aircraft model force and attitude instrumentation by integrating statistical methods. *Journal of Aircraft*, **44**(2), 436–443

Parker, P. A., Morton, M., Draper, N. R., and Line, W. P. (2001). A single-vector force calibration method featuring the modern design of experiments. *Proceedings of the American Institute of Aeronautics and Astronautics 39th Aerospace Sciences Meeting and Exhibit*, Reno, NV, USA.

Williams, J. D., Woodall, W. H., and Birch, J.B. (2007). Statistical monitoring of nonlinear product and process quality profiles. *Quality and Reliability Engineering International*, **23**(8), 925–941.

Zou, C., Tsung, F., and Wang, Z. (2007). Monitoring general linear profiles using multivariate exponentially weighted moving average schemes. *Technometrics*, **49**(4), 395–408.

CHAPTER 4

Binary Response Profiles

Arthur Yeh

Department of Applied Statistics and Operations Research, Bowling Green State University, Bowling Green, OH, USA

Longcheen Huwang

Institute of Statistics, National Tsing Hua University, Hsinchu, Taiwan

INTRODUCTION

In the previous two chapters, profile monitoring methodologies are presented and discussed for cases in which the response can be reasonably modeled as a linear function of either one explanatory variable (simple linear profile or polynomial profile) or multiple explanatory variables (multiple linear profile). Regardless of the type of profile functional form used, all the methodologies presented assume that the univariate response variable is continuous, or simply a variable in the statistical process control (SPC) terminology. Examples include the calibration curve in the photometric determination of Fe^{3+} with sulfosalicylic acid (Example 2.1) and the axial force calibration in wind tunnel experiments (Example 2.2). However, in many manufacturing as well as nonmanufacturing applications, the primary response variable of interest is of discrete or categorical nature. For example, the compressive strength of an alloy fastener used in aircraft construction is critical to ensure the quality of the aircraft built. However, when samples of alloy fasteners are put to test at different loading strengths (in pounds), the observed response is that a given alloy fastener fails at a certain loading strength. Another example deals with surgical outcomes. In a cardiac surgical center, patients are brought in for different types of cardiac surgeries. In order to sensibly measure the surgical outcomes, a commonly used measure is the patients' postsurgical mortality rate, i.e., whether a patient has died within a fixed time frame

Statistical Analysis of Profile Monitoring, First Edition. Edited by Rassoul Noorossana, Abbas Saghaei and Amirhossein Amiri.

after the surgery. These two seemingly different applications have a response variable that is binary.

While it is certainly no less important to control and monitor the quality of products or services whose quality characteristics are of discrete or categorical nature, there exist many control charts (the so-called attributes control charts) specifically designed to detect changes in discrete or categorical quality characteristics. However, in the context of statistical profile monitoring, very little attention has been paid to discrete or categorical responses. Yeh et al. (2009) develop profile monitoring methodologies for Phase I control for binary response. The methods hinge largely on fitting a logistic regression model of the binary response on a set of explanatory variables. Statistical control charts are developed to monitor possible changes in the parameters of the fitted models. These methods are extended in Noorossana et al. (2010) to Phase II monitoring. Therefore, the goal of this chapter is to discuss in more detail these relatively few and recent developments in statistical profile monitoring techniques for binary response. Section 4.1 deals with model formulation and parameter estimation. The Phase I methodologies will be discussed in Section 4.2, while the profile monitoring techniques for Phase II monitoring will be discussed in Section 4.3. A number of real applications, both in Phase I and Phase II, will be presented in Section 4.4. The chapter concludes in Section 4.5 with additional discussion on potential future research directions.

4.1 MODEL SETTING AND PARAMETER ESTIMATION

In general, when the response variable is categorical, the profile function can be reasonably handled by a generalized linear model. Particularly, when the response is binary, a commonly used model in practice is the logistic regression model (see, e.g., Montgomery et al. (2006)). Suppose there are $\boldsymbol{n}$ independent levels at which the outcomes of a binary response, $z_i, \boldsymbol{i} = \mathbf{1, 2, \ldots,} \boldsymbol{n}$, are to be collected. For the ith level, there corresponds a vector of $\boldsymbol{p}$ explanatory variables denoted by $(\boldsymbol{X}_i = \boldsymbol{X}_{i1}, \boldsymbol{X}_{i2}, \ldots, \boldsymbol{X}_{ip})^T, \boldsymbol{i} = \mathbf{1, 2, \ldots,} \boldsymbol{n}$. It is assumed that each z_i follows a Bernoulli distribution with success probability $\boldsymbol{\pi}_i, \boldsymbol{i} = \mathbf{1, 2, \ldots,} \boldsymbol{n}$, and thus $\boldsymbol{E}(z_i) = \boldsymbol{\pi}_i$ and $\boldsymbol{Var}(z_i) = \boldsymbol{\pi}_i \times (\mathbf{1} - \boldsymbol{\pi}_i)$. In the context of logistic regression model, the probability $\boldsymbol{\pi}_i$ is assumed to be a function of $\boldsymbol{X}_i$ through a link function $\boldsymbol{g}(\boldsymbol{\pi}_i)$ defined as:

$$g(\pi_i) = \log\left(\frac{\pi_i}{1-\pi_i}\right) = X_i^T \beta = \beta_1 X_{i1} + \beta_2 X_{i2} + \cdots + \beta_p X_{ip}, \tag{4.1}$$

where $\boldsymbol{\beta} = (\beta_1, \beta_2, \ldots, \beta_p)^T$ is the model parameter. In most practices, $X_{i1} \equiv 1$ and β_1 is considered to be the intercept of the model. The success probability for the ith level can be expressed as:

$$\pi_i = \frac{\exp\left(X_i^T \beta\right)}{1 + \exp\left(X_i^T \beta\right)^*} \tag{4.2}$$

We assume that there are m_i independent observations collected at the ith level of the explanatory variables, $i = 1, 2, \ldots, n$ and $M = \sum_{i=1}^{n} m_i$. Denote by $y_i = \sum_{j=1}^{m_i} z_{ij}$ the total number of successes at ith explanatory variables, where z_{ij} is the jth observation at ith level. It follows that $y_i \sim \text{binomial}\,(m_i, \pi_i)$ with $E\,(y_i) = m_i \pi_i$ and

$$\text{Var}\,(y_i) = m_i \pi_i\,(1 - \pi_i) = m_i \times \frac{\exp\left(X_i^T \beta\right)}{1 + \exp\left(X_i^T \beta\right)} \times \frac{1}{1 + \exp\left(X_i^T \beta\right)}. \tag{4.3}$$

The joint likelihood function of $y_1, y_2, \ldots, y_n$ is

$$L(\pi, y) = \prod_{i=1}^{n} \binom{m_i}{y_i} \pi_i^{y_i} (1 - \pi_i)^{m_i - y_i}, \tag{4.4}$$

where $\boldsymbol{y} = (y_1, y_2, \ldots, y_n)^T$ and $\boldsymbol{\pi} = (\pi_1, \pi_1, \ldots, \pi_n)^T$ are the vectors of n observed responses and n success probabilities, respectively. Taking the logarithm of the joint likelihood function, one gets

$$\log\,(L\,(\pi, y)) = \sum_{i=1}^{n} \log \binom{m_i}{y_i} + \sum_{i=1}^{n} y_i X_i^T \beta - \sum_{i=1}^{n} m_i \log\left(1 + \exp X_i^T \beta\right). \tag{4.5}$$

Taking the derivative of $\log\,(L\,(\boldsymbol{\pi}, y))$ with respect to $\boldsymbol{\beta}$ leads to

$$\frac{\partial \log\,(L\,(\pi, y))}{\partial \beta} = X^T\,(y - \mu), \tag{4.6}$$

where $\boldsymbol{\mu} = (\mu_1, \mu_2, \ldots, \mu_n)^T = E\,(\boldsymbol{y}) = (m_1\pi_1, m_2\pi_2, \ldots, m_n\pi_n)^T$, and $\boldsymbol{X} = (\boldsymbol{X}_1, \boldsymbol{X}_2, \ldots, \boldsymbol{X}_n)^T$ is an $n \times p$ design matrix.

The maximum likelihood estimator (MLE) of $\boldsymbol{\beta}$ is the solution to the score function $\boldsymbol{X}^T(\boldsymbol{y} - \boldsymbol{\mu}) = \boldsymbol{0}_p$, where $\boldsymbol{0}_p = (0, 0, \ldots, 0)^T$ is the p-dimensional 0 vector. In most practical applications, rather than solving the score function directly, the MLE of $\boldsymbol{\beta}$ can be approximated using the so-called iterative weighted least square (IWLS) estimation (McCullagh and Nelder (1989)). Let $\widehat{\boldsymbol{\beta}}_{ini} = (\widehat{\beta}_{1,ini}, \widehat{\beta}_{2,ini}, \ldots, \widehat{\beta}_{p,ini})^T$ be an estimate of $\boldsymbol{\beta}$, and $\hat{\boldsymbol{\eta}} = (\hat{\eta}_1, \hat{\eta}_2, \ldots, \hat{\eta}_n)^T = \boldsymbol{X}\hat{\boldsymbol{\beta}}_{ini}$. Define $i = 1, 2, \ldots, n$, $\hat{\pi}_i = \exp(\hat{\eta}_i)/[1 + \exp(\hat{\eta}_i)]$, $\hat{\boldsymbol{\mu}} = (m_1\pi_1, m_2\pi_2, \ldots, m_n\pi_n)^T$, and $\widehat{\boldsymbol{W}} = \text{diag}\{m_1\hat{\pi}_1(1 - \hat{\pi}_1), m_2\hat{\pi}_2(1 - \hat{\pi}_2), \ldots, m_n\hat{\pi}_n(1 - \hat{\pi}_n)\}$, where $\widehat{\boldsymbol{W}}$ is an $n \times n$ diagonal matrix. Further, we define

$$q = \hat{\eta} + \widehat{W}^{-1}\,(y - \hat{\mu}) = X\hat{\beta}_{ini} + \widehat{W}^{-1}\,(y - \hat{\mu}), \tag{4.7}$$

where $q = (q_1, q_1, \ldots, q_n)^T$ and $q_i = \hat{\eta}_i + (y_i - m_i\hat{\pi}_i) / [m_i\hat{\pi}_i(1 - \hat{\pi}_i)], i = 1, 2, \ldots, n$. Multiplying both sides of Equation (4.7) by $\boldsymbol{X}^T\widehat{\boldsymbol{W}}$ results in

$$X^T\widehat{W}q = X^T\widehat{W}X\hat{\beta}_{ini} + X^T\widehat{W}\widehat{W}^{-1}(y - \hat{\mu}) = X^T\widehat{W}X\hat{\beta}_{ini} + X^T(y - \hat{\mu}).$$

If one assumes that $\boldsymbol{X}^T(\boldsymbol{y} - \hat{\boldsymbol{\mu}}) = \boldsymbol{0}$, then the last equation leads to an updated version of the estimate of $\boldsymbol{\beta}$:

$$\hat{\beta}_{\text{updated}} = \left(X^T\widehat{W}X\right)^{-1} X^T\widehat{W}q \tag{4.8}$$

The IWLS procedure is now described more formally as follows:

Step1: Set $i = 0$, and let $\hat{\boldsymbol{\beta}}^{(0)}$ be an initial estimator of $\boldsymbol{\beta}$. Note that $\hat{\boldsymbol{\beta}}^{(0)}$ can be obtained from the ordinary least square estimation.

Step 2: On the basis of $\hat{\boldsymbol{\beta}}^{(i)}$, compute $\hat{\boldsymbol{\eta}}^{(i)}$, $\hat{\boldsymbol{\pi}}^{(i)}$, $\hat{\boldsymbol{\mu}}^{(i)}$, and $\widehat{\boldsymbol{W}}^{(i)}$.

Step 3: Compute $\boldsymbol{q}^{(i)} = \boldsymbol{X}\widehat{\boldsymbol{\beta}}^{(i)} + (\widehat{\boldsymbol{W}}^{(i)})^{-1}(\boldsymbol{y} - \hat{\boldsymbol{\mu}}^{(i)})$.

Step 4: The estimate of $\boldsymbol{\beta}$ is updated as $\hat{\boldsymbol{\beta}}^{(i+1)} = \left(\boldsymbol{X}^T\widehat{\boldsymbol{W}}^{(i)}\boldsymbol{X}\right)^{-1}\boldsymbol{X}^T\widehat{\boldsymbol{W}}^{(i)}\boldsymbol{q}^{(i)}$. Set $i = i + 1$.

Step 5: Repeat steps 2 through 4, say, k times until the $\boldsymbol{\beta}$ estimate converges. The convergence condition can be evaluated based on $\frac{\|\hat{\boldsymbol{\beta}}^{(k)} - \hat{\boldsymbol{\beta}}^{(k-1)}\|}{\|\hat{\boldsymbol{\beta}}^{(k-1)}\|} < \alpha$, where $\|\upsilon\|$ is the Euclidean distance of a vector υ, and the constant α is chosen to be sufficiently small (e.g., $\alpha = 10^{-5}$). If the convergence condition is met, then $\hat{\boldsymbol{\beta}} = \hat{\boldsymbol{\beta}}^{(k)}$ is the desired estimator.

Note that, as discussed in McCullagh and Nelder (1989), under regularity conditions, as n becomes large or as each m_i becomes large for fixed n, the $\widehat{\boldsymbol{\beta}}$ is asymptotically distributed as a p-dimensional normal distribution with mean vector equal to $\boldsymbol{\beta}$ and covariance matrix equal to $\left(\boldsymbol{X}^T\boldsymbol{W}\boldsymbol{X}\right)^{-1}$, where $\boldsymbol{W} = \text{diag}\{m_1\pi_1(1 - \pi_1), m_2\pi_2(1 - \pi_2), \ldots, m_n\pi_n(1 - \pi_n)\}$, i.e., $\widehat{\boldsymbol{\beta}}$ is asymptotically approximately distributed as a p-variate normal distribution $\boldsymbol{N}_p(\boldsymbol{\beta}, (\boldsymbol{X}^T\boldsymbol{W}\boldsymbol{X})^{-1})$.

4.2 PHASE I CONTROL

We assume that in Phase I control k independent samples, each with size $M = \sum_{i=1}^{n} m_i$, are taken, and that the design matrix $\boldsymbol{X}$ is the same for each sample. When the process is in-control, the model parameter is denoted by $\boldsymbol{\beta_0} = (\beta_{10}, \beta_{20}, \ldots, \beta_{p0})^T$. Let $\widehat{\boldsymbol{\beta}}_j$, $j = 1, 2, \ldots, k$ be the IWLS estimate of the model parameter for the jth sample based on the procedures described in previous section. The most commonly used control chart in Phase I monitoring is the Hotelling T^2-chart. The Hotelling T^2 charting statistic at jth sample is equal to

$$T_j^2 = \left(\widehat{\beta}_j - \beta_0\right)^T \boldsymbol{\Sigma}_0^{-1}\left(\widehat{\beta}_j - \beta_0\right), \tag{4.9}$$

where $\boldsymbol{\Sigma}_0$ is the population covariance matrix of $\widehat{\boldsymbol{\beta}}_j$ when the process is in-control. In Phase I control, both $\boldsymbol{\beta}_0$ and $\boldsymbol{\Sigma}_0$ need to be estimated. Yeh et al. (2009) propose and study several T^2 charting statistics, each of which is based on a different way of estimating both $\boldsymbol{\beta}_0$ and $\boldsymbol{\Sigma}_0$. These T^2-charts have been demonstrated to be effective for detecting various out-of-control scenarios, including the presence of a single or multiple out-of-control samples, process step shift, and process drift. The performance of a control chart is evaluated on the basis of the signal probability, which is defined as the probability of having at least one out-of-control sample.

4.2.1 T^2-Chart Based on Sample Mean and Sample Covariance Matrix

The most commonly used Phase I Hotelling T^2-chart is based on estimating the population mean and covariance matrix by their sample version. In the context of binary profile monitoring, let

$$\bar{\beta} = \frac{1}{k}\sum_{j=1}^{k} \hat{\beta}_j,\ S_H = \frac{1}{k-1}\sum_{j=1}^{k} \left(\hat{\beta}_j - \bar{\beta}\right)\left(\hat{\beta}_j - \bar{\beta}\right)^T \qquad (4.10)$$

be the sample mean and sample covariance matrix with respect to the k estimates of $\boldsymbol{\beta}$. The charting statistic is defined, for $j = 1, 2, \ldots, k$, as

$$T_{H,j}^2 = \left(\hat{\beta}_j - \bar{\beta}\right)^T S_H^{-1} \left(\hat{\beta}_j - \bar{\beta}\right). \qquad (4.11)$$

The T_H^2-chart signals when the charting statistic exceeds a predetermined upper control limit (UCL), i.e., $T_H^2 > \text{UCL}_H$. The UCL_H is determined on the basis of predetermined desirable signal probability when the process is assumed to be in-control.

It should be noted that this conventional T_H^2-chart is only effective when there is a single out-of-control sample. It is very ineffective in detecting other out-of-control scenarios such as multiple out-of-control samples, process step shift, and process drift. Despite its familiarity with most practitioners in the context of statistical control charts, it is not recommended (Yeh et al. (2009)) to be used in Phase I control in the context of monitoring binary response profile.

4.2.2 T^2-Chart Based on Sample Mean and Successive Differences Estimator

Instead of using the sample covariance matrix, another approach is to estimate the population covariance matrix of $\hat{\boldsymbol{\beta}}$ by the average of the successive differences of two $\boldsymbol{\beta}$ estimates calculated from two consecutive samples. Specifically, let

$$S_R = \frac{1}{2(k-1)}\sum_{j=1}^{k-1} \left(\widehat{\beta}_{j+1} - \widehat{\beta}_j\right)\left(\widehat{\beta}_{j+1} - \widehat{\beta}_j\right)^T, \qquad (4.12)$$

and, for $j = 1, 2, \ldots, k-1$,

$$T^2_{R,j} = \left(\widehat{\beta}_j - \bar{\beta}\right)^T S_R^{-1} \left(\widehat{\beta}_j - \bar{\beta}\right). \quad (4.13)$$

The T^2_R-chart signals when $T^2_R > \text{UCL}_R$.

The T^2_R-chart is generally more effective than the T^2_H-chart in detecting process step shift or process drift, but not as effective in detecting the presence of a single or multiple out-of-control samples.

4.2.3 T^2-Chart Based on Sample Mean and Within-Sample Covariance Estimate

Since the IWLS estimator $\widehat{\boldsymbol{\beta}}$ of the model parameter has an asymptotic covariance matrix $\left(\boldsymbol{X}^T \boldsymbol{W} \boldsymbol{X}\right)^{-1}$, one can estimate the covariance matrix $Var\left(\widehat{\boldsymbol{\beta}}_j\right)$, for each given sample, by $\left(\boldsymbol{X}^T \widehat{\boldsymbol{W}}_j \boldsymbol{X}\right)^{-1}$, for $j = 1, 2, \ldots, k$. Therefore, an estimate of $\boldsymbol{\Sigma}_0$ can be obtained by taking the average of these sample covariance estimates. Specifically, define

$$S_I = \frac{1}{k} \sum_{j=1}^{k} \left(X^T \widehat{W}_j X\right)^{-1}, \quad (4.14)$$

and the charting statistic is computed, for $j = 1, 2, \ldots, k$, as

$$T^2_{I,j} = \left(\widehat{\beta}_j - \bar{\beta}\right)^T S_I^{-1} \left(\widehat{\beta}_j - \bar{\beta}\right). \quad (4.15)$$

The T^2_I-chart signals when $T^2_I > \text{UCL}_I$.

The T^2_I-chart is most effective overall than both the T^2_H-chart and T^2_R-chart in detecting changes due to various out-of-control scenarios. It outperforms both the T^2_H-chart and T^2_R-chart when a single or multiple out-of-control samples are present. On the other hand, the T^2_I-chart performs competitively with the T^2_R-chart when there is a process step shift or process drift.

In Table 4.1, the UCL for the T^2_H-, T^2_R-, and T^2_I-charts are listed for different values of $m_i \equiv m$ and k. These UCLs were obtained based on simulations assuming that: (1) $\boldsymbol{\beta}_0 = (3, 2)^T$, (2) there are $n = 9$ levels of the explanatory variables at 0.1, 0.2, ..., 0.9, and (3) the signal probability when the process is in-control is 5%. As noted in Yeh et al. (2009), for a fixed $n \geq 8$ and different values of $\boldsymbol{\beta}_0$ and $\boldsymbol{X}$, the UCLs will be similar to those summarized in Table 4.1, provided that $m \geq 30$.

4.3 PHASE II MONITORING

We assume that in Phase I control once the out-of-control samples have been identified, all the corrected measures have been taken to find the root causes and the process has been corrected, if necessary, and brought back to be in-control. The out-of-control

Table 4.1 The Upper Control Limits of the T_H^2-, T_R^2-, and T_I^2-Charts

Replicates (m)	Total samples (k)	T_H^2-chart	T_R^2-chart	T_I^2-chart
	30	11.8166	13.4951	15.0796
30	60	15.2535	16.2155	17.9765
	90	17.2741	18.0524	19.9142
	30	11.3024	12.9983	13.9345
50	60	14.2448	15.2633	16.3131
	90	15.9921	16.6650	17.6259
	30	10.9355	12.6422	13.1126
100	60	13.5559	14.5646	15.1055
	90	14.9688	15.6884	16.2439

samples were excluded and the model parameters and covariance matrix were then estimated from the remaining samples. Here, we use the techniques discussed in Section 4.2.3 to estimate $\boldsymbol{\beta}_0$ and $\boldsymbol{\Sigma}_0$. Assuming that $\boldsymbol{k}$ in-control Phase I samples were used, then

$$\widehat{\beta}_0 = \frac{1}{k}\sum_{j=1}^{k} \widehat{\beta}_j, \text{ and } \widehat{\Sigma}_0 = \frac{1}{k}\sum_{j=1}^{k} \left(X^T \widehat{W}_j X\right)^{-1}. \tag{4.16}$$

When the Phase II monitoring begins, independent samples will be collected at a certain sampling interval, and at each sample, a total number of $M = \sum_{i=1}^{n} m_i$ observations will be collected, where m_i is the total number of replicates at the ith level of the explanatory variables, $i = 1, 2, \ldots, n$. There are a number of different charting schemes one can use in Phase II monitoring, such as the Shewhart, cumulative sum (CUSUM) and exponentially weighted moving average (EWMA) charts. Here, we only discuss the Shewhart chart. Let $\widehat{\boldsymbol{\beta}}_t, t = 1, 2, \ldots,$ be the estimate of the model parameter at time t based on the IWLS technique discussed in Section 4.1; then the charting statistic is equal to

$$T_t^2 = \left(\widehat{\beta}_t - \widehat{\beta}_0\right)^T \widehat{\Sigma}_0^{-1} \left(\widehat{\beta}_t - \widehat{\beta}_0\right). \tag{4.17}$$

The T^2-chart signals an out-of-control sample when $T_t^2 > UCL$. While it is complicated to derive the exact distribution of T_t^2, its asymptotic distribution, when the process is in-control, is equal to a chi-square distribution with p degrees of freedom, provided that n is large, or for a fixed n, m_i are large.

4.4 APPLICATIONS

In this section, we discuss two real applications of the Phase I and Phase II techniques discussed in previous sections.

Example 4.1 Compressive strength of alloy fastener used in aircraft construction (Phase I control). The original example came from Montgomery et al. (2006). It was analyzed in Yeh et al. (2009). The compressive strength of an alloy fastener used in aircraft construction is a very important quality characteristic. The explanatory variable (X) is the loading strength, as measured in pounds per square inch (psi). Nevertheless, the test result is whether or not each tested fastener fails at a certain load. Therefore, the response variable of main interest is binary. Ten levels of loading strength ranging from 2500 to 4300 psi were chosen and at each level a number of alloy fasteners were tested and the number of alloy fasteners (y) that failed at each level was observed.

A first-order logistic model was fit of the logit of failure rate on the logarithm of the loading strength. On the basis of the fitted model, 30 samples were generated to simulate the Phase I data. Of the 30 generated samples, samples 5, 12, 18, 23, and 27 were out-of-control due to a change in the value of the intercept of the fitted logistic model. The T_H^2-, T_R^2-, and T_I^2-charts were applied to the 30 generated samples, and these charts are shown in Figure 4.1. In this particular example, the T_H^2-chart (top panel) and T_R^2-chart (middle panel) are ineffective in detecting the out-of-control sample, while the T_I^2-chart (bottom panel) identifies all five out-of-control samples.

Example 4.2 Performance of cardiac surgeries (Phase II monitoring). The example was first discussed and analyzed in Steiner et al. (2000) in the context of constructing risk-adjusted CUSUM charts for monitoring surgical performance. The data set, gathered from a UK center for cardiac surgery, consists of 6994 operations from a single surgical center from 1992 to 1998. The information of a number of variables was collected from each patient. These variables include preoperative variables such as age, gender, diabetes status, and Parsonnet score (Parsonnet et al. (1989)) as well as operative variables such as surgeon, type of procedure, and reason for operation. The Parsonnet score is basically a score obtained on the basis of a combination of many other explanatory variables thought to be important to cardiac surgery, such as hypertension, morbid obesity, the status of left ventricular function, etc. The higher the Parsonnet score a patient has prior to operation, the higher the risk the patient is associated with. The outcome variable is the 30-day postsurgery mortality of each patient. The data from the first 2 years (1992 and 1993 for about 2218 observations) were used as Phase I data to fit a logistic regression model of 30-day mortality on patients' Parsonnet scores. The fitted model was

$$\log\left(\frac{\pi_i}{1-\pi_i}\right) = -3.68 + .077X_i,$$

where π_i is the 30-mortality rate of a patient whose Parsonnet score is X_i. The fitted model was then used to adjust the CUSUM chart for monitoring the 30-day mortality of the patients from 1994 to 1998. Please see Steiner et al. (2000) for a detailed account.

In the analysis of the surgical data set and the construction of the risk-adjusted CUSUM charts, the logistic regression model was assumed to remain the same in

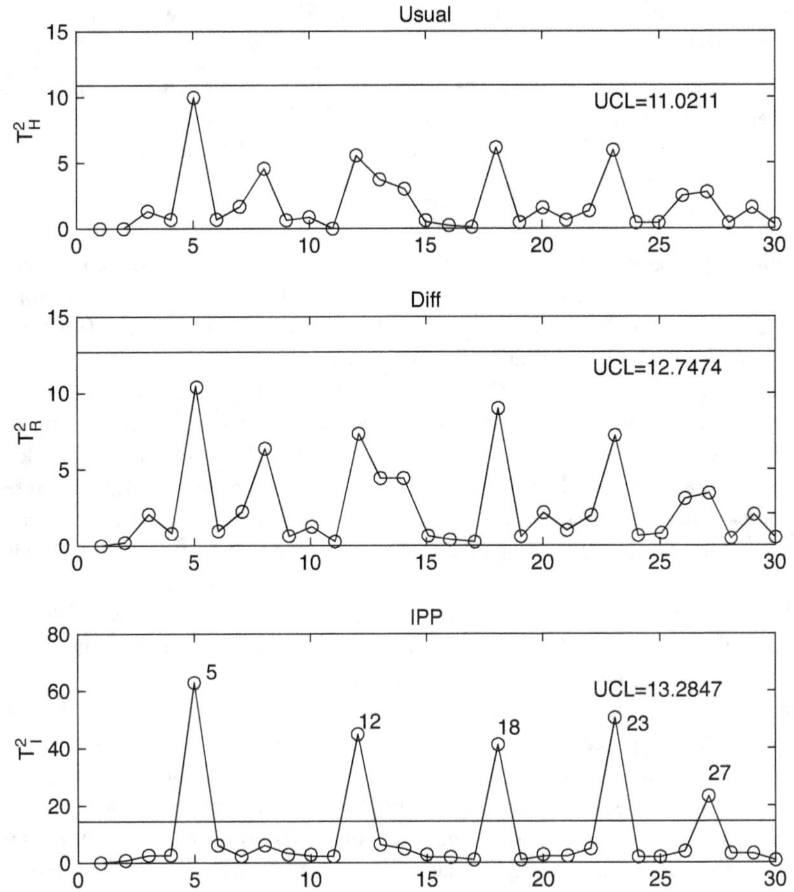

Figure 4.1 The T_H^2-, T_R^2-, and T_I^2-charts of the alloy fastener example.

Phase II monitoring of the 30-day mortality. If we consider the fitted logistic model based on 1992–1993 data as the in-control binary response profile characterizing the relationship between patients' Parsonnet scores and 30-day mortality rates, then the question naturally arises: Has the binary profile changed from 1994 to 1998?

In order to estimate the covariance matrix of the coefficient estimates based on 1992–1993 data, the binomial random observations, for each given X_i, were repeatedly generated. For each generated data set, the coefficient estimates were calculated, and the estimate of the covariance matrix of $\widehat{\boldsymbol{\beta}}$ was calculated as the sample covariance matrix of the calculated $\widehat{\boldsymbol{\beta}}$. Note that repeated generations of random observations in this example were necessary since the original observations were individual observations not grouped in rational subgroups. Furthermore, for each generated $\widehat{\boldsymbol{\beta}}$, the T^2 statistic was also calculated. These generated T^2 statistics were used to determine the UCL of the binary profile control chart. Using a Type-I error probability of 0.0027, we found the UCL to be 14.4858. The original data were collected from

Table 4.2 The Intercept, Slope Estimates, and T^2 Statistics of the Cardiac Surgeries Data

Year	94.1	94.2	95.1	95.2	96.1	96.2	97.1	97.2	98.1	98.2
Intercept	−3.84	−3.78	−3.43	−3.19	−3.11	−3.86	−3.99	−3.67	−3.79	−4.08
Slope	.082	.074	.075	.062	.050	.069	.072	.073	.090	.091
T^2 statistic	11.92	.32	18.66	26.31	28.42	3.45	6.42	1.97	14.04	5.27

individual patients. In order to apply the Phase II control chart for binary response profile monitoring for 1994–1998 data, we grouped the data by the operation dates with every six months constituting one subgroup, averaging about 470 patients per group. It should be noted that the grouping was rather arbitrary and was primarily done to ensure enough observations in each subgroup. In practical applications, other factors such as domain knowledge should be taken into account in forming the appropriate subgroup. For each of the subgroups from 1994 to 1998, we calculated the $\widehat{\boldsymbol{\beta}}$ and the corresponding T^2 statistics as discussed in Section 4.3. The intercept and slope estimates and the corresponding T^2 statistics are summarized in Table 4.2. The T^2 statistics are plotted against the sampling sequence along with the UCL in Figure 4.2.

It is evident from Figure 4.2 that the binary profile, as characterized by the logistic regression model of 30-day mortality on patient's Parsonnet score, is out-of-control for periods including 1995 and the first half of 1996. It is worth noting that the $\widehat{\boldsymbol{\beta}}$ in each of the three out-of-control periods sees significant decreases in both the intercept and slope parameters (see Table 4.2). The reduction in those parameters translates into reduction in 30-day mortality for similar Parsonnet score, as compared to the 1992–1993 baseline model. It should be noted that Biswas and Kalbfleisch (2008), Sego et al. (2009) and Steiner and Jones (2010) developed various types of risk-adjusted CUSUM or EWMA charts and applied the developed charts to the same data set. They all observed out-of-control samples in 1995 (but not in 1996). Because these risk-adjusted control charts mainly focus on the response variable, observing out-of-control samples on these charts would likely lead one to conclude that the 30-day mortality is out-of-control during 1995. By focusing on the binary profile, our analysis further reveals that the profile that governs how the patient's Parsonnet score affects the 30-day mortality has undergone some fundamental changes during the same period.

4.5 CONCLUSIONS

In this chapter, we discussed the Phase I and Phase II control charting techniques for detecting and monitoring binary response profiles. Two very different examples, one manufacturing and the other nonmanufacturing, were presented for demonstrating how the control charting techniques developed can be used in practice.

In many applications, there exists some dependency between the explanatory variables within the subgroup or some clustering due to a particular explanatory variable.

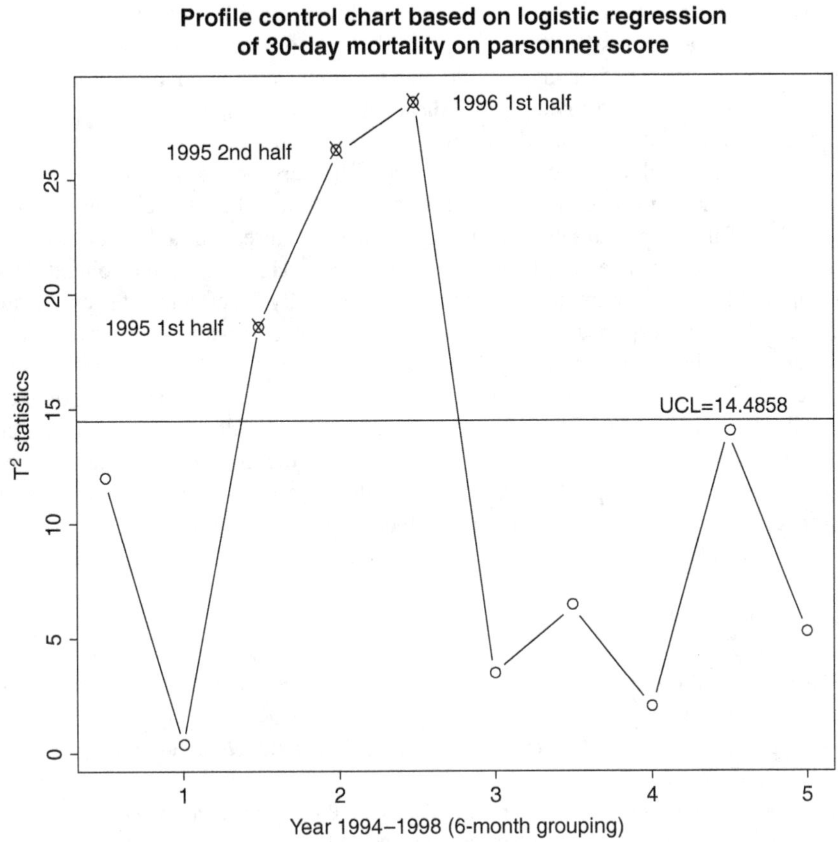

Figure 4.2 The Phase II binary profile control chart for cardiac surgery example.

For instance, in the surgical data, there appears to be some clustering of 30-day mortality due to different surgeons as well as different types of procedure. These potentially dependent or clustering structures can be reasonably captured by generalized linear mixed models. It would be worthwhile to study how statistical profile monitoring can be accomplished when the binary profile function is characterized by a generalized linear mixed model. Furthermore, for many practical applications in various industries, the response variable could be nominal with more than two categories, such as failure mode, or could be of discrete nature, such as the number of defective parts on printed circuit boards. These examples require different types of profile function modeling, and are therefore worthy of further investigation as to how the binary profile monitoring techniques discussed in this chapter can be extended to these response variables. For response that can be modeled as Poisson distribution, Roostaei (2008) studied control charting techniques for Phase II applications.

Other nonconventional modeling techniques such as Bayesian network and wavelets have been applied to profile monitoring for continuous responses. There exist other types of modeling techniques for dealing with binary or categorical responses such as regression tree, machine learning, or neural network. It would be interesting to see whether these modeling techniques can be adapted to monitoring profile, and what are the advantages or disadvantages of using these techniques versus the more traditional generalized linear model. Another important direction of potential future research is diagnostic techniques. From the perspective of using the T^2-chart to monitor the changes of parameters of a profile, one may be tempted, when the chart signals, to follow up with a number of diagnostic procedures available in the context of monitoring the mean vector of a multivariate process. However, in the context of profile monitoring, the parameters correspond to essentially the effects of the explanatory variables on the response. It is conceivable that, say, two parameters have changed to cause the chart to signal, and yet the parameters could change in such a way as to cancel each other out the effects of the corresponding explanatory variables on the response. In such a scenario, it is less likely to observe an out-of-control signal on a control chart applied solely on the response. In addition, the interpretations and diagnostics of out-of-control signals on a profile monitoring control chart will inevitably get more complicated if the profile model includes two-way or higher order interactions.

REFERENCES

Biswas, P. and Kalbfleisch, J. D. (2008) A risk-adjusted CUSUM in continuous time based on the Cox model. *Statistics in Medicine*, **27**(17), 3382–3406.

McCullagh, P. and Nelder, J. A. (1989) *Generalized Linear Models*, 2nd Edition, Chapman & Hall, London.

Montgomery, D. C., Peck, E. A., and Vining, G. G. (2006) *Introduction to Linear Regression Analysis*, 4th Edition, John Wiley & Sons, Inc., NJ.

Noorossana, R., Dorri, M., and Paynabar, K. (2010) Phase II monitoring of binary response profiles. Under review.

Parsonnet, V., Dean, D., and Bernstein, A. D. (1989) A model of uniform stratification of risks for evaluating the results of surgery in acquired adult heart disease. *Circulation*, **779**(1), 1–12.

Roostaei, A. (2008) *Monitoring Profiles Based on Poisson Observations* (in Farsi). M.S. Thesis in Industrial Engineering, Iran University of Science and Technology.

Sego, L. H., Reynolds Jr., M. R., and Woodall, W. H. (2009) Risk-adjusted monitoring of survival time. *Statistics in Medicine*, **28**(9), 1386–1401.

Steiner, S. H., Cook, R. J., Farewell, V. T., and Treasure, T. (2000) Monitoring surgical performance using risk-adjusted cumulative sum charts. *Biostatistics*, **1**, 441–452.

Steiner, S. H. and Jones, M. (2010) Risk-adjusted survival time monitoring with an updating exponentially weighted moving average (EWMA) control chart. *Statistics in Medicine*, **29**(4), 444–454.

Yeh, A. B., Huwang, L., and Li, Y. -M. (2009) Profile monitoring for a binary response. *IIE Transactions*, **41**(11), 931–941.

CHAPTER 5

Parametric Nonlinear Profiles

James D. Williams

INTRODUCTION

Profiles can take on several different functional forms, depending on the specific application. In some applications, the functional form of the profile cannot be adequately characterized using linear regression methods. Such profiles can be modeled either using parametric nonlinear regression or by nonparametric smoothing methods. Nonparametric regression techniques provide great flexibility in modeling the response. However, one disadvantage of nonparametric smoothing methods is that the subject-specific interpretation of the estimated nonparametric curve may be more difficult, and may not lead the user to discover as easily assignable causes that lead to an out-of-control signal.

Often, however, scientific theory or subject-matter knowledge leads to a natural parametric nonlinear function that describes the profiles well. Hence, one method is to model each profile by a parametric nonlinear regression function. The mathematical formulation for a *nonlinear profile* of an item is given by

$$y_{ij} = f(\mathbf{x}_{ij}, \boldsymbol{\beta}_i) + \epsilon_{ij}, \tag{5.1}$$

where $\mathbf{x}_{ij}$ is a $k \times 1$ vector of regressors for the jth observation of the ith profile, ϵ_{ij} is the random error, $\boldsymbol{\beta}_i$ is a $p \times 1$ vector of parameters for profile i, and f is a nonlinear function in the parameters $\boldsymbol{\beta}_i$. The random errors ϵ_{ij} are assumed to be i.i.d. normal random variates with mean zero and variance σ^2. In many applications, there is only one regressor ($k = 1$), but there are multiple parameters to monitor ($p > 1$). Williams et al. (2007), henceforth denoted WWB, give an example of a the

Statistical Analysis of Profile Monitoring, First Edition. Edited by Rassoul Noorossana, Abbas Saghaei and Amirhossein Amiri.

four-parameter logistic model often used to model dose–response profiles of a drug. The model is given by

$$y_{ij} = A_i + \frac{D_i - A_i}{1 + \left(\dfrac{x_{ij}}{C_i}\right)^{B_i}} + \epsilon_{ij}, \tag{5.2}$$

where y_{ij} is the measured response of the subject exposed to dose x_{ij} for batch i, $i = 1, \ldots, m,\ \ j = 1, \ldots, n$. In Equation (5.2), $k = 1$ and $p = 4$, giving four parameters to monitor, each parameter having a specific interpretation: A_i is the upper asymptote parameter, D_i is the lower asymptote parameter, B_i is the rate parameter for the ith batch, and C_i is the ED_{50} parameter. Another example is the "bathtub" function described in WWB where the density of particleboard is measured across the vertical profile. Note that for any given application, the specific form of the nonlinear function, f, in Equation (5.1) must be specified by the user.

The basic idea behind parametric nonlinear profile monitoring is to reduce the complex nonlinear profile into a few parameters through the nonlinear model estimation, then form a control chart scheme on the estimated parameters of each individual profile. The methods are usually multivariate in nature, since in most cases $p > 1$.

In a Phase I analysis, the concern is distinguishing between in-control conditions and the presence of assignable causes so that in-control parameters may be estimated for further product or process monitoring in Phase II analysis. If out-of-control observations are included in the estimation of in-control parameters, then the subsequent monitoring procedure will be less effective. Therefore, it is imperative in Phase I that outliers be identified and excluded from further analysis. Further, step or ramp shifts (if any) must be detected in the mean profile, so that in-control parameters may be estimated to reflect what would be expected from a stable process. In order to accomplish this a historical data set (HDS) consisting of m items sampled over time must be obtained. For each item i a response y_{ij} is observed and a set of predictor variables $\mathbf{x}_{ij}$, $i = 1, \ldots, m,\ \ j = 1, \ldots, n$, resulting in the quality profile for item i, i.e., $(y_{i1}, \mathbf{x}_{i1}), (y_{i2}, \mathbf{x}_{i2}), \ldots, (y_{in}, \mathbf{x}_{in})$.

5.1 NONLINEAR MODEL ESTIMATION

The n observations within each profile of the scalar model in Equation (5.1) are stacked into matrix form as $\mathbf{y}_i = (y_{i1}, y_{i2}, \ldots, y_{in})'$, $\mathbf{f}(\mathbf{X}_i, \boldsymbol{\beta}_i) = (f(\mathbf{x}_{i1}, \boldsymbol{\beta}_i), f(\mathbf{x}_{i2}, \boldsymbol{\beta}_i), \ldots, f(\mathbf{x}_{in}, \boldsymbol{\beta}_i))'$, and $\epsilon_i = (\epsilon_{i1}, \epsilon_{i2}, \ldots, \epsilon_{in})'$. The vector form of the nonlinear model is then given by

$$\mathbf{y}_i = \mathbf{f}(\mathbf{X}_i, \boldsymbol{\beta}_i) + \epsilon_i, \quad i = 1, \ldots, m. \tag{5.3}$$

Estimates of $\boldsymbol{\beta}_i$ must be obtained for each sample of the HDS. This can be accomplished by employing the iterative Gauss–Newton procedure to obtain the maximum likelihood estimates. The Gauss–Newton algorithm requires the $m \times p$

matrix of derivatives of $f(\mathbf{X}_i, \boldsymbol{\beta}_i)$ with respect to $\boldsymbol{\beta}_i$, which is given by

$$\mathrm{D}_i = \frac{\partial \mathbf{f}(\boldsymbol{X}_i, \boldsymbol{\beta}_i)}{\partial \boldsymbol{\beta}_i} = \begin{bmatrix} \frac{\partial f(\mathrm{x}_{i1},\beta_i)}{\partial \beta_{i1}} & \frac{\partial f(\mathrm{x}_{i1},\beta_i)}{\partial \beta_{i2}} & \cdots & \frac{\partial f(\mathrm{x}_{i1},\beta_i)}{\partial \beta_{ip}} \\ \frac{\partial f(\mathrm{x}_{i2},\beta_i)}{\partial \beta_{i1}} & \frac{\partial f(\mathrm{x}_{i2},\beta_i)}{\partial \beta_{i2}} & \cdots & \frac{\partial f(\mathrm{x}_{i2},\beta_i)}{\partial \beta_{ip}} \\ \vdots & \vdots & \ddots & \vdots \\ \frac{\partial f(\mathrm{x}_{in},\beta_i)}{\partial \beta_{i1}} & \frac{\partial f(\mathrm{x}_{in},\beta_i)}{\partial \beta_{i2}} & \cdots & \frac{\partial f(\mathrm{x}_{in},\beta_i)}{\partial \beta_{ip}} \end{bmatrix}. \tag{5.4}$$

Define $\mathbf{f}\left(\mathbf{X}_i, \hat{\boldsymbol{\beta}}_i^{(h)}\right) = f\left((\mathbf{x}_{i1}, \hat{\boldsymbol{\beta}}_i^{(h)}), f(\mathbf{x}_{i2}, \hat{\boldsymbol{\beta}}_i^{(h)}), \ldots, f(\mathrm{x}_{in}, \hat{\boldsymbol{\beta}}_i^{(h)})\right)'$, where $\hat{\boldsymbol{\beta}}_i^{(h)}$ is the estimate of $\boldsymbol{\beta}$ at iteration h, and define $\hat{\mathbf{D}}_i^{(h)}$ as the matrix of derivatives of f given in Equation (5.4) evaluated at $\hat{\boldsymbol{\beta}}_i^{(h)}$. Then the Gauss–Newton iterative solution for $\hat{\boldsymbol{\beta}}_i$ is given by

$$\hat{\boldsymbol{\beta}}_i^{(h+1)} = \hat{\boldsymbol{\beta}}_i^{(h)} + \left(\hat{\mathbf{D}}_i^{\prime(h)}\hat{\mathbf{D}}_i^{(h)}\right)^{-1}\hat{\mathbf{D}}_i^{\prime(h)}\left(\mathbf{y}_i - \mathbf{f}(\mathbf{X}_i, \hat{\boldsymbol{\beta}}_i^{(h)})\right).$$

Upon convergence of the algorithm, the estimated covariance matrix of $\hat{\boldsymbol{\beta}}_i$ is the estimated Fisher information matrix given by

$$\hat{Var}\left(\hat{\boldsymbol{\beta}}_i\right) = \hat{\sigma}_i^2(\hat{\mathbf{D}}_i'\hat{\mathbf{D}}_i)^{-1},$$

where $\hat{\sigma}_i^2 = \sum_{j=1}^{m}(y_{ij} - f(\mathbf{x}_{ij}, \hat{\boldsymbol{\beta}}_i))^2/(n-p)$ and $\hat{\mathbf{D}}_i$ is the derivative matrix in Equation (5.4) evaluated at the converged parameter vector estimate $\hat{\boldsymbol{\beta}}_i$. Myers (1990, Chapter 9) and Schabenberger and Pierce (2002, Chapter 5) give a concise discussion of nonlinear regression model estimation. A more detailed treatment can be found in Gallant (1987) and Seber and Wild (1989).

Unlike linear regression, the exact distribution of parameter estimators in nonlinear regression is unobtainable, even if the errors ϵ_{ij} are assumed to be i.i.d. normal random variables. Instead, asymptotic results must be applied. Seber and Wild (1989, Chapter 12) give the asymptotic distribution of $\hat{\boldsymbol{\beta}}_i$ and the necessary assumptions and regularity conditions for the asymptotic distribution to be obtained. Given their regularity conditions hold and assuming that $n^{-1}\hat{\mathbf{D}}'_i\hat{\mathbf{D}}_i$ converges to some nonsingular matrix Ω_i as $n \to \infty$, then the asymptotic distribution of $\hat{\boldsymbol{\beta}}_i$ is given by

$$\sqrt{n}(\hat{\boldsymbol{\beta}}_i - \boldsymbol{\beta}_i) \xrightarrow{D} N_p(0, \sigma^2\Omega_i^{-1}), \tag{5.5}$$

where $\boldsymbol{\beta}_i$ is the asymptotic expected value of $\hat{\boldsymbol{\beta}}_i$ and N_p indicates a p-dimensional multivariate normal distribution. For practical purposes, the distribution given by Equation (5.5) is incalculable since the matrix Ω_i is unknown. Instead the approximate asymptotic distribution of $\hat{\boldsymbol{\beta}}_i$ is commonly used, given by

$$\hat{\boldsymbol{\beta}}_i \dot{\sim} N_p(\boldsymbol{\beta}_i, \sigma_i^2(\mathbf{D}_i'\mathbf{D}_i)^{-1}). \tag{5.6}$$

This asymptotic distribution result will be exploited to determine approximate control limits in the multivariate T^2 control chart.

The standard estimator of $Var(\hat{\boldsymbol{\beta}}_i)$ is $\hat{\sigma}_i^2\hat{\mathbf{D}}_i'\hat{\mathbf{D}}_i$. For the most traditional "in-control" case, $\boldsymbol{\beta}_i = \boldsymbol{\beta}$ for all profiles $i = 1, \ldots, m$, where $\boldsymbol{\beta}$ is the in-control parameter vector. Consequently, the $\boldsymbol{\Omega}_i$ and $\mathbf{D}_i$ matrices are the same across all m items since all items have the same underlying model, f, the same x-values are observed, and the same values of $\boldsymbol{\beta}_i$. However, the $\hat{\mathbf{D}}_i$ matrices are not equal since the $\hat{\boldsymbol{\beta}}_i$ values vary from profile to profile.

5.2 PHASE I METHODS

The methods presented in this section are designed to detect out-of-control conditions in the HDS. Out-of-control cases are subsequently removed from the HDS, and the final in-control model parameters are estimated from the in-control HDS. The in-control estimates are then used for ongoing process monitoring in a Phase II analysis.

5.2.1 Multivariate T^2 Control Chart

In this section, the general framework of the multivariate T^2 statistic as well as its usefulness in a Phase I analysis is discussed. From the HDS there are m independent sample observation vectors, denoted, $\mathbf{w}_i$ $(i = 1, \ldots, m)$, each of dimension p. The general form of the T^2 statistic for sample i is

$$T_i^2 = (\mathbf{w}_i - \bar{\mathbf{w}})' \mathbf{S}^{-1} (\mathbf{w}_i - \bar{\mathbf{w}}), \tag{5.7}$$

where $\bar{\mathbf{w}} = \frac{1}{m}\sum_{j=1}^{m} \mathbf{w}_i$ and $\mathbf{S}$ is some estimator of the variance–covariance matrix of $\mathbf{w}_i$ (Mason and Young 2002). The T_i^2 statistics, $i = 1, \ldots, m$, are plotted against i, and out-of-control signals will be given for any T_i^2 value exceeding an upper control limit (UCL). It is usually assumed that each of the $\mathbf{w}_i$ vectors follows a multivariate normal distribution with common mean vector $\boldsymbol{\mu}$ and covariance matrix $\boldsymbol{\Sigma}$. This assumption is critical to finding the marginal distribution of T_i^2, as discussed in Section 5.2.1.1.

The vector of parameters $\boldsymbol{\beta}_i$ from the nonlinear regression model given in Equation (5.1) determines the shape of the curve $f(\mathbf{X}_i, \boldsymbol{\beta}_i)$. The multivariate T^2 statistic is employed to assess stability of the p parameters simultaneously. Individual control charts are not used for each of the p nonlinear regression parameters because of the correlation structure of the parameter estimators in nonlinear regression. The correlation structure can be accounted for in the multivariate T^2 statistic.

After calculating $\hat{\boldsymbol{\beta}}_i$ from each sample in the HDS, the average of the $\hat{\boldsymbol{\beta}}_i$ is calculated, which is denoted as $\bar{\hat{\boldsymbol{\beta}}}$. Further, some estimate of the covariance matrix is also calculated. Then $\mathbf{w}_i$ is replaced with $\hat{\boldsymbol{\beta}}_i$ and $\bar{\mathbf{w}}$ is replaced with $\bar{\hat{\boldsymbol{\beta}}}$ in Equation (5.7) to obtain the T^2 statistic

$$T_i^2 = \left(\hat{\boldsymbol{\beta}}_i - \bar{\hat{\boldsymbol{\beta}}}\right)' \mathbf{S}^{-1} \left(\hat{\boldsymbol{\beta}}_i - \bar{\hat{\boldsymbol{\beta}}}\right).$$

Large values of T_i^2 indicate abnormal $\hat{\boldsymbol{\beta}}_i$, suggesting that the profile for item i might be out-of-control.

There are several choices for the estimator $\mathbf{S}$. In their paper, WWB discuss three choices, which are given here.

The first choice for $\mathbf{S}$ is the *sample covariance* matrix, which is calculated by

$$\mathbf{S}_C = \frac{1}{m-1}\sum_{i=1}^{m}\left(\hat{\boldsymbol{\beta}}_i - \bar{\hat{\boldsymbol{\beta}}}\right)\left(\hat{\boldsymbol{\beta}}_i - \bar{\hat{\boldsymbol{\beta}}}\right)'. \tag{5.8}$$

Subsequently, the T_i^2 statistics are calculated as

$$T_{C,i}^2 = \left(\hat{\boldsymbol{\beta}}_i - \bar{\hat{\boldsymbol{\beta}}}\right)' \mathbf{S}_C^{-1}\left(\hat{\boldsymbol{\beta}}_i - \bar{\hat{\boldsymbol{\beta}}}\right). \tag{5.9}$$

Brill (2001) suggested use of the T_C^2 statistics in the context of monitoring nonlinear profiles of a chemical product. The advantage of this statistic is that it is very well understood and widely used. However, Sullivan and Woodall (1996) and Vargas (2003) demonstrated that T_C^2 is ineffective in detecting sustained shifts in the mean vector in a Phase I analysis. In fact, it was shown that as the step shift size increased, the power to detect the shift actually decreased.

An second choice of $\mathbf{S}$ is one based on *successive differences*, proposed originally by Hawkins and Merriam (1974) and again later by Holmes and Mergen (1993). To obtain the estimator, one first calculates the difference vector $\hat{\mathbf{v}}_i = \hat{\boldsymbol{\beta}}_{i+1} - \hat{\boldsymbol{\beta}}_i$ for $i = 1, \ldots, m-1$ and then stacks the transpose of these $m-1$ difference vectors into the matrix $\hat{\mathbf{V}}$ as

$$\hat{\mathbf{V}} = \begin{bmatrix} \hat{\mathbf{v}}_1' \\ \hat{\mathbf{v}}_2' \\ \vdots \\ \hat{\mathbf{v}}_{m-1}' \end{bmatrix}.$$

The estimator of the variance–covariance matrix is then calculated as

$$\mathbf{S}_D = \frac{\hat{\mathbf{V}}'\hat{\mathbf{V}}}{2(m-1)} \tag{5.10}$$

Sullivan and Woodall (1996) showed that $\mathbf{S}_D$ is an unbiased estimator of the true covariance matrix if the process is stable in Phase I. The resulting T_i^2 statistics are given by

$$T_{D,i}^2 = \left(\hat{\boldsymbol{\beta}}_i - \bar{\hat{\boldsymbol{\beta}}}\right)' \mathbf{S}_D^{-1}\left(\hat{\boldsymbol{\beta}}_i - \bar{\hat{\boldsymbol{\beta}}}\right). \tag{5.11}$$

Sullivan and Woodall (1996) and Vargas (2003) showed that a T^2 chart based on values of $T_{D,i}^2$ was effective in detecting both a step and ramp shift in the mean vector during Phase I. Sullivan and Woodall (1996) also showed that the T_D^2 values are invariant to a full-rank linear transformation on the observations.

The third choice for $\mathbf{S}$ is a robust estimator of the variance–covariance matrix known as the minimum volume ellipsoid (MVE) estimator, first proposed by Rousseeuw (1984) and studied in profile monitoring for Phase I analysis by Jensen et al. (2007). To apply the MVE method, one finds outlier-robust estimates for both the in-control parameter vector and the variance–covariance matrix. The MVE estimator is based on finding the ellipsoid with the smallest volume that contains at least half of the $\hat{\boldsymbol{\beta}}_i$ vectors, $i = 1, \ldots, m$. The MVE estimator of $\boldsymbol{\beta}$ is the mean vector of the smallest ellipsoid, and the estimator of the variance–covariance matrix is the sample covariance matrix of the observations within the smallest ellipsoid multiplied by a constant to make the estimator unbiased for multivariate normal data. In a simulation study, Vargas (2003) studied the power properties of several different choices of $\mathbf{S}$ in the context of the T^2 statistic given in Equation (5.7) and found that the T^2 statistic based on the MVE estimators of $\boldsymbol{\beta}$ and the variance–covariance matrix was very powerful in detecting multivariate outliers. The MVE estimators of $\boldsymbol{\beta}$ and the covariance matrix are denoted by $\hat{\boldsymbol{\beta}}_{MVE}$ and $\mathbf{S}_{MVE}$, respectively. Hence, the third choice of T^2 is

$$T^2_{MVE,i} = \left(\hat{\boldsymbol{\beta}}_i - \hat{\boldsymbol{\beta}}_{MVE}\right)' \mathbf{S}^{-1}_{MVE} \left(\hat{\boldsymbol{\beta}}_i - \hat{\boldsymbol{\beta}}_{MVE}\right), \quad i = 1, \ldots, m. \tag{5.12}$$

5.2.1.1 *Control Limits*

The distribution of the T_i^2 statistics for monitoring nonlinear profiles is more complex than in the case of linear profiles. The asymptotic distribution of $\hat{\boldsymbol{\beta}}_i$, $i = 1, \ldots, m$, must be employed instead. In order to determine the marginal distribution of T_i^2 in this case, it is assumed that the sample size, n, from each item in the baseline data set is of sufficient size such that the distributions of $\hat{\boldsymbol{\beta}}_i$, $i = 1, \ldots, m$, are approximately multivariate normal. The subsequent UCLs for the multivariate T^2 control charts are determined on the basis of this normality assumption.

In order to control the overall probability of a false alarm based on some appropriate UCL, the joint distribution of the T_i^2 values is required. However, these statistics are correlated since $\bar{\hat{\boldsymbol{\beta}}}$ and $\mathbf{S}$ are used in all T_i^2 statistics ($i = 1, \ldots, m$), thus making the joint distribution of the T_i^2 values difficult to obtain. As an alternative, Mahmoud and Woodall (2004) suggested using an approximate joint distribution assuming that the T_i^2 statistics are independent. Let α be the probability of a false alarm for any individual T_i^2 statistic. Then the approximate overall probability of a false alarm for a sample of m items is given by $\alpha_{overall} = 1 - (1 - \alpha)^m$. Thus, for a given overall probability of a false alarm, $\alpha = 1 - (1 - \alpha_{overall})^{1/m}$ is used in the calculation of UCLs. In their simulation study, Mahmoud and Woodall (2004) found that this approximation used to determine the UCLs performed well.

Tracy et al. (1992) noted in their paper that Gnanadesikan and Kettenring (1972) proved that for a stable process the marginal distribution of $T_{C,i}^2$ is proportional to a

β distribution, i.e.,

$$T_{C,i}^2 \frac{m}{(m-1)^2} \sim B\left(\frac{p}{2}, \frac{m-p-1}{2}\right).$$

Chou et al. (1999) give a formal proof. One key assumption is that the distribution of $\hat{\boldsymbol{\beta}}_i$ is approximately normal. Hence, an approximate UCL is

$$UCL_C = \frac{(m-1)^2}{m} B_{1-\alpha, p/2, (m-p-1)/2}, \tag{5.13}$$

where $B_{1-\alpha, p/2, (m-p-1)/2}$ is the $1-\alpha$ quantile of a β distribution with shape parameters $p/2$ and $(m-p-1)/2$.

The marginal distribution of the $T_{D,i}^2$ statistic is unknown. However, Williams et al. (2006) and Williams et al. (2009) gave an approximate distribution based on the chi-squared distribution for large sample sizes and an approximate distribution based on the $\boldsymbol{\beta}$ distribution for small sample sizes. For large sample sizes where $m > p^2 + 3p$, the UCL is

$$UCL_D = \chi^2(1-\alpha, p).$$

For small sample sizes where $m \leq p^2 + 3p$ and $p < 10$, the UCL is a vector given by

$$\mathbf{UCL}_D = (UCL_1, UCL_2, \ldots, UCL_m), \tag{5.14}$$

where

$$UCL_i = MV(m, i)\, BETA_{1-\alpha, \beta(m,p,i), \gamma(m,p,i)}, \quad i = 1, \ldots, m,$$

and $\beta(m, p, i)$ and $\gamma(m, p, i)$ are functions of m, p, and i that define the two shape parameters for the β distribution. The equations for $\beta(m, p, i)$ and $\gamma(m, p, i)$ given by Williams et al. (2006) are included in the Appendix.

The exact marginal distribution of $T_{MVE,i}^2$ is also unknown and intractable. Hence, in order to find the UCL for $T_{MVE,i}^2$ simulation must be employed.

5.2.2 Lack-of-Fit Control Chart

Another type of out-of-control profile is one where the functional form is different from the hypothesized model. Hence, in addition to monitoring the mean profile for abnormalities, it also may be of interest to check for changes in the profile function over time. This can be done by employing the lack-of-fit (LOF) control chart proposed by Williams et al. (2007), henceforth denoted WBWF. In order to compute the LOF control chart, the observations taken at each point must be replicated. The replicated observations are denoted as $\mathbf{x}_{ijk}$, where $k = 1, \ldots, r$ and $r > 1$. Replication is common for dose–response monitoring schemes in the agricultural products and pharmaceutical industries, among others.

The LOF statistic proposed in WBWF compares the error sum of squares of a saturated model, sometimes called the means or full model, to the error sum of squares of the specified nonlinear regression model. The sum of squared errors for the saturated model is

$$SSE_i^{full} = \sum_{j=1}^{n} \sum_{k=1}^{r} (y_{ijk} - \bar{y}_{ij.})^2,$$

where $\bar{y}_{ij.} = r^{-1} \sum_{k=1}^{r} y_{ijk}$ is the mean of the r replications at the point j in profile i. Further, let $f(\mathbf{x}_{ijk}, \hat{\boldsymbol{\beta}}_i)$ represent the estimated profile replication k of profile i, then the sum of square errors for the nonlinear regression model is given by

$$SSE_i^{reg} = \sum_{j=1}^{n} \sum_{k=1}^{r} (y_{ijk} - f(\mathbf{x}_{ijk}, \hat{\boldsymbol{\beta}}_i))^2.$$

The LOF statistic compares the nonlinear model fit to the saturated model. The LOF statistic for profile i is

$$LOF_i = \frac{(SSE_i^{reg} - SSE_i^{full})/df^{LOF}}{SSE_i^{full}/df^{full}}, \tag{5.15}$$

where $df^{LOF} = n - p$ is the degrees of freedom for the LOF and $df^{full} = n(r-1)$ is the degrees of freedom of the full model. Neill (1988) demonstrated that the LOF_i statistic for the nonlinear regression approximately follows an $\mathcal{F}$ distribution with df^{LOF} and df^{full} numerator and denominator degrees of freedom, respectively.

To assess LOF for profile i, one plots all LOF_i statistics versus i in an LOF chart. The UCL for the LOF chart is based on the $\mathcal{F}$ distribution, and is given by

$$UCL_{LOF} = \mathcal{F}(1-\alpha, df^{LOF}, df^{full}), \tag{5.16}$$

where $\mathcal{F}(q, df_1, df_2)$ is the qth quantile from an $\mathcal{F}$ distribution with df_1 and df_2 numerator and denominator degrees of freedom, respectively. In those cases where LOF_i exceeds UCL_{LOF}, the proposed parametric nonlinear model is determined to be inadequate for profile i. Those profiles that exhibit LOF are subject to being removed from the HDS.

5.2.3 Monitoring the Variance σ^2

In addition to checking abnormal mean profiles in the HDS, it is also important to check for abnormal variance about each profile. This is analogous to monitoring the process variance in the standard univariate case. In nonlinear profile monitoring, it is of interest to monitor the within-profile variability. The measure of within-profile variability is the mean square error (MSE) defined as $MSE_i = \sum_{j=1}^{n}(y_{ij} - \hat{y}_{ij})^2/(n-p)$, where $\hat{y}_{ij}$ is the predicted value of y_{ij} based on the nonlinear regression model in Equation (5.1). Wludyka and Nelson (1997) recommended a method to monitor variances based on an analysis-of-means-type test utilizing $S_i^2 = MSE_i$. In their paper, S_i^2 is plotted against i with associated lower- and upper-control limits equal to $(L_{\alpha,m,n-p})m\overline{S}^2$ and $(U_{\alpha,m,n-p})m\overline{S}^2$, respectively, where L and U are critical values given in their paper and $\overline{S^2}$ is the average of the S_i^2 values, $i = 1, \ldots, m$. For large n, their approximate upper and lower control limits are $\overline{S}^2 \pm h_{\alpha,m,\infty}\hat{\sigma}$, where h is a critical value given in Nelson (1983) and $\hat{\sigma} = \overline{S}^2\sqrt{2(m-1)/m(n-p)}$. The S_i^2 statistics are plotted on a separate control chart to monitor the variance of the error terms and LOF simultaneously with a T^2 control chart for the nonlinear regression parameters. However, use of this method is recommended only when within-profile error terms are independent.

5.2.4 In-Control Profile Estimate

The main purpose of a Phase I analysis is to estimate in-control parameters, including parameters for the mean profile and parameters for the variance. Then in ongoing process monitoring, new profiles are then monitored in a Phase II analysis against the baseline profile determined in Phase I. In Phase I, out-of-control profiles are removed from the HDS and in-control parameters are estimated from the remaining in-control profiles in the HDS. Let m^* be the number of in-control profiles remaining in the HDS. Then the in-control mean profile is estimated as $\mathbf{f}(\mathbf{X}, \boldsymbol{\beta})$ from Equation (5.3), where

$$\boldsymbol{\beta} = \frac{1}{m^*}\sum_{i=1}^{m^*}\hat{\boldsymbol{\beta}}_i. \tag{5.17}$$

The final estimate of the variance–covariance matrix of $\hat{\boldsymbol{\beta}}$, denoted by $\boldsymbol{\Sigma}_\beta$, is given by

$$\boldsymbol{\Sigma}_\beta = \frac{1}{m^*-1}\sum_{i=1}^{m^*}\left(\hat{\boldsymbol{\beta}}_i - \bar{\hat{\boldsymbol{\beta}}}\right)\left(\hat{\boldsymbol{\beta}}_i - \bar{\hat{\boldsymbol{\beta}}}\right)'. \tag{5.18}$$

Finally, the in-control estimate of σ^2 is given by

$$\sigma^2 = \frac{1}{m^*}\sum_{i=1}^{m^*}\sum_{j=1}^{n}\frac{(y_{ij} - f(\mathbf{x}_{ij}, \hat{\boldsymbol{\beta}}_i))^2}{n-p}. \tag{5.19}$$

One basic assumption is that the estimated parameters from Equations (5.17)–(5.19) are known and exact. This assumption affects the distributional properties of the control chart statistics in Phase II. One should also check that $\boldsymbol{\Sigma}_\beta$ is positive definite, since this is a requirement in further stages of the analysis. One can check this assumption by calculating the eigenvalues of the $\boldsymbol{\Sigma}_\beta$ matrix and checking for any non-positive eigenvalues.

5.2.5 VDP Data Example

WWB gave the following example using the vertical density profile (VDP) data from Walker and Wright (2002) to illustrate the nonlinear profile monitoring methods of this section. The original data is available at the Web site http://filebox.vt.edu/users/bwoodall/VDP%20nonlinear%20profile%20data.txt. In the manufacture of particleboard, the quality characteristic monitored over time is the density properties of the finished boards. In manufactured particleboard, the density (in pounds per ft^3) near the core or center of a particleboard is much less than the density at the top and bottom faces of a board (see Young et al. 1999). The industry standard sampling procedure calls for a laser-aided density measuring device that scans fixed vertical depths of a board and records the density at each depth. Since the scan depths are fixed for each sample, the depth x_{ij} is denoted by simply x_j. Density measurements for this HDS were taken at depths of $x_j = (0.002)j$ inches, $j = 0, 1, 2, \ldots, 313$, giving a sequence of ordered pairs, (x_j, y_{ij}), $j = 1, \ldots, n$. The set of order pairs forms a *VDP* of the board. The HDS contains 24 particleboards measured in this way, and the 24 profiles are illustrated in Figure 5.1.

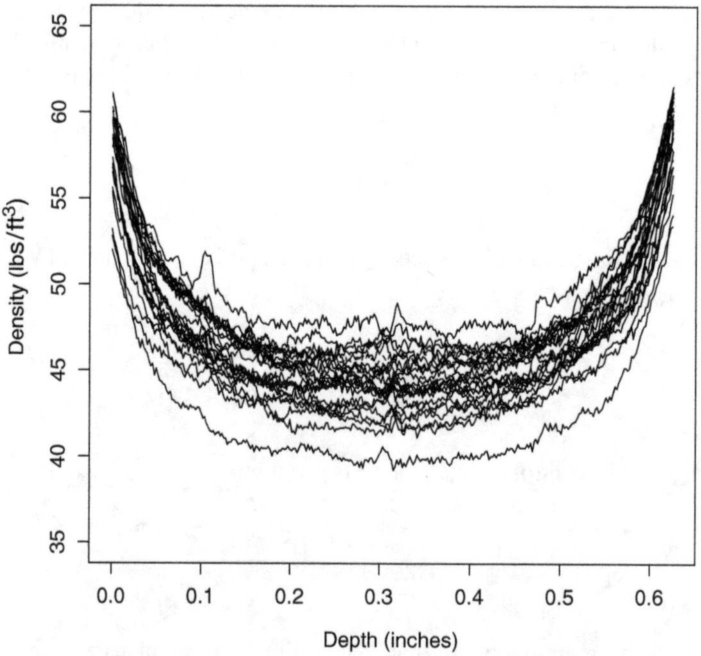

Figure 5.1 Vertical density profile (VDP) of 24 particleboards.

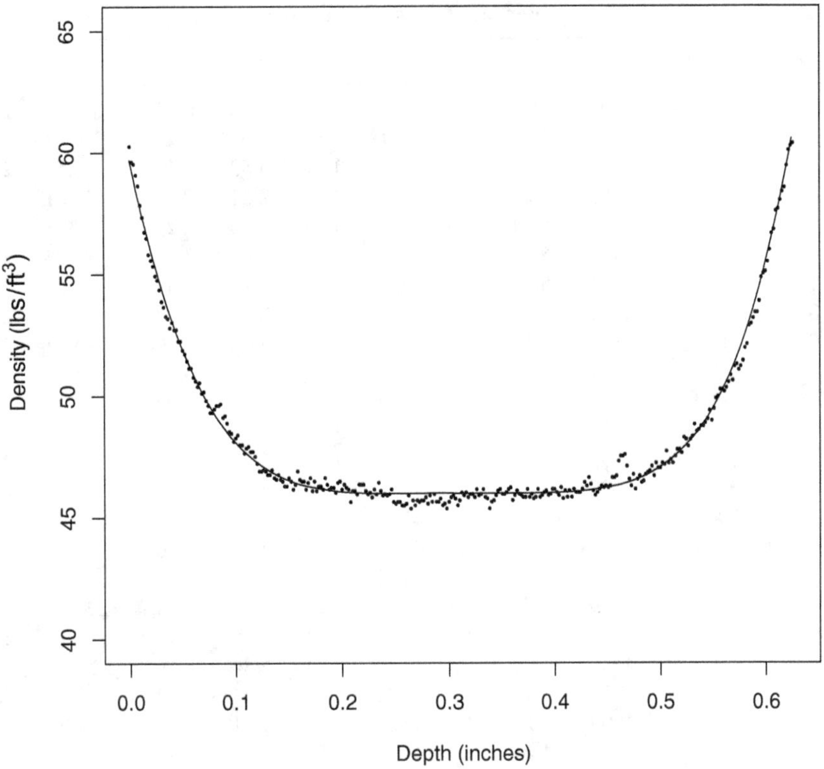

Figure 5.2 "Bathtub" function fit to board 1.

The nonlinear function used to model particleboard profile i is a "bathtub" function given by

$$f(x_{ij}, \boldsymbol{\beta}) = \begin{cases} a_1(x_{ij} - c)^{b_1} + d & x_j > c \\ a_2(-x_{ij} + c)^{b_2} + d & x_j \le c \end{cases}, \quad i = 1, \ldots, m; \quad j = 1, \ldots, n, \tag{5.20}$$

where $\boldsymbol{\beta} = (a_1, a_2, b_1, b_2, c, d)$. One advantage of this nonlinear model is the interpretability of the model parameters. For example, a_1, a_2, b_1, and b_2 determine the "flatness," c is the center, and d is the bottom, or the "level" of the curve. Differing values of a_1 and a_2 or different values of b_1 and b_2 allow for an asymmetric curve about the center c. Figure 5.2 contains the "bathtub" function fit to board 1 from the VDP data.

Although this parametric nonlinear regression model does not explain all of the variability in the VDP of board 1, the bathtub function model has $R^2 > 0.9999$, and is acceptable for the purposes of this analysis. The bathtub function of Equation (5.20) was fit to each of the 24 boards in the HDS, and the T_C^2, T_D^2, and T_{MVE}^2 statistics of Equations (5.9), (5.11), and (5.12), respectively, were calculated based on the $\hat{\boldsymbol{\beta}}_i$

Table 5.1 Estimated Parameter Values and T^2 Statistics for the VDP Data

Board	$\hat{a}_1$	$\hat{a}_2$	$\hat{b}_1$	$\hat{b}_2$	$\hat{c}$	$\hat{d}$	T_C^2	T_D^2	T_{MVE}^2	UCL_D
1	6560	3259	5.63	4.40	45.98	0.29	2.65	1.91	6.00	27.88
2	470	291	3.01	2.74	42.08	0.32	7.56	5.27	6.97	22.29
3	1812	2871	3.99	5.02	47.66	0.34	5.83	7.17	8.64	22.27
4	6171	15,009	4.25	7.39	46.63	0.39	12.21	17.28	1131.81	22.24
5	4963	2251	5.14	4.20	43.43	0.30	1.65	2.27	2.88	22.21
6	4556	3758	5.28	4.72	40.13	0.30	8.49	13.03	9.83	22.17
7	5542	3815	5.25	5.00	44.15	0.31	2.15	3.49	3.58	22.12
8	3664	2979	4.89	4.41	44.06	0.30	0.79	0.97	2.69	22.07
9	28,041	8872	7.58	4.95	43.22	0.26	4.62	7.10	385.03	22.01
10	1640	1207	4.17	3.39	41.84	0.28	4.30	5.05	4.61	21.95
11	3492	1031	5.82	3.17	46.06	0.25	8.66	8.95	10.00	21.91
12	915	750	3.45	3.52	44.37	0.32	1.80	1.99	2.22	21.88
13	989	1392	3.58	4.05	45.47	0.32	3.42	4.42	5.18	21.88
14	1474	620	4.82	3.29	42.52	0.27	3.28	4.50	7.04	21.91
15	129,068	5420	12.40	3.33	45.90	0.15	21.45	22.18	17018.91	21.95
16	10,166	3822	5.83	4.86	44.19	0.30	3.83	5.60	12.93	22.01
17	1483	603	4.07	3.26	44.83	0.30	2.30	2.53	2.36	22.07
18	31,156	31,069	7.70	5.94	46.46	0.27	14.55	19.75	8221.00	22.12
19	418	198	3.22	2.67	42.84	0.30	4.58	3.90	5.16	22.17
20	3207	4741	4.88	5.02	44.45	0.30	5.34	5.59	34.00	22.21
21	672	773	3.37	3.37	44.46	0.31	2.64	3.42	2.79	22.24
22	3520	1807	5.10	4.01	45.52	0.29	1.71	1.37	1.73	22.27
23	1979	845	4.24	3.66	45.53	0.32	4.45	4.85	7.38	22.29
24	6095	26,778	5.41	6.67	44.46	0.31	9.75	10.55	6676.21	27.88

values. Parameter estimates for each of the twenty-four boards and the corresponding T^2 statistics are given in Table 5.1.

The control limits for the T_C^2 and T_D^2 statistics were calculated from Equations (5.13) and (5.14), respectively. The UCL for the T_{MVE}^2 statistic was simulated to achieve an overall probability of a signal equal to 0.05 for $m = 24$ boards. In this simulation, samples were drawn from a multivariate normal distribution of dimension six, mean vector zero, and variance–covariance matrix I. The identity covariance matrix was used since the in-control performance of the methods does not depend on the assumed in-control parameter vector or the variance–covariance matrix. The simulation was repeated 200,000 times, which gives a standard error for the estimated control limits of less than 0.0005. The UCL values are 14.72 and 65.37, for the T_C^2 and T_{MVE}^2 control charts, respectively. The control limit vector for the T_D^2 control chart is given in Table 5.1.

The purpose of a Phase I analysis is to identify "outlying" or out-of-control boards or a shift in the process that might affect the estimation of in-control parameters. Subsequently, the three T^2 control charts were compared for assessing process stability and identifying outlying profiles. Figure 5.3 below is a plot of all three T^2 control charts for the VDP data.

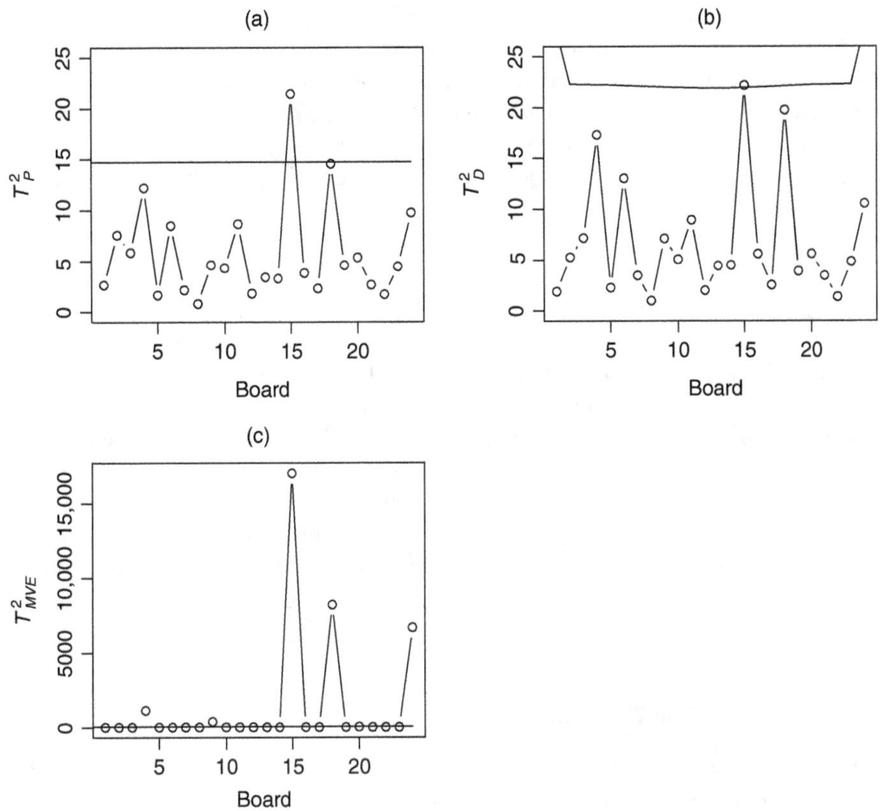

Figure 5.3 The T^2 control charts for the VDP data. (a) The T_C^2 control chart based on the sample covariance matrix, (b) T_D^2 control chart based on the successive differences estimator, and (c) T_{MVE}^2 control chart based on the minimum volume ellipsoid.

Both the T_C^2 and the T_D^2 control charts signal that board 15 is a potential out-of-control profile, and the profile for board 18 produces a signal in the T_C^2 control chart. Note that the T_D^2 statistic accentuates the same outlying observations of the T_C^2 chart, but has a larger UCL. Sullivan and Woodall (1996) found that the T_C^2 control chart has greater power to detect isolated outlying observations than the T_D^2 control chart based on the successive differences variance–covariance matrix estimator; however, the T_D^2 chart is better for detecting a sustained shift in the mean vector. For this HDS, there is no apparent sustained shift in the regression parameter vector.

The T_{MVE}^2 control chart based on the minimum variance ellipsoid estimator indicates that boards 4, 9, 15, 18, and 24 have outlying profiles. The most pronounced outlier is board 15, which both the T_C^2 and the T_D^2 charts also indicated as the most severe outlier. As shown in Vargas (2003), the T_{MVE}^2 control chart is very powerful in detecting multivariate outliers. Investigating the table of parameter estimates for these boards, given in Table 5.1, it seems reasonable that the boards 15 and 18 are outliers, with boards 4, 9, and 24 worthy of further investigation.

In addition to monitoring the regression parameter vectors of the profiles in a Phase I analysis, the variation about the profiles should be monitored to check for stability. In their paper, WWB recommended using the methods of Wludyka and Nelson (1997) to monitor the variance σ^2. Use of their method is appropriate when the error terms within a profile are independent. In this VDP example, however, the within-profile density measurements are spatially correlated so the variance chart is not used.

5.3 PHASE II METHODS

In-control model parameters are estimated in a Phase I analysis. A Phase II analysis consists of continuous monitoring of ongoing profiles from the product or process, denoted by $\mathbf{y}_i, i = 1, \ldots$. The methods presented in this section are designed to detect out-of-control conditions of new profiles compared to the baseline profile estimated from the Phase I analysis. For each new profile, $\mathbf{y}_i$, the nonlinear model estimation methods of Section 5.1 are applied to obtain the estimated parameter vector $\hat{\boldsymbol{\beta}}_i$, the fitted sample profile $\mathrm{f}(\mathbf{X}_i, \hat{\boldsymbol{\beta}}_i)$, and the estimate of the profile variance S_i^2. Phase II control charts are then applied on $\hat{\boldsymbol{\beta}}_i$ and S_i^2 to detect changes in the mean profile, changes in the variance of the profile, or both.

5.3.1 Multivariate T^2 Control Chart

Analogous to the multivariate T^2 control chart given in Equation (5.7) for a Phase I application, the multivariate T^2 control chart to monitor $\hat{\boldsymbol{\beta}}_i$ for Phase II applications is given by

$$T_i^2 = (\hat{\boldsymbol{\beta}}_i - \boldsymbol{\beta})' \boldsymbol{\Sigma}_{\beta}^{-1} (\hat{\boldsymbol{\beta}}_i - \boldsymbol{\beta}), \tag{5.21}$$

where $\boldsymbol{\beta}$ and $\boldsymbol{\Sigma}_{\beta}$ are given in Equations (5.17) and (5.18), respectively. Since $\boldsymbol{\beta}$ and $\boldsymbol{\Sigma}_{\beta}$ are assumed to be known, then T_i^2 of Equation (5.21) has a chi-square distribution with p degrees of freedom. Hence, a UCL for this control chart is given by $UCL_{T^2} = \chi^2(1 - \alpha, p)$.

In their paper, WBWF give a more conservative UCL for T_i^2 by relaxing the assumption that the parameters in Equations (5.17) and (5.18) are known and exact. This UCL is based on the fact that these parameters are estimated rather than known. This UCL is based on a results in Mason et al. (2001), and is given by

$$UCL_{T^2} = \frac{p(m^* + 1)(m^* - 1)}{m^*(m^* - p)} \mathcal{F}(1 - \alpha, p, m^* - p),$$

where $\mathcal{F}(\alpha, df_1, df_2)$ is the α quantile of an F distribution with df_1 numerator and df_2 denominator degrees of freedom. Whenever T_i^2 exceeds UCL_{T^2} then profile i is declared to be out of control, and is subject to further investigation.

5.3.2 EWMA Control Chart

Vaghefi et al. (2009) extended the EWMA control chart methods of the linear profile given in Kang and Albin (2000) to the nonlinear profile case. This method is based on monitoring the residuals between the reference and sample profiles, and can be applied to either parametric nonlinear profiles $\mathbf{f}(\mathbf{X}_{ij}, \boldsymbol{\beta}_i)$ or nonparametric nonlinear profiles, that is, mean profile curves estimated using nonparametric regression techniques. The EWMA chart is based on monitoring the expected value of

$$e_{ij} = y_{ij} - f(\mathbf{x}_{ij}, \boldsymbol{\beta}), \tag{5.22}$$

which is the error between the reference profile and the sample profile. Recall that $\boldsymbol{\beta}$ was calculated from the in-control profiles of the HDS in a Phase I analysis, as described in Section 5.2.4. The EWMA statistics are calculated as $z_i, i = 1, 2, 3, \ldots,$ where $z_i = \lambda \bar{e}_i + (1 - \lambda) z_{i-1}$, $\bar{e}_i$ is the sample mean of the residuals from profile i, $0 < \lambda \leq 1$ is the smoothing parameter, and z_0 is assumed to equal zero. The upper and lower control limits of the EWMA control chart are symmetric around zero, and are given by

$$\pm c_1 \sigma \sqrt{\frac{\lambda}{(2 - \lambda) n}},$$

where c_1 is a constant chosen to produce the desired average run length (ARL) and $\sigma = \sqrt{\sigma^2}$ from Equation (5.19).

5.3.3 MCUSUM Control Chart

Vaghefi et al. (2009) proposed a multivariate cumulative sum (MCUSUM) control chart to monitor the parameter vector $\hat{\boldsymbol{\beta}}_i$. Their method assumes that $\hat{\boldsymbol{\beta}}_i$ comes from a multivariate normal distribution with either an in-control mean denoted by $\boldsymbol{\beta}_0$ or an out-of-control mean denoted by $\boldsymbol{\beta}_\delta$, where $\boldsymbol{\beta}_\delta = \boldsymbol{\beta} + \boldsymbol{\delta}$ and $\boldsymbol{\delta}$ is the magnitude of the shift one is interested in detecting. The MCUSUM statistic is given by

$$S_i = max(S_{i-1} + \mathbf{a}'(\hat{\boldsymbol{\beta}}_i - \boldsymbol{\beta}) - 0.5D, 0), \tag{5.23}$$

where $S_0 = 0$,

$$D = \sqrt{\boldsymbol{\delta}' \boldsymbol{\Sigma}_\beta^{-1} \boldsymbol{\delta}}$$

and $\mathbf{a}$ is a $p \times 1$ vector given by

$$\mathbf{a}' = \frac{\boldsymbol{\delta}' \boldsymbol{\Sigma}_\beta^{-1}}{D}.$$

A signal is given whenever S_i exceeds the UCL, which can set to obtain an overall ARL.

5.3.4 Control Charts Based on Metrics

Other Phase II control charts that can be applied to either parametric or nonparametric linear profiles are those given in Vaghefi et al. (2009) based on metrics. These charts, like the EWMA chart in Section 5.3.2, are based on the residual error e_{ij} of Equation (5.22). Each metric is designed to measure a different aspect of the "distance" between the reference profile and the sample profile. The metrics are defined as

$$
\begin{aligned}
M_{i1} &= \max_{1 \leq j \leq n} |e_{ij}| \\
M_{i2} &= \sum_{j=1}^{n} |e_{ij}| \\
M_{i3} &= \sum_{j=1}^{n} n^{-1} |e_{ij}| \\
M_{i4} &= \sum_{j=1}^{n} e_{ij}^2 \\
M_{i5} &= \left| S(\mathbf{y}_i) - \int_{x_1}^{x_n} f(\mathbf{x}_i, \boldsymbol{\beta}) dx \right|,
\end{aligned}
$$

where $S(\mathbf{y}_i)$ approximates the integral of y between the range of the x_{ij} by the trapezoidal method. The four metrics M_{i1} through M_{i4} represent the maximum deviation, sum of absolute deviations, the mean absolute deviation, and the sum of square deviations between the reference profile and the sample profile, respectively. The metric M_{i5} represents the absolute difference between the area under the reference profile and the area under the sample profile.

The UCL for the control charts based on metrics $M_{i1} - M_{i5}$ must be obtained via simulation in order to obtain the desired ARL.

5.3.5 Lack-of-Fit Control Chart

If there exists replication in the observed profile at each point $\mathbf{x}_{ij}$, then LOF can be monitored at time i using methods analogous to those in Section 5.2.2. From the sample profile obtained at time i, the LOF statistic is calculated as

$$LOF_i = \frac{(SSE_i^{reg} - SSE_i^{full})/df_i^{LOF}}{SSE_i^{full}/df^{full}}.$$

As mentioned in Section 5.2.2, the LOF_i statistic follows an approximate $\mathcal{F}$ distribution with df^{LOF} and df^{full} numerator and denominator degrees of freedom, respectively. The LOF_i statistics are plotted by i in a LOF chart. The associated UCL is given by

$$UCL_{LOF} = \mathcal{F}(1 - \alpha, df^{LOF}, df^{full}).$$

Whenever LOF_i exceeds UCL_{LOF}, then it is determined that the proposed dose–response model has changed for profile i, and is subject to further investigation.

5.3.6 Monitoring the Variance

In addition to monitoring the mean of the profiles, it is also important to check for changes in the residual variance of the profiles. From a Phase I analysis, the in-control residual variance is estimated to be σ^2 of Equation (5.19). Similar to the Phase I control chart for the variance given in Section 5.2.3, the Phase II control statistic to monitor the variance is based on the estimated within-profile error variance, estimated by the MSE. The chart statistic is calculated as

$$\chi^2 = \frac{(n-p)S_i^2}{\sigma^2},$$

which has an asymptotic χ^2 distribution with $n-p$ degrees of freedom. Hence, an approximate UCL for the chart is given by $UCL = \chi^2(1-\alpha, n-p)$.

5.4 VARIANCE PROFILES

In some applications where $\mathbf{x}_{ij}$ is replicated r times, such as is common in dose–response profiles of a drug or chemical, one may find that the variance is not homoscedastic throughout the range of $\mathbf{x}_{ij}$. One would expect that the variability in the response of an organism to large doses of a chemical would be much smaller than the variability in the response to small doses. In this case, the variance can be described as some function of dose. In their paper, WBWF give an example of a heteroscedastic dose–response relationship of a chemical in the agricultural crop protection industry.

The body of literature on variance function modeling is extensive. A thorough treatment of variance function estimation is given in Davidian and Carroll (1987), Carroll and Ruppert (1988), and Arbogast and Bedrick (2004). Davidian and Carroll (1987) characterized the general variance function model as

$$\text{Var}(y_{ijk}) = \sigma_{ij}^2 = \sigma_i^2 g(\mathbf{z}_{ij}, \boldsymbol{\beta}_i, \boldsymbol{\theta}_i),$$

where in this case σ_i^2 is a scale parameter for profile i and $g(\cdot)$ is some function of regressor variables $\mathbf{z}_{ij}$, the parameter vector $\boldsymbol{\beta}_i$, and other parameters $\boldsymbol{\theta}_i$. The variance predictor variables $\mathbf{z}_{ij}$ can be $\mathbf{x}_{ij}$, but not necessarily so. The form of $g(\cdot)$ will depend on the specific application.

When there are replications at every $\mathbf{x}_{ij}$, then an unbiased estimator of σ_{ij}^2 can be obtained. The estimator is

$$\hat{\sigma}_{ij}^2 = S_{ij}^2 = \frac{1}{r-1}\sum_{k=1}^{r}(y_{ijk} - \bar{y}_{ij.})^2. \tag{5.24}$$

If it is assumed that y_{ijk}, $k = 1, \ldots, r$, are i.i.d. normal random variables (i.e., the r replications within a given dose are independent), then

$$\frac{(r-1)S_{ij}^2}{\sigma_{ij}^2} \sim \chi^2(r-1). \tag{5.25}$$

A useful variance model proposed by Bellio et al. (2000) is the so-called power of x (POX) model, given by

$$\sigma_{ij}^2 = \sigma_i^2 g(\mathbf{z}_{ij}, \boldsymbol{\beta}_i, \boldsymbol{\theta}_i) = a_{0,i} x_{ij}^{\theta_{1,i}}. \tag{5.26}$$

Then, as suggested by Aitkin (1987), using the distributional result of Equation (5.25) the S_{ij}^2 may be modeled using a generalized linear model (GLIM) framework with the natural logarithm link function and gamma errors. Specifically, the GLIM model used is S_{ij}^2 following a gamma distribution with scale parameter equal to $\frac{r-1}{2}$ and mean function equal to

$$\sigma_{ij}^2 = \exp\{\theta_{0,i} + \theta_{1,i} \log(x_{ij})\}. \tag{5.27}$$

Note that $\theta_{0,i} = \log(a_{0,i})$. This model is based on the assumption that $\log(S_{ij}^2)$ has a simple linear relationship with $\log(x_{ij})$. Subject-matter theory and experience often give rise to an appropriate variance model.

Once the model parameters in Equation (5.27) are estimated using GLIM techniques, then estimates of σ_{ij}^2 may be obtained. Let $\hat{\theta}_{0,i}$ and $\hat{\theta}_{1,i}$ be the GLIM estimators of $\theta_{0,i}$ and $\theta_{1,i}$, respectively. These estimators have an asymptotic normal distribution, and the estimator of σ_{ij}^2 is

$$(\hat{\sigma}_{ij}^2)^{GLIM} = \exp\{\hat{\theta}_{0,i} + \hat{\theta}_{1,i} \log(x_{ij})\}. \tag{5.28}$$

The expression in Equation (5.28) is referred to as the estimated *variance profile* for sampling period i.

In order to check that the estimated variance profiles are in-control, one looks for unusual values of

$$\hat{\boldsymbol{\theta}}_i = \begin{bmatrix} \hat{\theta}_{0,i} \\ \hat{\theta}_{1,i} \end{bmatrix}.$$

For example, an extremely large value of $\theta_{1,i}$ could indicate that the overall variability for variance profile i is too large. On the other hand, an out-of-control value of $\theta_{1,i}$ indicates that the rate of increase or decrease of heterogeneity of variance is different for variance profile i. For positive values of $\theta_{1,i}$, the variability increases with x_{ij}, whereas for negative values the variability decreases with x_{ij}.

Monitoring the variance profile in a Phase I application can be done using the multivariate T^2 control chart methods described in Section 5.2.1 applied to $\hat{\boldsymbol{\theta}}_i, i = 1, \ldots, m$. The associated UCL values for the T^2 control charts are based on the assumption that the $\hat{\boldsymbol{\theta}}_i$ have an asymptotic multivariate normal distribution with mean vector $\boldsymbol{\theta}$ and covariance matrix $\boldsymbol{\Sigma}_\theta$. A common estimator of the mean is $\bar{\hat{\boldsymbol{\theta}}} = m^{-1} \sum_{i=1}^{m} \hat{\boldsymbol{\theta}}_i$. The estimator of the covariance matrix can be based on the sample covariance matrix estimator of Equation (5.8), the successive differences estimator of Equation (5.10), or the MVE estimator of Section 5.2.1. The associated UCL values of the T^2 charts are calculated in the same manner as described in Section 5.2.1.1.

It is important to note that in the case of heteroscedasticity, a more appropriate mean profile estimation method is to use a weighted Gauss–Newton estimation method, where the weights are given by the $n \times n$ diagonal weight matrix $\mathbf{W}_i = \text{diag}\{(\hat{\sigma}_{ij}^2)^{GLIM}\}$. This will affect both the estimate of $\boldsymbol{\beta}$ and the associated estimate of the variance–covariance matrix $\boldsymbol{\Sigma}_\beta$.

With the aid of the T^2 charts and the associated UCLs, the out-of-control variance profiles are investigated an possibly removed from the HDS. With the remaining m^* profiles, the in-control mean vector $\boldsymbol{\mu}_\theta$ and covariance matrix $\boldsymbol{\Sigma}_\theta$ for use in continuous process monitoring are calculated as

$$\boldsymbol{\mu}_\theta = \frac{1}{m^*} \sum_{i=1}^{m^*} \hat{\boldsymbol{\theta}}_i \tag{5.29}$$

and

$$\boldsymbol{\Sigma}_\theta = \frac{1}{m^* - 1} \sum_{i=1}^{m^*} \left(\hat{\boldsymbol{\theta}}_i - \boldsymbol{\mu}_\theta\right) \left(\hat{\boldsymbol{\theta}}_i - \boldsymbol{\mu}_\theta\right)'. \tag{5.30}$$

5.4.1 Phase II Variance Profile Monitoring

To monitor the variance profile collected at time t, the chosen variance profile model from the Phase I analysis is employed to estimate the value of $\boldsymbol{\theta}$ from the profile generated at time t. The estimated mean vector and variance–covariance matrix of $\hat{\boldsymbol{\theta}}$, given in Equations (5.29) and (5.30), are then used to construct a T^2 control chart. The T_t^2, $t = 1, 2, 3, \ldots$, statistics are calculated as

$$T_t^2 = \left(\hat{\boldsymbol{\theta}}_t - \boldsymbol{\mu}_\theta\right)' \boldsymbol{\Sigma}_\theta^{-1} \left(\hat{\boldsymbol{\theta}}_t - \boldsymbol{\mu}_\theta\right).$$

The UCL for this chart is given by

$$UCL_{T_\theta^2} = \frac{2(m^* + 1)(m^* - 1)}{m^*(m^* - 2)} \mathcal{F}(1 - \alpha_{II}, p, m^* - 2)$$

(Mason et al. 2001). Whenever $T_t^2 > UCL_{T_\theta^2}$ then the chart signals, and hence profile t is subject to further investigation.

5.4.2 Dose–Response Data Example

To illustrate the profile monitoring methods for heteroscedastic data, the HDS from DuPont Crop Protection given in WBWF is analyzed. The data consist of 44 weeks ($m = 44$) of in vivo bioassay results for standards (commercial crop protection products) run alongside experimental compounds over a 1-year time period. The bioassay procedure was employed using a commercial crop protection product and a test organism. Because of the proprietary nature of the bioassay, the commercial compound and test organism used are undisclosed.

The commercial compound was diluted to eight doses (0.003, 0.009, 0.028, 0.084, 0.25, 0.76, 2.27, and 6.8) and replicated four times at each dose ($r = 4$) in 96-well microtiter plates for each sampling period i. A spectrophotometer measured the optical density (OD) of the plant organism after the inoculation period. Both treated and untreated wells were measured for growth inhibition. The PC_{ijk} values were calculated using the median OD (M_i) from 96 replications of untreated wells.

Based on subject-matter theory of the biological process, the proposed variance model is the POX model given in Equation (5.26). Equation (5.24) was employed to calculate S_{ij}^2 for $i = 1, \ldots, 44$ and $j = 1, \ldots, 4$, used the distributional properties of the S_{ij}^2 statistics given in Equation (5.25), and employed a GLIM model from Equation (5.27). Using this model values of the estimators, $\hat{\theta}_{0,i}$ and $\hat{\theta}_{1,i}$ were obtained for all 44 weeks of the DuPont data.

To graphically visualize the goodness-of-fit of this model, the $\log(S_{ij}^2)$ versus the logarithm of dose is plotted, superimpose with the logarithm of the predicted values of the variance from Equation (5.28), given by $\log((\hat{\sigma}_{ij}^2)^{GLIM})$. The plot is given in Figure 5.4.

To check for unusual variance profiles, $T_{MVE,i}^2$ and $T_{D,i}^2$ statistics were calculated based on the values of $\hat{\boldsymbol{\theta}}_i$, as given in Equations (5.12) and (5.11), respectively. The overall α was chosen to be $\alpha_{overall} = 0.05$. The UCL for the chart based on $T_{D,i}^2$ is 13.51 and the UCL for the chart based on the MVE (obtained via simulation) is approximately 23. The two charts are given in Figure 5.5.

The T^2 chart based on the $T_{D,i}^2$ statistic does not signal, whereas observations 6, 20, 22, 24, 26, and 45 produced a signal in the chart based on $T_{MVE,i}^2$. Recall that the T^2 chart based on $T_{D,i}^2$ statistics is not as powerful in detecting multivariate outliers as the chart based on the $T_{MVE,i}^2$ statistics. The observations that signal are extreme in both $\theta_{0,i}$ and $\theta_{1,i}$. In Figure 5.6, the 44 ordered pairs $\hat{\theta}_{0,i}$ and $\hat{\theta}_{1,i}$ are plotted. The 6 weeks that produced the signal are extreme in both $\theta_{0,i}$ and $\theta_{1,i}$. All are associated with variance profiles with a positive slope value $\hat{\theta}_{1,i}$ Hence, these observations are removed from the HDS, resulting in $m' = 38$.

WNLS was used to estimate the mean profiles, where the estimated weights are $(\hat{\sigma}_{ij}^{-2})^{GLIM}$ from Equation (5.28). The estimated mean profiles for all 44 weeks are given in Figure 5.7.

To check for appropriateness of the mean profile model, the LOF statistic of Equation (5.15) were calculated based on the weighted sums of squares. From Equation

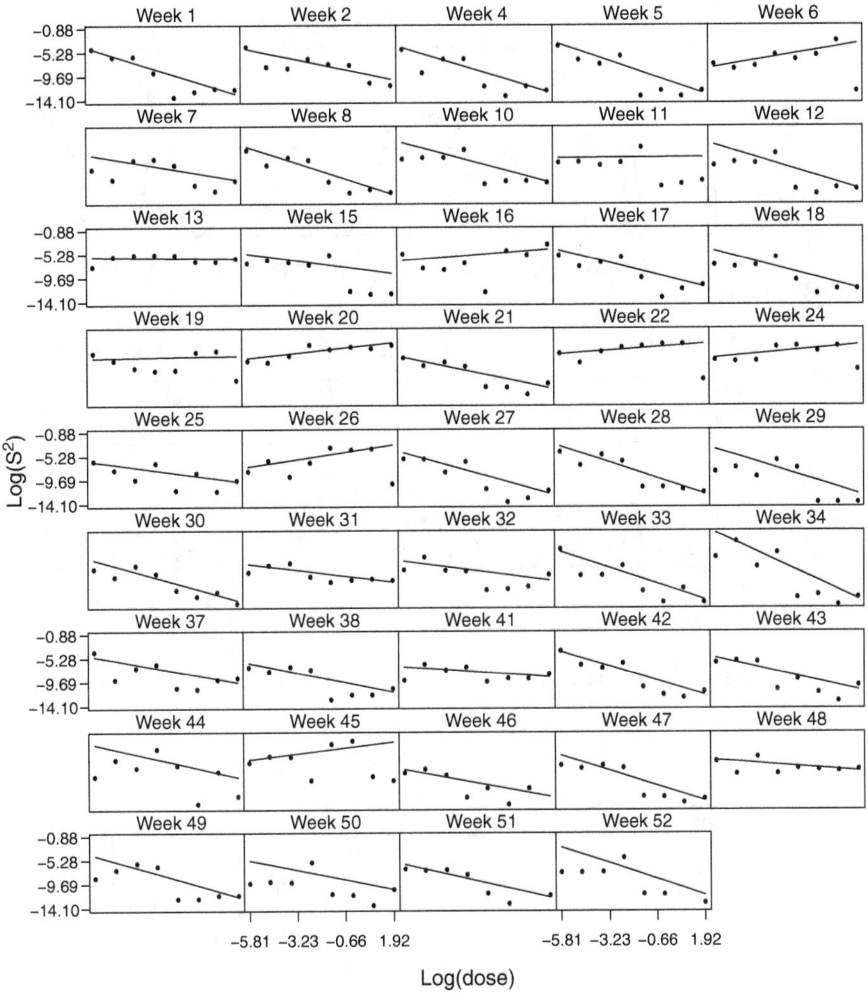

Figure 5.4 Estimated variance profiles for all 44 weeks.

(5.16) the approximate UCL associated with the chart is 5.26, and the chart is given in Figure 5.8.

Although weeks 21, 30, 32, and 33 produced a signal, only observations 21 and 32 were removed from the HDS. Weeks 30 and 33 exhibit only marginal LOF. This leaves $m'' = 36$ profiles remaining in the HDS.

Next unusual values of estimated parameters were examined using a T^2 chart based on the $T^2_{MVE,i}$ and $T^2_{D,i}$ statistics, as given in Equations (5.12) and (5.11), respectively. The UCLs associated with the charts are 38 and 17.68, respectively. The charts are given in Figure 5.9.

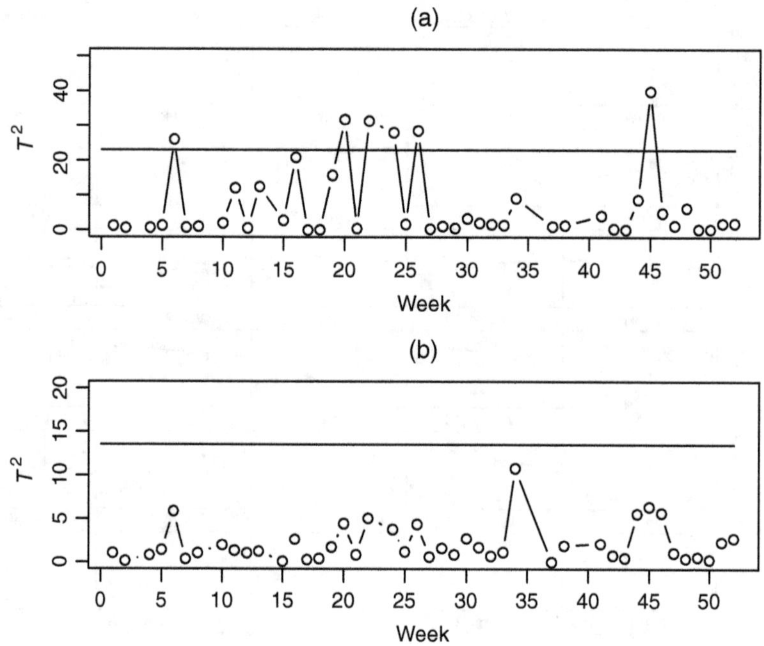

Figure 5.5 T^2 charts based on (a) $T^2_{MVE,i}$ and (b) $T^2_{D,i}$ to detect unusual variance profiles.

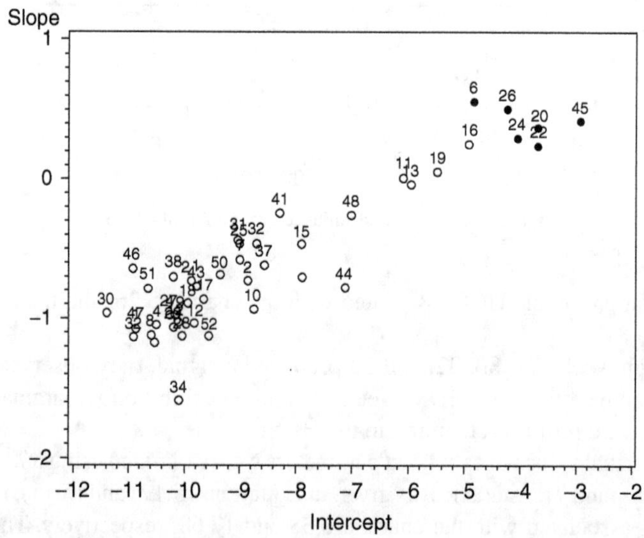

Figure 5.6 Scatterplot of values of $\hat{\theta}_{0,i}$ and $\hat{\theta}_{1,i}$ for all 44 weeks of the DuPont HDS. Those six observations that produced a signal in the T^2 chart are indicated with the solid black dot.

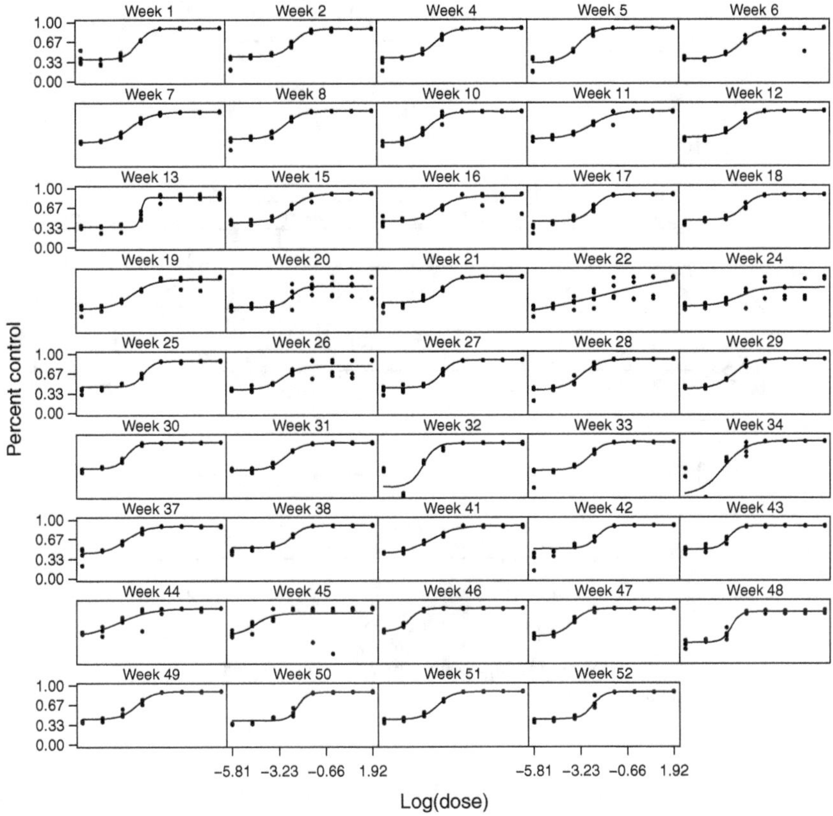

Figure 5.7 Estimated mean profiles based on estimated weights for all 44 weeks.

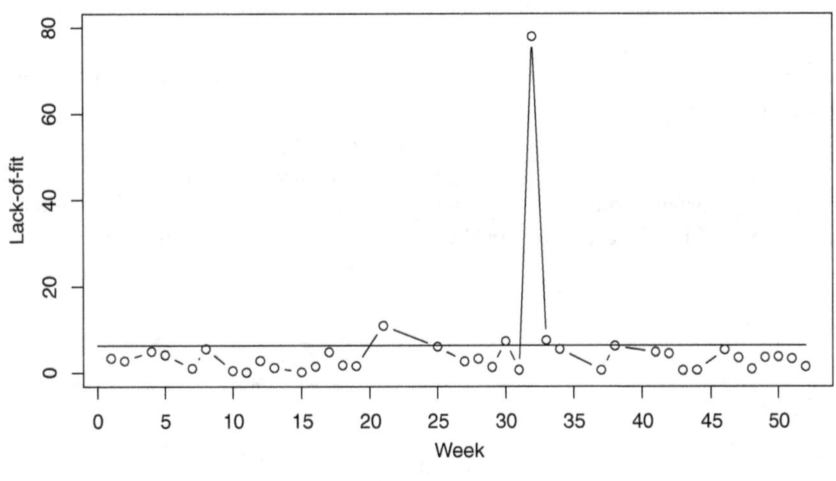

Figure 5.8 Lack-of-fit chart based on the weighted sums of squares.

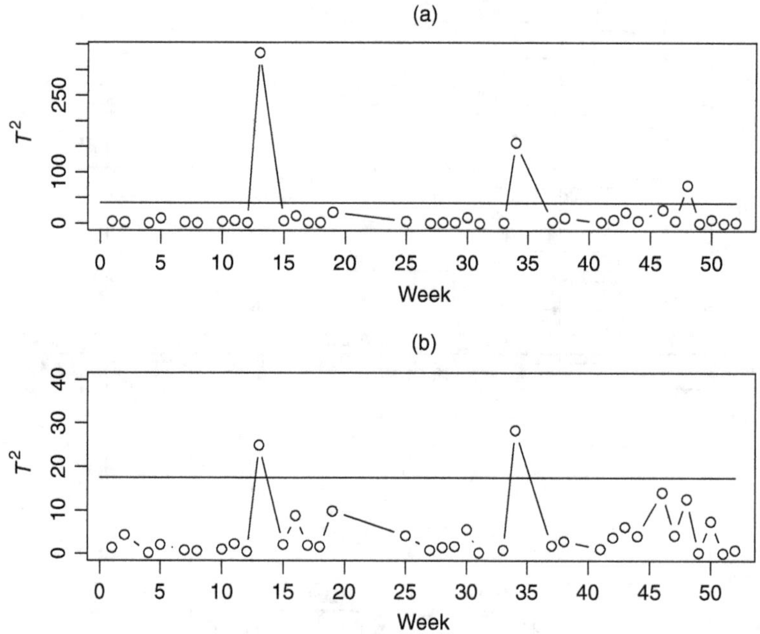

Figure 5.9 T^2 charts based on (a) $T^2_{MVE,i}$ and (b) $T^2_{D,i}$ to detect unusual values of the parameters estimates of the mean profiles.

Weeks 13, 34, and 48 produced a signal in the T^2 chart based on $T^2_{MVE,i}$, and observations 13 and 34 produced a signal in the chart based on $T^2_{D,i}$. All three of these weeks were removed from the HDS. Subsequently, a second T^2 chart was created to further check for out-of-control profiles after removing these weeks. The new UCLs for the two charts are 40 and 17.49, respectively. The charts are given in Figure 5.10.

Week 46 was identified as being out of control and was removed from the HDS. The remaining weeks were presumed to be in-control. A plot of the in-control estimated mean profiles is given in Figure 5.11.

The remaining $m^* = 32$ profiles are now used to estimate the in-control mean vector and variance–covariance matrix of $\hat{\boldsymbol{\theta}}_i$ and the in-control mean vector and variance–covariance matrix of $\hat{\boldsymbol{\beta}}_i$ for future process monitoring. From Equations (5.29) and (5.30), the estimates for the variance profile are calculated as

$$\boldsymbol{\mu}_\theta = \begin{bmatrix} -9.326028 \\ -0.765682 \end{bmatrix}$$

and

$$\boldsymbol{\Sigma}_\theta = \begin{bmatrix} 2.4730289 & 0.5147257 \\ 0.5147257 & 0.1396993 \end{bmatrix}.$$

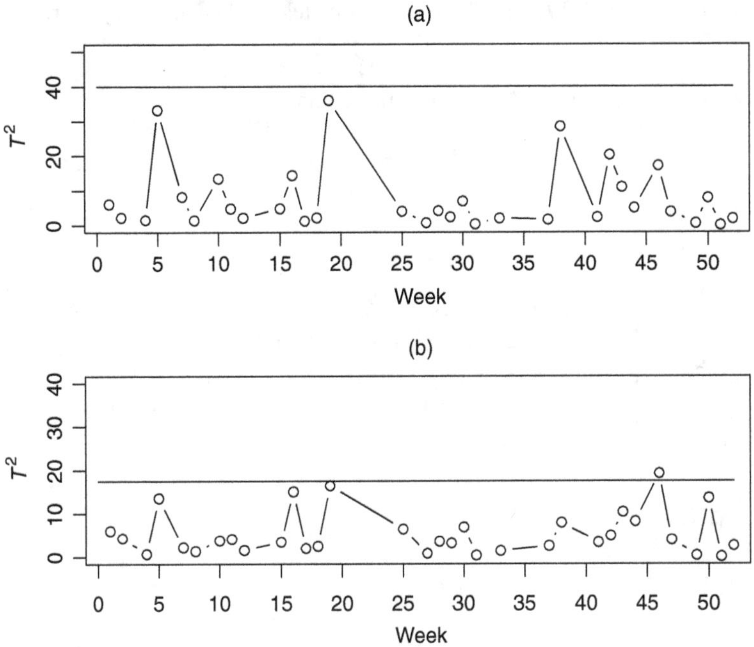

Figure 5.10 Second set of T^2 charts based on (a) $T^2_{MVE,i}$ and (b) $T^2_{D,i}$ to detect unusual values of the parameters estimates of the mean profiles.

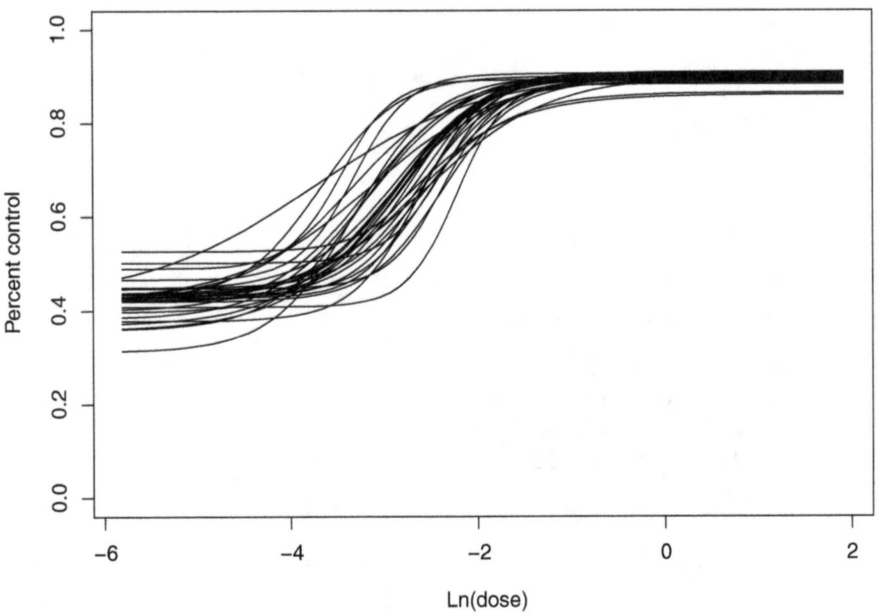

Figure 5.11 Estimated in-control mean profiles.

From Equations (5.17) and (5.19), the estimates for the mean profile are calculated as

$$\boldsymbol{\mu}_\beta = \begin{bmatrix} 0.8959855 \\ 2.3857821 \\ 0.0608633 \\ 0.4227484 \end{bmatrix}.$$

and

$$\boldsymbol{\Sigma}_\beta = \begin{bmatrix} 0.0001282 & -0.000134 & -0.000055 & 0.0000786 \\ -0.000134 & 0.4280911 & 0.0067914 & 0.0120498 \\ -0.000055 & 0.0067914 & 0.0004831 & 0.0002597 \\ 0.0000786 & 0.0120498 & 0.0002597 & 0.0017581 \end{bmatrix}.$$

APPENDIX

The function for $\beta(m, p, i)$ is

$$\beta(m, p, i) = I_{\{i=1,m\}}\left(\frac{p}{2} - \frac{1}{a_{11}(m - b_{11})}\right) + I_{\{i=2,\ldots,m-1\}}\left(a_{12}p + b_{12}\right),$$

and the function for $\gamma(m, p, i)$ is given by

$$\gamma(m, p, i) = I_{\{i=1,m\}}a_{21} + I_{\{i=2,\ldots,m-1\}}\left[a_{22}\left(i - \frac{m+1}{2}\right)^2 + b_{22}\right],$$

where

$$I_{\{i=1,m\}} = \begin{cases} 1 & \text{if } i = 1 \text{ or } i = m \\ 0 & \text{otherwise} \end{cases}$$

$$I_{\{i=2,\ldots,m-1\}} = \begin{cases} 1 & \text{if } 2 \le i \le m - 1 \\ 0 & \text{otherwise} \end{cases}$$

$$\begin{aligned}
a_{11} &= 6.356e^{-0.825p} + 0.06 \\
b_{11} &= 0.5564p + 0.9723 \\
a_{12} &= 0.54 - 0.25e^{-0.25(m-15)} \\
b_{12} &= -0.085 + 0.2e^{-0.2(m-22)} \\
a_{21} &= (-0.5m + 2)p + \frac{1}{3}(m + 3)(m - 5) \\
a_{22} &= 0.99 + 0.38e^{0.38(p-13.5)} - \frac{1}{0.25e^{-0.25(p-10)}\left(m - 11 + \frac{(p-7)^2}{3}\right)} \\
b_{22} &= (0.07e^{-0.07(m-42)} - 1.95)p + 0.0833m^2
\end{aligned}$$

REFERENCES

Aitkin, M. (1987) Modelling variance heterogeneity in normal regression using GLIM. *Applied Statistics*, **36**, 332–339.

Arbogast, P. G. and Bedrick, E. J. (2004) Model-checking techniques for linear models with parametric variance functions. *Technometrics*, **46**, 404–410.

Bellio, R., Jensen, J. E., and Seiden, P. (2000) Applications of likelihood asymptotics for nonlinear regression in herbicide bioassays. *Biometrics*, **56**, 1204–1212.

Brill, R. V. (2001) A case study for control charting a product quality measure that is a continuous function over time. Presentation at the 47th Annual Fall Technical Conference, Toronto, Ontario.

Carroll, R. J. and Ruppert, D. (1988) *Transformation and Weighting in Regression*. Chapman and Hall, New York.

Chou, Y. -M., Mason, R. L., and Young J. C. (1999) Power comparisons for a hotelling's T^2 statistic. *Communications in Statistics, Part B—Simulation and Computation*, **28**, 1031–1050.

Davidian, M. and Carroll, R. J. (1987) Variance function estimation. *Journal of the American Statistical Association*, **82**, 1079–1091.

Gallant, A. R. (1987) *Nonlinear Statistical Models*. John Wiley & Sons, Inc., New York.

Gnanadesikan, R. and Kettenring, J. R. (1972) Robust estimates, residuals, and outlier detection with multiresponse data. *Biometrics*, **28**, 81–124.

Hawkins, D. M. and Merriam, D. F. (1974) Zonation of multivariate sequences of digitized geologic data. *Mathematical Geology*, **6**, 263–269.

Holmes, D. S and Mergen, A. E. (1993) Improving the performance of the T^2 control chart. *Quality Engineering*, **5**, 619–625.

Jensen, W. A., Birch, J. B., and Woodall, W. H. (2007) High breakdown estimation methods for phase I multivariate control charts. *Quality and Reliability Engineering International*, **23** (5), 615–629.

Kang, L. and Albin, S. L. (2000) Online monitoring when the process yields a linear profile. *Journal of Quality Technology*, **32**, 418–426.

Mahmoud, M. A. and Woodall, W. H. (2004) Phase I analysis of Linear Profiles with Calibration Applications. *Technometrics*, **46**, 380–391.

Mason, R. L., Chou, Y. M., and Young, J. C. (2001), Applying Hotelling's T^2 Statistic to Batch Processes, *Journal of Quality Technology* **33**, pp. 466–479.

Mason, R. L. and Young, J. C. (2002) *Multivariate Statistical Process Control with Industrial Applications*. SIAM, Philadelphia.

Myers, R. H. (1990) *Classical and Modern Regression with Applications*. 2nd Edition. Duxbury Press, Belmont, California.

Neill, J. W. (1988) Testing for lack-of-fit in nonlinear regression. *The Annals of Statistics*, **16**, 733–740.

Nelson, L. S. (1983) Exact critical values for the analysis of means. *Journal of Quality Technology*, 15, 40–44.

Rousseeuw, P. J. (1984) Least median of squares regression. *Journal of the American Statistical Association*, **79**, 871–880.

Schabenberger, O. and Pierce F. J. (2002) *Contemporary Statistics for the Plant and Soil Sciences*. CRC Press, Boca Raton, Florida.

Seber, G. A. F. and Wild, C. J. (1989) *Nonlinear Regression*. John Wiley & Sons, Inc., New York.

Sullivan, J. H. and Woodall, W. H. (1996) A comparison of multivariate control charts for individual observations. *Journal of Quality Technology*, **28**, 398–408.

Tracy, N. D., Young, J. C., and Mason, R. L. (1992) Multivariate control charts for individual observations. *Journal of Quality Technology*, **24**, 88–95.

Vaghefi, A., Tajbakhsh, S. D., and Noorossana, R. (2009) Phase II monitoring of nonlinear profiles. *Communications in Statistics—Theory and Methods*, **38**, 1834–1851.

Vargas J. A. (2003) Robust estimation in multivariate control charts for individual observations, *Journal of Quality Technology*, **35**, 367–376.

Walker, E. and Wright, S. (2002) Comparing curves using additive models. *Journal of Quality Technology*, **34**, 118–129.

Williams, J. D., Woodall, W. H., Birch, J. B. and Sullivan, J. H. (2006) Distribution of hotelling's T^2 statistic based on successive differences. *Journal of Quality Technology*, **38**,217–229.

Williams, J. D., Birch, J. B., Woodall, W. H., and Ferry, N. M. (2007) Statistical monitoring of heteroscedastic dose-response profiles from high-throughput screening. *Journal of Agricultural Environmental, and Biological Statistics*, **12**, 216–235.

Williams, J. D., Sullivan, J. H., and Birch, J. B. (2009) Maximum value of hotelling's T^2 statistic based on the successive differences covariance matrix estimator. *Communications in Statistics: Theory and Methods*, **38** (4), 471–483.

Williams, J. D., Woodall, W. H., and Birch, J. B. (2007) Statistical monitoring of nonlinear product and process quality profiles. *Quality & Reliability Engineering International*, **23**, 925–941.

Wludyka, P. S. and Nelson, P. R. (1997) An analysis-of-means-type test for variances from normal populations. *Technometrics*, **39**, 274–285.

Young, T. M., Winistorfer, P. M., and Wang, S. (1999) Multivariate control charts of MDF and OSB vertical density profile attributes. *Forest Products Journal*, **49**(5), 79–86.

CHAPTER 6

Nonparametric Nonlinear Profiles

Eric Chicken

Department of Statistics, Florida State University, Tallahassee, FL, USA

INTRODUCTION

The initial chapters of this book discussed modeling and analyzing profiles that possessed linear structure. The class of profiles under consideration was extended in Chapter 5 by allowing nonlinear structure for the response curves. Such methods are inherently parametric in that the methods developed depend upon the assumption that there exist parameters relating the structure of observed data x_i to observed responses y_i. This reliance on parameters is evident in the relation

$$y_i = f(x_i, \tilde{\beta}) + \varepsilon_i, \quad i = 1, 2, \ldots, n, \tag{6.1}$$

where $\tilde{\beta}$ is a vector of parameters, x_i is the explanatory variable, y_i is the response, and ε_i is the error term. In the case of simple linear profiles, $\tilde{\beta} = (\beta_0, \beta_1)$ is the usual intercept and slope,

$$y_i = \beta_0 + \beta_1 x_i + \varepsilon_i, \quad i = 1, 2, \ldots, n. \tag{6.2}$$

The formulation in Equation (6.1) allows for greater freedom in modeling profiles beyond the simple linear relation. For example, an exponential relation may be parameterized by a three-dimensional $\tilde{\beta} = (\beta_0, \beta_1, \beta_2)$,

$$y_i = \beta_0 + \beta_1 e^{-\beta_2 x_i} + \varepsilon_i, \quad i = 1, 2, \ldots, n. \tag{6.3}$$

Statistical Analysis of Profile Monitoring, First Edition. Edited by Rassoul Noorossana, Abbas Saghaei and Amirhossein Amiri.

However, the problem with using Equation (6.1) is that if the data does not follow a particular assumed parametric form, or if the parametric model is incorrectly specified, the estimated models are not useful.

Not all profiles lend themselves well to parametric forms. There are occasions when the observed profile f does not readily admit to a parameterization using Equation (6.1). This may be due to difficulty in determining the correct parametric form of f or because the function is, in fact, nonparametric, i.e., it has no finite parametric representation. In this case, a more general form of the profile is used,

$$y_i = f(x_i) + \varepsilon_i, \quad i = 1, 2, \ldots, n. \tag{6.4}$$

There are many examples of profiles that may only be modeled using Equation (6.4). Several will be examined in the remainder of the chapter. One example that has been used extensively is the vertical density profile (VDP) of wood panel products. Such data was considered for nonparametric analysis in Winistorfer et al. (1996) and analyzed in a statistical process control (SPC) setting in Williams et al. (2007), Chang and Yadama (2010), and many others. Twenty-four of these profiles are shown in Figure 6.1. Each profile is measured at 314 positions 0.002 inches apart. Note that it is not clear what parametric form would be appropriate for the profiles shown.

Nonparametric methods, like parametric methods, are employed to reduce the high dimensionality of the data. The parametric methods described in earlier chapters accomplished this by collapsing each of the m observed profiles of length n into a collection of estimated parameters. In Phase I, it is desired to determine what an in-control profile looks like, while in Phase II the goal is to detect an occurrence of an out-of-control profile. For parametric methods, this is equivalent to describing the parameters and noting when they vary from nominal values. An advantage of parametric methods, when such models are appropriate, is that no matter how large n

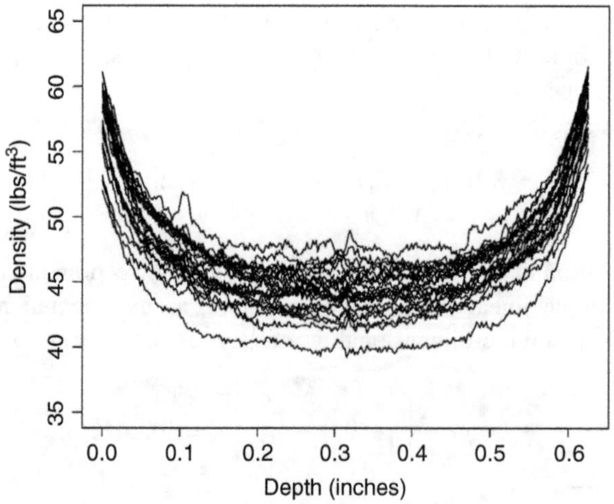

Figure 6.1 VDP particleboard data. (Adapted from Williams et al. 2007.)

is, it does not affect the number of parameters to be estimated. In fact, the larger the value of n, the more accurate the estimated parameters are. The disadvantage is that if the parametric model is not valid, then the accuracy of the methods is questionable, regardless of how much data is collected.

Another disadvantage of parametric methods in profile estimation is that information about localized changes in a profile is difficult to detect unless parameters are included in the model that specifically address localization. Unfortunately, including such locality parameters is usually not efficient. First, they increase the dimensionality of the problem. Second, and more importantly, it is typically not known a priori where to model localized differences. In Figure 6.1, for example, it appears that a localized phenomenon occurs in several of the profiles at 0.3 on the horizontal axis. Modeling such a localized feature would complicate the parametric model considerably. Parametric models may be thought of as global descriptions of the profiles. In contrast, some of the nonparametric methods to be discussed in this chapter include locality as well as global measures.

Nonparametric methods are considered robust. While a parametric method is generally only useful when the assumed model is correct, nonparametric methods are applicable whether or not the profiles follow a parametric form.

Many methods have been proposed for profile monitoring when the situation at Equation (6.4) is the most reasonable assumption. Some of these include the use of spline estimators, component analysis, wavelet methods, and simple data-driven metrics. The remainder of this chapter will discuss several nonparametric methods of profile monitoring. These methods are similar in that as few assumptions as possible are made on the nature of the relation between the explanatory variable x and the response y. Section 6.1 provides the model formulation for these methods and gives an introductory example. Section 6.2 covers spline methods for profiles, component analysis is discussed in Section 6.3, and an overview of wavelet-based SPC methods is given in Section 6.4.

6.1 MODEL FORMULATION AND NONPARAMETRIC EXAMPLE

The usual model for nonparametric nonlinear profile analysis is

$$y_{ij} = f_j(x_i) + \varepsilon_i, \quad i = 1, 2, \ldots, n, \quad j = 1, 2, \ldots, m. \tag{6.5}$$

For each of the m profiles, the response y is measured at the same n fixed locations x_i, typically equally spaced apart over the interval of the profile's support. No restrictive parametric assumptions are placed on the structure of f. Assumptions are usually placed on the error terms. The errors ε_{ij} are generally assumed to be independently distributed normal random variables with mean 0 and variance σ_{ij}^2.

A median-based nonparametric method may be found in Zhang and Albin (2009). The model used is Equation (6.5) with no assumptions on f, and the errors are independent identically distributed normal random variables with mean 0 and variance σ_j^2. Note that here the variance of the errors is assumed constant for a particular observed profile j. It is assumed that less than $m/2$ of the profiles are outliers or out-of-control. The remaining profiles are in-control. When in-control, the unknown form of the profile function is at a nominal value, say $f_j = f_0$. Additionally, the variance is constant for these same profiles, say $\sigma_j^2 = \sigma_0^2$. For the outlier profiles, $f_j \neq f_0$ or $\sigma_j^2 > \sigma_0^2$, or both.

The authors propose a control chart control chart for this problem that may be used in both Phase I and Phase II analysis. From a sequence of m profiles, the outliers will be noted and removed. This will allow a more precise modeling of the in-control condition. With regards to Phase II monitoring, the control chart may be used to classify a newly observed profile as either in-control or out-of-control with respect to the baseline profile estimated in Phase I.

The sequence of profiles contains both in-control observations and outlier observations. Therefore, robust measures of center and variability are called for. For the center,

$$\hat{\mu} = [\hat{y}_1, \hat{y}_2, \ldots, \hat{y}_n]', \tag{6.6}$$

where $\hat{y}_i$ is the median of $\{y_{ij} | j = 1, 2, \ldots, m\}$. To estimate the variability, for each $x_i, i = 1, 2, \ldots, n$, the authors find all $m(m-1)/2$ pairwise differences over the profiles:

$$\delta_{ijk} = y_{ij} - y_{ik}, \quad j, k = 1, 2, \ldots m, \quad j \neq k. \tag{6.7}$$

Taking the average of the squared differences at each location x_i, define

$$\hat{\sigma}_{jk}^2 = \frac{1}{2n} \sum_{i=1}^{n} \delta_{ijk}^2. \tag{6.8}$$

Finally, estimate the in-control variance σ^2 as the median of the $m(m-1)/2$ estimates $\hat{\sigma}_{jk}^2$. These two median-based estimators are nonparametric estimates of the center and the spread of the data. Both estimates are robust to outliers. A test statistic is formed using these two estimators and the data. For profile j, set

$$\Delta_j = \sum_{i=1}^{n} \frac{(y_{ij} - \hat{y}_i)^2}{\hat{\sigma}^2 (m-1)/m}. \tag{6.9}$$

If profile j is not an outlier, then Δ_j is approximately χ^2 with n degrees of freedom. A profile j is considered an outlier if $\Delta_j > UCL = \chi_{n,\alpha}^2$. Plotting Δ_j against j provides a chi-square control chart.

The authors provide extensions to this control chart. In particular, they allow for the variance to differ at the locations x_i within a profile.

Example 6.1 This nonparametric method is compared to the parametric nonlinear regression method given in Williams et al. (2007) on both simulated and real data.

The simulated data consisted of a specified parametric nonlinear function with independent normal errors. This function may be modeled with only one parameter, but is sufficiently complex so that nonlinear regression models perform poorly. In these simulations, the chi-square control chart always had a much lower Type I error (in-control outlier is incorrectly classified as out-of-control) than the parametric methods.

The out-of-control condition is defined by modifying two settings: (1) the parameter used to define the test function and (2) the level of the variance. When the level of the change in the parameter is small, the chi-square control chart had a higher Type II error (out-of-control profile not classified as such) than the parametric method. As σ increases, the Type II error for the chi-square control chart usually has a smaller Type II error than the parametric method. Increases in variance are more likely to be detected by the chi-square control chart than the nonlinear regression method. See Zhang and Albin (2009) for more detailed discussions on the simulation study comparing these two methods.

The two methods are applied to a real data set, the VDP data. For the chi-square control chart, a modification was made to account for changing variances at the x_i within a profile. The control chart is shown in the left panel of Figure 6.2. Five profiles (3, 6, 9, 10, and 14) are labeled as outliers. The right panel of Figure 6.2 shows these out-of-control profiles. For comparison, Williams et al. (2007) designated profiles 15 and 18 as out-of-control and 4, 9, and 24 as in need of further investigation.

Once outliers have been detected in the sequence of profiles, they may be removed in order to form a standard in-control profile to be used in Phase II analysis. For Phase II, a new observation is observed and a statistic similar to Equation (6.9) is calculated:

$$\Delta_{new} = \sum_{i=1}^{n} \frac{\left(y_{i,new} - \hat{y}_i\right)^2}{\hat{\sigma}^2 (m-1)/m}. \tag{6.10}$$

As before, Δ_{new} has a χ_n^2 distribution if profile *new* is in-control.

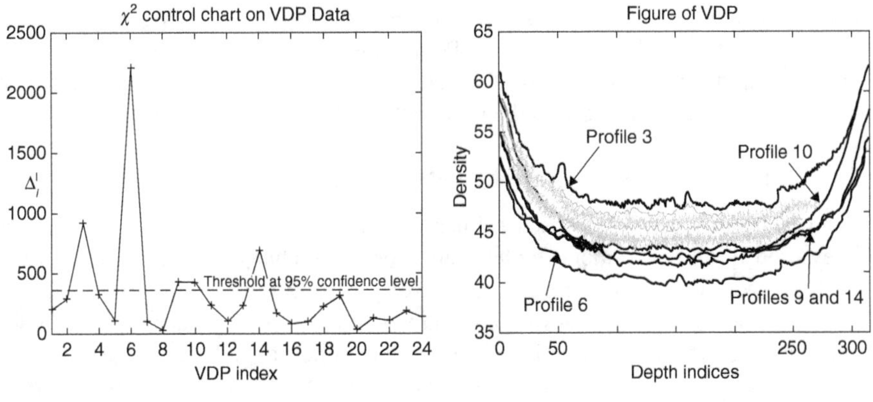

Figure 6.2 VDP particleboard analysis. (Adapted from Zhang and Albin 2009.)

This introductory example highlights some of the basics in nonparametric methods. No assumptions are placed on the shape of the profile f. In particular, there are no parameters associated with the structure of f. The Hotelling-like statistic in Equation (6.9) is based not on the means and covariances estimated in a parametric fashion (e.g., with maximum likelihood estimation), but using robust nonparametric estimates of center and spread. These types of estimation methods will be seen in the proposed profiling methods discussed in the remaining sections.

6.2 SPLINES

Splines are a part of a large class of function analysis methods referred to as smoothers. Rather than estimating parameters that describe a function, smoothers provide a fitted curve to the observed data (x_i, y_i). As the name implies, the result is a smoothed curve through the observed points, removing unwanted noise. Fitting a spline to data is a computationally complex problem, which is not generally performed manually. It requires the use of specialized software routines for implementation. Most commercial statistical software packages have the ability to implement spline fitting. In the SPC context, the observed profiles are smoothed with splines, then monitoring is done using statistics based on the resultant smoothed profiles.

An early use of splines is given in Winistorfer et al. (1996). They found spline estimates of profiles of VDP data across differing experimental conditions. They then proceeded to perform multiple comparisons of the splines. Thus, this is not Phase I or II profile monitoring.

The spline method used in Winistorfer et al. is a cubic spline. Each profile is partitioned into disjoint intervals, then cubic polynomials are fitted locally to each interval such that the resulting spline estimate is continuous and differentiable at the join points. The spline estimate is found by balancing a trade-off between smoothness and fit to the observed data. The smoothed spline curves were compared to each other via test statistics and inference methods developed by King et al. (1991).

Walker and Wright (2002) also used splines to model profile data. Specifically, they employed a common type of spline known as the B-spline, which is available widely in statistical and mathematical software packages. As with Winistorfer et al., they did not analyze the profiles in an SPC-monitoring setting. In this case, they were interested in determining sources of variation among the profiles. They applied their method to the VDP data used in the introductory example in the previous section.

6.2.1 Monitoring Spline Control Points

Chang and Yadama (2010) used splines to develop a profile monitoring method. They used B-splines to identify mean shifts or shape changes in the profiles. Their method's goal is not only to detect an out-of-control profile, but to also be able to specify the location of the out-of-control area within the profile. As mentioned in the introduction, such localization is difficult with parametric models.

When estimating a function from observed data with B-splines, it is necessary to specify a set of control points and a set of knots. The B-spline fit is created from piecing together several polynomial segments. These polynomial segments meet at the knots and the fitted B-spline passes near the control points. The number of control points, the number of knots, and the degree of the polynomials are all related. Chang and Yadama used uniform cubic B-splines. This type of B-spline uses third degree polynomials and the knots are uniformly spaced apart. They create a statistic for profile monitoring based on the control points.

To get to their statistic, a profile is initially smoothed to remove noise. Chang and Yadama perform this via wavelet thresholding, to be described later in this chapter. For now, it is enough to assume that the noisy profile has had the noise filtered from it. In accordance with Equation (6.5), the noise is assumed to be multivariate normal, the profiles are all of the same length, and the x_i are equally spaced apart.

The next step is to segment the data. This is done through user input. Segments may be chosen to highlight certain regions of the profile that may be of interest. Specifying the entire profile to consist of just one segment will provide no localized information. The choice of the number and locations of the segments is based on the expertise of the practitioner. There is no automated method provided.

After the profile is segmented, a B-spline is fit to each segment. Specifying the number of control points is all that is needed. The x-axis values of the control points are placed uniformly over the segment. The y-axis locations for a segment are found by minimizing the sum of squared distances between the data points and the corresponding spline fit. The complexity of the fitted spline depends on the number of control points specified. Since the B-spline will pass near the control points, to model a profile that is irregular one would specify a large number of control points. To accurately model a profile that is relatively smooth, only a few control points would be necessary.

Consider the VDP profile in Figure 6.3. This profile of 540 equally spaced measurements has a noticeable localized fluctuation near x_{300}. It seems reasonable to

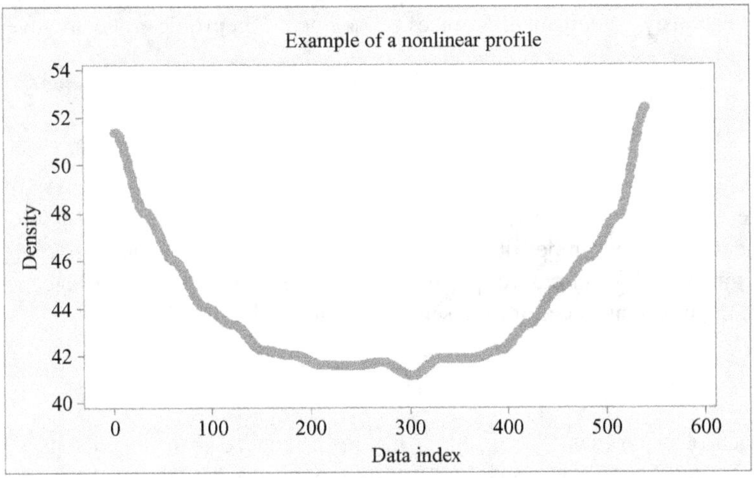

Figure 6.3 VDP particleboard. (Adapted from Chang and Yadama 2010.)

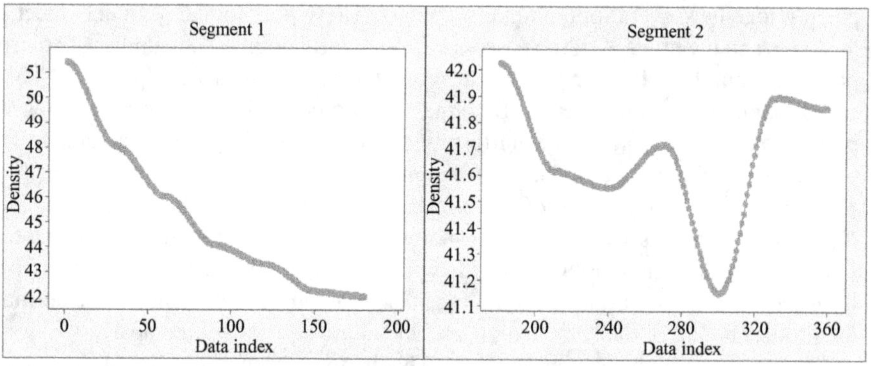

Figure 6.4 VDP particleboard segments. (Adapted from Chang and Yadama 2010.)

segment this data to localize this artifact. Figure 6.4 shows two of the three segments. The left panel is the downward sloping, smooth portion of the VDP from Figure 6.3, the right panel shows the central portion near the artifact of interest. The third segment is not shown, but is approximately a reflection of the first segment.

The number of control points in each segment must still be specified. Choosing more control points allows the B-spline to model functions that are more complex. So, in Figure 6.4, fewer control points would be needed for segment 1 than segment 2. Once the number of control points per segment is determined, then the entire profile may be fit with the B-spline. This results in values for the control points for each profile. Chang and Yadama's method selects the same segments and number of control points for every profile. This allows for a direct comparison of the control points from one profile to the next.

Mean differences between a profile j and the estimated Phase I in-control profile or a known in-control profile are calculated. If neither of these is available, an average profile may be used, the average found by fitting a B-spline to all m profiles simultaneously. The mean differences for segment k in profile j and are given by

$$D_{jk} = \frac{\sum_{l=1}^{c_k} d\left(c_{jkl}, \bar{c}_{jkl}\right)}{c_k}, \quad j = 1, 2, \ldots, m, \quad k = 1, 2, \ldots, p, \tag{6.11}$$

where c_k is the number of control points in segment k, c_{jkl} is control point l in the segment, p is the number of segments, and $\bar{c}_{jkl}$ is the corresponding control point from the average or in-control profile. The measure d may be any distance measure. Then, the mean distance for each segment within a profile j is

$$Y_j = \left[D_{j1}, \ldots, D_{jp}\right]' \tag{6.12}$$

for each of m profiles. The profile j has been reduced to p dimensions, the user-specified number of segments. The profile monitoring is performed on the Y_j, and not on the original observed profiles.

The vectors Y_j are now used in a Hotelling T^2 control chart. Set

$$T^2 = \left(Y_i - \bar{Y}\right)' S^{-1} \left(Y_i - \bar{Y}\right), \tag{6.13}$$

where $\bar{Y}$ and S must be estimated if they are not already known. If estimating them from m' in-control Phase I profiles,

$$\bar{Y} = \left[\bar{D}_1, \ldots, \bar{D}_p\right]', \quad \bar{D}_k = \frac{1}{m'} \sum_{j=1}^{m'} D_{jk}, \quad S = \frac{V'V}{2(n-1)}, \tag{6.14}$$

and, from Sullivan and Woodall (1995),

$$V = \left[(Y_2 - Y_1)', (Y_3 - Y_2)', \ldots, (Y_n - Y_{n-1})'\right]'. \tag{6.15}$$

Tracy et al. (1992) determine the UCL for Phase I monitoring to be

$$\frac{(n-1)^2}{n} \beta_{\alpha, p/2, (n-p-1)/2}, \tag{6.16}$$

while for Phase II it is

$$\frac{p(n-1)(n+1)}{n(n-p)} F_{\alpha, p, n-p}, \tag{6.17}$$

where β and F refer here to critical values of the beta and F-distributions. If an out-of-control profile is detected, Mason and Young (1995, 2001) provide a method that can be used to break the T^2 statistic into components that correspond to the different segments. This provides the ability to localize the fault that causes the out-of-control condition.

Example 6.2 Chang and Yadama provide a simulation example to examine the effect of the number of control points per segment. In the simulation setup the profiles are of length 240, each profile is divided into four segments, the number of control points c_k is the same in each segment and varies from 5 to 20, and the mean is shifted in different locations within the profile. For comparison purposes, the UCL was adjusted so that the in-control ARL is about 500 for each simulation configuration. A portion of the results are given in Table 6.1. In this table, the mean shift is located near the center of the profile. As expected, the ARLs decrease as the magnitude of the shift increases. Additionally, increasing the number of control points for each segment further reduces the ARL, although the improvement is not linear. In particular, doubling the number of control points p from 10 to 20 increases the dimension of the problem with reduced improvement in the ARL values.

Table 6.1 Select B-Spline Simulation (Results from Chang and Yadama 2010.)

Mean Shift	5 Control Points	10 Control Points	20 Control Points
0.0	505.47	498.26	495.15
0.5	280.19	161.06	149.19
1.0	107.50	40.07	12.06
2.0	64.60	5.03	1.85
3.0	1.37	1.00	1.00

6.2.2 Spline Metrics

Another use of profile monitoring using splines is found in Williams et al. (2007). Like Chang and Yadama, they use a B-spline. However, there are several differences. In Chang and Yadama, the user specifies segments within a profile. This is not done here. The test statistic in Chang and Yadama is based on comparing the control points. Here, the statistics used for profile monitoring are based on the fitted values of the spline.

Let $\hat{y}_{ij}$ be the spline fit of x_i in profile j and set $\tilde{y}_i = \sum_{j=1}^{m} \hat{y}_{ij}/m$. Then $\tilde{y}_i$ is the average estimated profile over all m profiles. Williams et al. (2007) provide five metrics on the basis of the following values:

1. $M_{j1} = s_0 \cdot \max_i \left|\hat{y}_{ij} - \tilde{y}_i\right|$, maximum deviation
2. $M_{j2} = \sum_{i=1}^{n} \left|\hat{y}_{ij} - \tilde{y}_i\right|$, sum of absolute deviations
3. $M_{j3} = \frac{1}{m} \sum_{i=1}^{n} \left|\hat{y}_{ij} - \tilde{y}_i\right|$, mean absolute deviation
4. $M_{j4} = \left|M_{j1}\right|$, maximum absolute deviation
5. $M_{j5} = \sum_{i=1}^{n} \left(\hat{y}_{ij} - \tilde{y}_i\right)^2$, sum of squared differences

where $s_0 = \operatorname{sgn}\left(\hat{y}_{i'j} - \tilde{y}_{i'}\right)$ with $i' = \arg\max_i \left|\hat{y}_{ij} - \tilde{y}_i\right|$. Metrics 2 and 3 differ only by a constant and both are not useful when used together. These five metrics are calculated for each of the m profiles and charted on an I-chart with control limits determined by the moving range.

Example 6.3 Figure 6.5 shows the average spline $\tilde{y}$ and the spline fit to a single VDP profile. The line through the points is the spline fit to just one of the 24 VDP profiles, the line beneath it is the average spline.

For the VDP data, metrics 1, 3, 4, and 5 are plotted in Figure 6.6. The top row contains the charts for metrics 1 and 4, the bottom row displays the charts for metrics 3 and 5. Metric 2 is proportional to metric 3 and provides the same results. It is not plotted. Metrics 1 and 4 agree on the 24 VDP boards, stating all 24 profiles are in-control. Note that metric 4 is the absolute value of metric 1, so this agreement is not unexpected. The largest values for these two metrics occur on profile 14, $M_{14,4} = 5.79 = -M_{14,1}$. This measures the largest deviation for a single point in profile 14

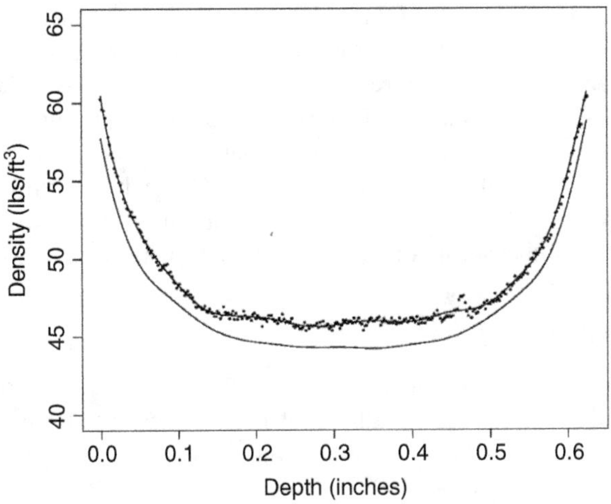

Figure 6.5 Spline fits to VDP particleboard data from Williams et al. (2007).

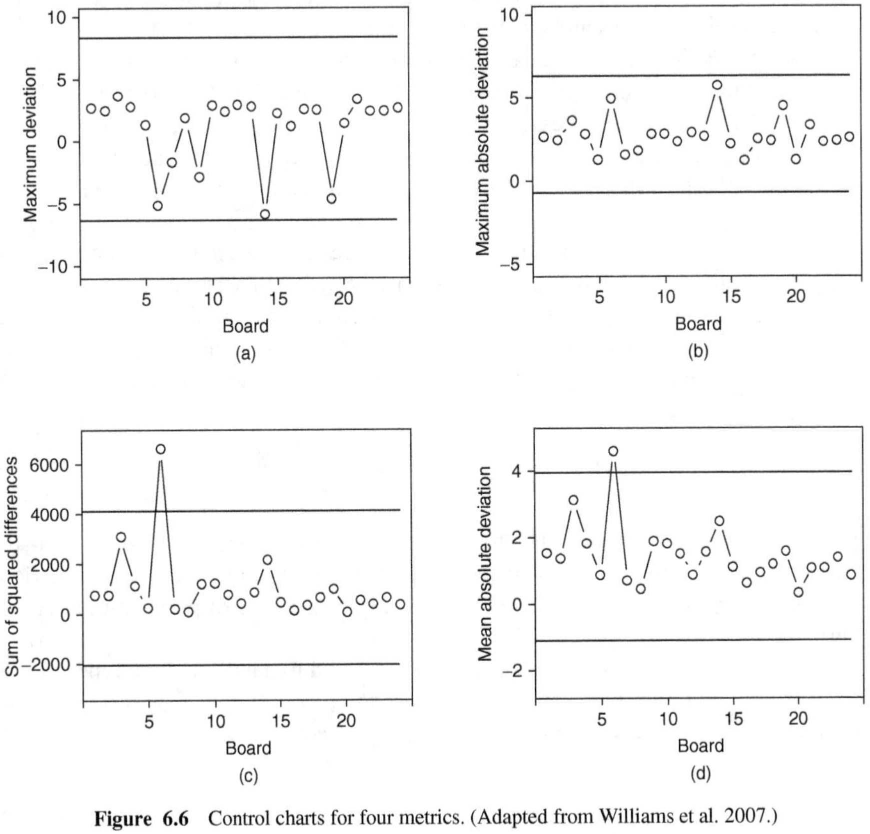

Figure 6.6 Control charts for four metrics. (Adapted from Williams et al. 2007.)

from the corresponding point in the average profile $\bar{y}$. In a sense, these two metrics are detecting differences between profiles at a particular point within the profile.

Metrics 3 and 5 are also in agreement. Both charts have classified VDP 6 as out-of-control. This profile is shown in Figure 6.1 as the lowest profile. For each of these two metrics, profile 3 is the in-control profile that is nearest to being out-of-control. In Figure 6.1 this is the highest profile. While metrics 1 and 4 look for large difference between a profile and the average profile at particular points within the profiles, metrics 3 and 5 are measuring the overall difference between the two profiles.

6.2.3 Bivariate Splines

The splines considered so far in Winistorfer et al. (1996), Walker and Wright (2002), Chang and Yadama (2010), and Williams et al. (2007) have all been univariate. There is only one explanatory variable x. Splines are easily extended to bivariate data. This is illustrated in Gardner et al. (1997). Their goal is to develop a Phase II method to detect and classify out-of-control surface profiles. They use thin-plate splines, an extension of cubic splines, to model surfaces nonparametrically. It is similar to the methods above in that it employs a smoothness-fit trade-off when fitting the spline. Like Williams et al., this method creates metrics based on differences between observed profile surfaces and an average profile.

Since the method is proposed as Phase II monitoring, the average profile is replaced with an assumed in-control profile response. This in-control profile is determined from a user-specified set of profiles. The in-control profile is the average of the selected Phase I profiles. The average is taken over the two-dimensional spline fits of the in-control profile surfaces, rather than fitting a spline to the average of the in-control profiles.

The metrics are taken off the spline estimate from each profile surface. The proposed metrics are all based on the difference between an observed profile response surface f_j and the in-control surface f_0. This difference is quantified by the difference between g_j and g_0, the thin-plate spline fits to the corresponding surfaces. Note that these are bivariate functions. The general form for the metrics is given by

$$M_{GEN} = \sum_{i=1}^{k} \int_{R_i} w_i h_i \left(g_j - g_0\right) dR_i, \tag{6.18}$$

where h_i is a penalty function for the difference, w_i are weights, and the R_i break the two-dimensional interval into k disjoint, nonoverlapping subintervals. This metric is a bivariate integral, but is evaluated numerically over a grid of points on the profile surfaces.

The h_i are selected by user expertise and are used to detect specific types of out-of-control conditions. The weights can be used to place emphasis on differing parts of the monitored surfaces. Some specific examples are

$$M_{SQ} = \int_R \left(g_j - g_0\right)^2 dR \tag{6.19}$$

and

$$M_{ABS} = \int_R \left| g_j - g_0 \right| dR. \tag{6.20}$$

Here $k = 1$, so the entire profile surface is considered as one piece. Each of these two metrics measures changes between the two profiles over the entire profile surface simultaneously. The penalty functions h are either absolute value or squared difference. The square penalty highlights large differences between a profile j and the in-control profile more than the absolute value penalty does. Metrics M_{j2} and M_{j5} from Williams et al. (2007) are equivalent to M_{ABS} and M_{SQ}, respectively: one need only reduce the dimension of the data from 2 to 1 and approximate the integrals.

A more complex metric is given by

$$M_{WT} = \begin{cases} \int_R w_1 h_1 \left(g_j - g_0 \right) dR, & g_j > g_0, \\ \int_R w_2 h_2 \left(g_j - g_0 \right) dR, & g_j \le g_0. \end{cases} \tag{6.21}$$

Here, weights can be used to detect a particular condition. For example, if h_1 and h_2 are any two penalty functions, setting $w_1 = 1$ and $w_2 = 0$ leads to a metric that signals only when a profile surface j is too thick.

To determine the in-control distributions for the metrics, the distribution of a spline surface g is estimated as a multivariate normal with the mean and covariance matrix estimated from profiles assumed to be in-control (Phase I). Repeat observations are generated from this distribution, and the metrics are calculated for each observation. Quantiles are then found and used to determine UCLs for Phase II.

Example 6.4 Gardner et al. apply this bivariate method to silicon wafer surfaces in an experimental 3^2 factorial design. The thickness of the wafers is the measurement of interest. Two factors, air flow rate and number of operating lamps, were varied. A total of 15 observations of wafer surfaces were gathered. Two of these observations were declared to be baseline Phase I observations and were used to construct an in-control surface and to generate UCLs for the metrics. In this example, five metrics were employed. In addition to M_{SQ} and M_{ABS}, there were three metrics based on M_{WT} as follows:

1. $$M_3 = \begin{cases} \int_R \left(g_j - g_0 \right)^2 dR, & \left| g_j - g_0 \right| > 1.5, \\ 0, & \left| g_j - g_0 \right| \le 1.5, \end{cases}$$

2. $$M_4 = \begin{cases} \int_R \left(g_j - g_0 \right)^2 dR, & g_j > g_0, \\ -\int_R \left| g_j - g_0 \right| dR, & g_j \le g_0, \end{cases}$$

3. $M_5 = \begin{cases} \int_R (g_j - g_0)^2 \, dR, & g_j \leq g_0, \\ -\int_R |g_j - g_0| \, dR, & g_j > g_0. \end{cases}$

Metrics M_4 and M_5 are designed to detect wafers with thicker and thinner surfaces, respectively, than the estimated in-control profile surface. Twelve of the thirteen Phase II profiles were out-of-control and one was in-control. With a UCL based on a Type I error rate of $\alpha = 0.01$, metrics M_{SQ}, M_{ABS}, and M_3 successfully detected all out-of-control profile surfaces. Metric M_4 detected the two out-of-control profiles where a thicker than expected surface was evident, and metric M_5 detected the nine profile surfaces where thinner than expected surfaces occurred. This is in line with the designed purpose of these two metrics. No metric classified the in-control profile as out-of-control.

All these spline methods monitor profiles by creating a smoothed fit to the data and then comparing the spline fits from one profile to another. Smoothing the observed profiles is also part of the methods discussed in Section 6.3.

6.3 COMPONENT ANALYSIS

Another nonparametric method for dealing with nonlinear profiles is through the use of component analysis. In this section, two types of component analysis are examined: (1) principal component analysis (PCA) and (2) independent component analysis (ICA). Both methods reduce the dimension of a problem by rotating the axes of the data with respect to some criteria.

6.3.1 Principal Component Analysis

In PCA, a rotation is found that places the maximum amount of data variability into the first component, the next highest amount of variability into the second component, and so on. Typically, many components are discarded since they explain little of the variability of the data.

Shiau et al. (2009) apply PCA to the problem of profile monitoring. They use charts of the principle component (PC) scores of profiles in both Phase I and II monitoring. Before finding the PCs, they smooth the data of the m profiles using splines to reduce the noise in the data. Figure 6.7 shows the results of applying splines to the 24 VDP profiles displayed in Figure 6.1.

Let y_j be the $n \times 1$ smoothed spline fit of profile j, and S be the $n \times n$ sample covariance matrix of the y_j, $j = 1, 2, \ldots, m$. In PCA, the components are found by determining the eigenvalues and eigenvectors of S. Let λ_r be the rth largest eigenvalue and ν_r the corresponding eigenvector. Then ν_r is the rth PC and $S_{jr} = \nu_r' y_j$ is the score of the rth PC of profile j. Define the score vector to be $s_j = \left[S_{j1}, S_{j2}, \ldots, S_{jk}\right]'$, where K is the number of PCs chosen to represent a specified amount of the variation.

In Phase I, a recursive use of a Hotelling T^2 chart is used to classify in-control profiles. For profile j,

$$T_j^2 = \left(s_j - \bar{s}\right)' B^{-1} \left(s_j - \bar{s}\right), \tag{6.22}$$

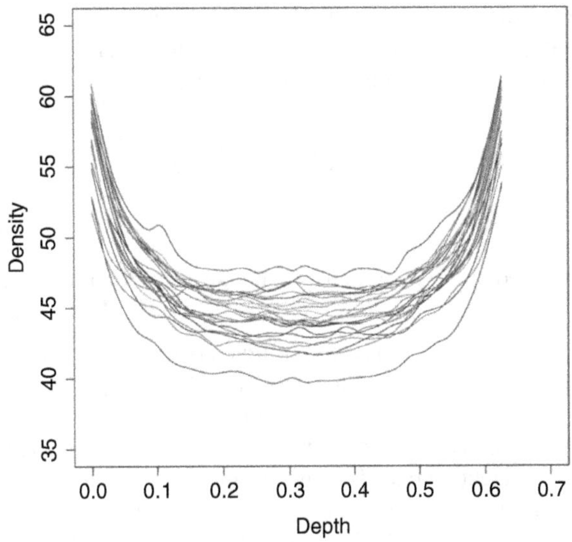

Figure 6.7 Smoothed VDP particleboard data. (Adapted from Shiau et al. 2009.)

where s and B are the sample mean and covariance of the score vectors. This statistic is approximately distributed as a beta,

$$\frac{n}{(n-1)^2}T_j^2 \sim \text{beta}\left(\frac{K}{2}, \frac{n-K-1}{2}\right). \tag{6.23}$$

Using a specified α-level UCL, recursive control charting of these statistics on a suitable historical control set will remove any outlier profiles and retain a set of in-control profiles.

In Phase II, T^2 charts are again used. The in-control distribution of a profile is multivariate normal with covariance Σ_0 and mean μ_0. Obtain the eigenvectors $\nu_1, \nu_2, \ldots, \nu_n$ and eigenvalues $\lambda_1 \geq \lambda_2 \geq \ldots \geq \lambda_n$ via PCA on the in-control covariance matrix and select the K largest components that represent the desired amount of variability in the data. A newly observed profile is smoothed and PCA is applied resulting in K PC scores $S_1, S_2, \ldots, S_K$. A control chart is constructed for each of these K PC scores. The ith PC score is normal with mean $\nu_i'\mu_0$ and variance λ_i. A control chart to monitor this PC is based on

$$\nu_i'\mu_0 \pm z_{\alpha/2}\sqrt{\lambda_i}. \tag{6.24}$$

It is not necessarily known which PC score to observe for detecting an out-of-control profile. An out-of-control condition may also be evident in multiple PC scores. In each of these cases, a combined control chart is called for. The relevant statistic is

$$J = \max_{1 \leq i \leq K} \left| \frac{S_i - \nu_i'\mu_0}{\sqrt{\lambda_i}} \right|. \tag{6.25}$$

An out-of-control profile is identified when J is larger than $z_{\alpha'/2}$, where α' has been adjusted for multiple PC scores to give an overall rate of α, i.e., $\alpha' = 1 - (1-\alpha)^{1/K}$.

An alternative to J is a combined control chart using

$$T^2 = \sum_{i=1}^{K} \frac{\left(S_i - \nu_i'\mu_0\right)^2}{\sqrt{\lambda_i}}. \tag{6.26}$$

This is distributed as a χ_K^2 random variable and may be monitored appropriately.

Example 6.5 Shiau et al. (2009) report the results of a simulation study investigating the number of PC scores needed to detect certain types of out-of-control profiles. An increased value of K will, by construction, explain more variation in the data. However, such an increase does not always coincide with an increase in the ability of the methods to detect an out-of-control profile. In fact, for the T^2 chart in Equation (6.26), the power will drop when K gets large.

Simulations comparing the Phase II methods (6.24), (6.25), and (6.26) are also examined. It is clear from the study that a single PC cannot be relied on to explain changes from an in-control to an out-of-control profile, even for simple deviations from in-control. The charts defined by Equations (6.25) and (6.26) give very similar results, and these two are close to the best of the individual PC charts at Equation (6.24). If a profile is classified as out-of-control by the combined chart in Equation (6.25) or the T^2 chart in Equation (6.26), the authors recommend examining the PC charts in Equation (6.24) to determine the source of the fault.

This method is applied in Phase I to the smoothed VDP data shown in Figure 6.7. The first three eigenvectors account for 85.26%, 10.83%, and 1.90% of the variability for a total of 97.99%, so K is chosen to be 3. The Phase I T^2 chart described above shows no out-of-control profiles. This is in agreement with the I-chart methods of Williams et al. (2007) for metrics M_{j1} and M_{j4} described in Section 6.2.

The authors point out that the smoothing step in the above methods has a large influence on the subsequent analysis. Increasing the level of smoothing applied to the profiles will concentrate the variation of the profiles into fewer PCs. Less smoothing will result in the need for more PCs and hence less data reduction is realized. This will make it more difficult to pinpoint source faults in out-of-control conditions.

6.3.2 Independent Component Analysis

Another example of using component analysis for profile monitoring is given by Ding et al. (2006). They examine charts on the basis of not only on PCA, but on ICA as well. PCA and ICA are used to reduce the dimension of the data.

PCA projects the data into a new space, where the dimensions are ordered by the amount of variability each explains. Geometrically, PCA is a rotation of the original axes into a new set of axes. The first axis points in direction of greatest variability, the second points in the direction of the next greatest amount of variability, and so on.

ICA projects the data into a new space, where the axes may be ordered by their divergence from normality. This is desirable if the in-control profiles are multivariate normal. This is a common assumption, and is made here. The first IC is the least normal, the second is the next least normal, etc. Taking a subset of the ICs provides a reduced dimension for the data.

For a continuous random variable Y, divergence from normality is measured by its entropy

$$H(Y) = -E\left(\log\left(f(Y)\right)\right), \tag{6.27}$$

where f is the density of Y. This measure is at its maximum when Y is normal, and strictly less than this maximum for other distributions. The ICs are determined with respect to this criterion in readily available software. The number of components (PC or IC) chosen is selected using the minimum description length as given in Apley and Shi (2001).

After the dimension reduction step using component analysis, either PCA or ICA, is complete, the profiles are classified as either in-control or out-of-control using the change-point detection (CPD) method described in Sullivan (2002). This CPD method works well at detecting multiple change points in a sequence of observations and detecting outliers. Sullivan's CPD is applied to each of the retained ICs separately. A profile is in-control if each of its ICs is in-control. This is referred to as ICA + CPD by the authors. Additionally, CPD is applied to the each of the selected PCs of a profile. This is the PCA + CPD method. For comparison purposes, theauthors also examine the efficiency of a T^2 chart applied recursively to the PCs. This is the PCA + T^2 method.

Example 6.6 A simulation study with $m = 1000$ profiles each of length $n = 20$ was run comparing the three methods on a variety of out-of-control scenarios. Clear distinctions among these three methods are evident. First, PCA + T^2 does well at classifying in-control profiles, but at the expense of classifying many out-of-control profiles as in-control, as well. A notable exception is when there are only a few out-of-control profiles (outliers). In this case, PCA + T^2 classifies nearly all profiles correctly. However, PCA + ICA and PCA + CPD both perform better than PCA + T^2 in this case.

Second, in a scenario specifically known where PCA would have trouble distinguishing between in- and out-of-control data (the difference between in- and out-of-control is modeled by a PC associated with a small eigenvalue), ICA + CPD is superior to PCA + CPD in classifying the profiles. This implies that ICA + CPD may be more robust than PCA + CPD.

Finally, in most scenarios ICA + CPD and PCA + CPD seem to be approximately equal in their Phase I profile classification performance. Overall, the authors use these conclusions as a basis for recommending ICA + CPD over the two alternatives.

6.4 WAVELETS

Wavelets present a way in which to model profiles in an efficient manner without losing important features of the data. A profile is projected into the wavelet domain where it is represented by a set of wavelet coefficients. The information about the profile is compressed into relatively few large magnitude coefficients, and these coefficients are readily interpreted and extracted in terms of differing scale-frequencies and locations of features within the profile.

Suppose a profile f is sampled at $n = 2^J$ points on the interval [0, 1]. The wavelet representation of f is

$$f(x) = \sum_{k=1}^{2^{j_0}} \xi_{j_0 k} \phi_{j_0 k}(x) + \sum_{j=j_0}^{J-1} \sum_{k=1}^{2^j} \theta_{jk} \psi_{jk}(x), \tag{6.28}$$

where $\xi_{j_0 k} = \langle f, \phi_{j_0 k} \rangle$ and $\theta_{jk} = \langle f, \psi_{jk} \rangle$ are the wavelet coefficients for f. The ξ are often called coarse coefficients, the θ are detail coefficients. The functions ϕ and ψ are specific to a particular wavelet family. The translations and dilations of these two functions,

$$\phi_{j_0 k}(x) = 2^{j_0/2} \phi\left(2^{j_0} x - k\right), \quad \psi_{jk}(x) = 2^{j/2} \psi\left(2^j x - k\right) \tag{6.29}$$

form an orthonormal basis for square integrable functions under certain readily met conditions. An important property resulting from using an orthonormal basis of wavelets is that it maintains the energy of the data: the sum of squared wavelet coefficients is the same as the sum of squared data values.

For the representation of f given above, it is assumed that the wavelets have been periodized to the interval [0, 1]:

$$\phi_{j_0 k}^p(x) = \sum_{l=-\infty}^{\infty} \phi_{j_0 k}(x - l), \quad \psi_{jk}^p(x) = \sum_{l=-\infty}^{\infty} \psi_{jk}(x - l). \tag{6.30}$$

With periodized orthonormal wavelets, a profile sampled at n points will have n coefficients. The superscript p will be dropped from here on.

The wavelet coefficients are computed from the sampled data points of f using the orthogonal transformation known as the discrete wavelet transform or DWT (Mallat (2009)). The DWT is available in many commercially available statistical software packages. The DWT may be represented as a matrix W operating on the sampled points of f,

$$Wf = \left[\xi_{j_0,1}, \xi_{j_0,2}, \ldots, \xi_{j_0,2^{j_0}}, \theta_{j_0,1}, \theta_{j_0 2}, \ldots, \theta_{J-1,2^{J-1}}\right]' = \theta. \tag{6.31}$$

The wavelet representation of f can be broken down into as many as J levels, called resolution levels. These levels correspond to the summation series for differing values of j in Equation (6.28). The lowest level of $j = j_0$ is the coarsest part of f, while increased levels of j correspond to increasingly detailed parts of f. Higher levels of j model those portions of f which show features at small scales or high frequencies. Lower levels of j are associated with large scale or low-frequency features of f. From Equation (6.28), it is evident that the original profile f is the sum of these individual projections of f into the different resolution levels. Additionally, the translations of the wavelet basis functions allow the wavelet representation in Equation (6.28) to also find features in f that are localized with respect to the x-axis. This ability of wavelets to analyze a function both locally and at differing scale-frequency levels is one of its most useful properties.

If $j_0 = 0$, then the function f is decomposed into the maximum possible number of resolution levels, J. In this case, there is only one wavelet coefficient for each of the lowest two summation series in Equation (6.28). The single coefficient ξ at this lowest resolution level represents the mean of the sampled points of f. If $j_0 > 0$, then the coefficients ξ at the lowest resolution level represent a smoothed version of f. As j_0 increases, the smoothness of this coarsest part decreases. The level of decomposition j_0 is a parameter selectable by the user.

Consider the profile displayed in Figure 6.8. This function is analyzed with wavelets using the orthonormal compactly supported wavelet basis derived from the Daubechies least asymmetric wavelet of length 8 (Daubechies (1992)). The profile "Blocks" is sampled at $n = 2^J = 2^8 = 256$ equally spaced points and is decomposed by wavelets down to resolution level $j_0 = 4$. The plots on the left of Figure 6.9 show the projection of "Blocks" into three different resolution levels: the highest possible detail level at $j = 7$, the detail portion of f at level $j = 5$, and the coarsest level at $j_0 = 4$. The detail resolution levels corresponding to $j = 4$ and $j = 6$ are not displayed. The sum of all the projections of f results in the original sampled profile in Figure 6.8. The plots on the right side of Figure 6.9 show the wavelet coefficients for the same resolution levels. There are 2^j coefficients at resolution level j. Note that the scales of these three coefficient plots are not the same. The wavelet coefficients correspond to features of f at particular locations and at particular scale-frequencies.

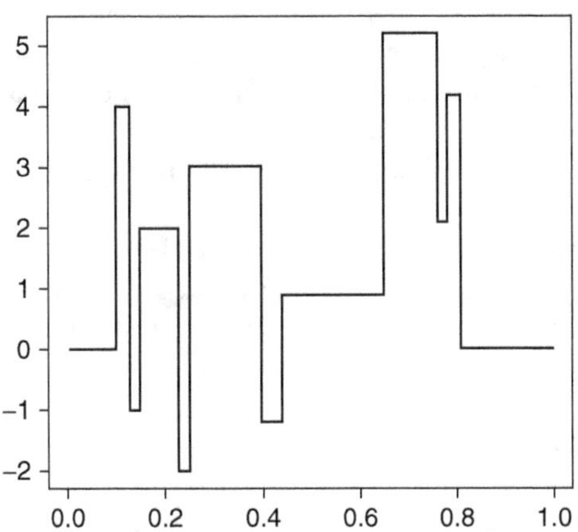

Figure 6.8 The common wavelet test function "Blocks."

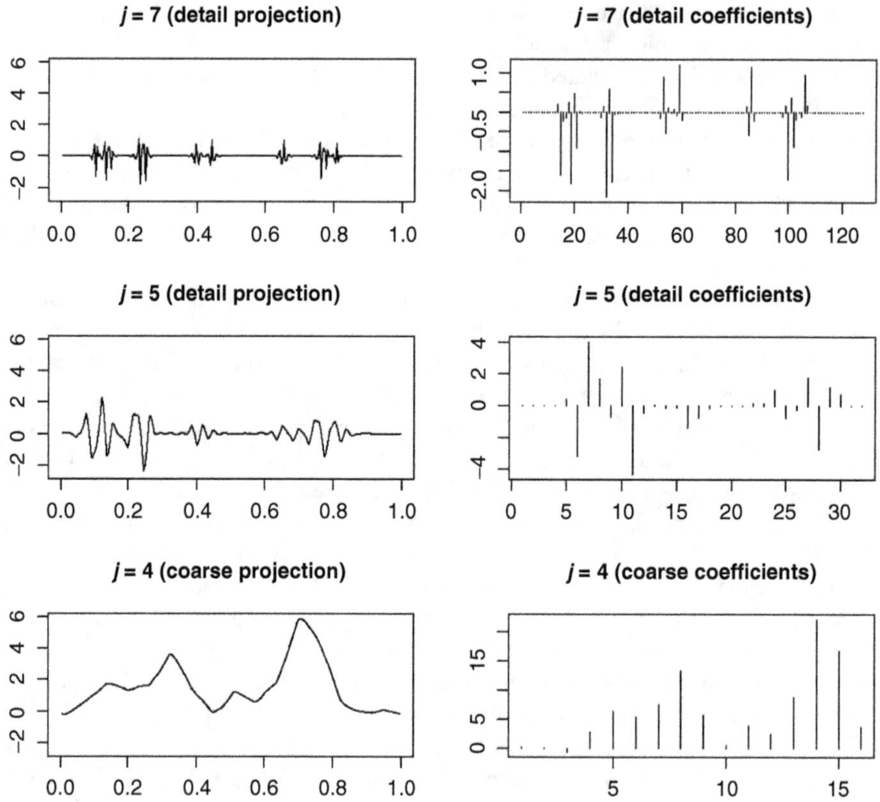

Figure 6.9 Analysis of "Blocks" by wavelets.

The coarse projection of f shown in the bottom left panel of Figure 6.9 is modeling the general shape of "Blocks", but is missing important details such as the number and steepness of the jumps. As each successive detail resolution level is added to this coarse projection, the approximation gets closer and closer to the original sample profile. The differing resolution levels model behavior in the profile that goes from global (smooth) approximation at low levels j to increasingly localized modeling at higher levels.

The wavelet representation in Equation (6.28) is for a profile f. What is usually observed is a noisy profile as in Equation (6.5). However, all that is needed is to replace the n sampled values of f with the n observed y values and apply the DWT transformation W,

$$Wy = W(f + \varepsilon) = Wf + W\varepsilon = \theta + \varepsilon_W = \tilde{\theta}. \tag{6.32}$$

If the errors ε are normally distributed, then the wavelet coefficients $\tilde{\theta}$ will also be normal since the DWT is orthogonal. Additionally, the DWT will preserve the covariance structure of the data when moving into the wavelet domain. Note that θ are the coefficients for the unobserved f, while $\tilde{\theta}$ are the coefficients for the observed values y of the noisy profile. If the ε are normal with mean 0 and variance σ^2, then the distribution of the $\tilde{\theta}$ is multivariate normal with mean θ and variance $\sigma^2 \cdot I_{n\times n}$.

Typically, many, if not most, of the wavelet coefficients θ are very small and may safely be discarded or shrunk further to zero. This sparseness property is useful in denoising a profile and reducing the dimension of the analysis. This process of removing coefficients is called thresholding. There are multiple methods of thresholding coefficients. A simple example is the hard threshold rule of Donoho and Johnstone (1998):

$$\hat{\theta} = \begin{cases} \tilde{\theta}, & |\tilde{\theta}| > \lambda, \\ 0, & |\tilde{\theta}| \le \lambda. \end{cases} \tag{6.33}$$

This threshold rule keeps all observed noisy coefficients larger than λ and sets the rest to zero. Usually, only detail coefficients from a user-specified resolution level j and up are thresholded. In the case of a profile contaminated with noise with variance σ^2 as in Equation (6.5), the value of λ used in Donoho and Johnstone is $\lambda = \sqrt{2\hat{\sigma}^2 \ln(n)}$, where $\hat{\sigma}$ is an estimate for σ. A similar threshold rule is

$$\hat{\theta} = \operatorname{sgn}(\tilde{\theta})\left(|\tilde{\theta}| - \lambda\right)_{+}. \tag{6.34}$$

This is the soft threshold rule. When used with λ above, it is referred to as the VisuShrink estimator.

The reconstruction $\hat{f}$ is made by replacing the coefficients in Equation (6.28) with the thresholded coefficients. Equivalently, the inverse of the DWT is applied to the thresholded coefficients to obtain a reconstruction. Many other threshold methods exist with differing values of λ and differing rules relating θ to λ.

The estimate of σ needed for thresholding is found using the wavelet coefficients. Since the DWT is an orthogonal transform, the wavelet coefficients are normal random variables with variance σ^2. The coefficients at the highest detail resolution level represent mainly noise. For most profiles, very little profile structure is found at this resolution level, so the majority of these coefficients are expected to have mean 0. The median absolute deviation (MAD) estimate of σ is used on these high level coefficients to determine $\hat{\sigma}$ for thresholding purposes. This estimate is desirable since it is robust to outliers. In this situation, outliers are those wavelet coefficients that are not purely noise, but are modeling some features of the profile.

Thresholding is a nonlinear operator that provides both smoothing and data reduction to profile monitoring. Smoothing is accomplished by identifying those coefficients that model noise in the profile and thresholding them by reducing their magnitude or removing them entirely. Reconstructing the profile with the modified coefficients provides a smoothed version of the original observed profile. Some of the nonparametric methods in the previous sections of this chapter used wavelets in this manner.

Data reduction may be achieved with wavelets by the choice of the threshold parameters. By selecting which coefficients to threshold and how much to modify them, one may reduce a profile of length n to a considerably smaller set of wavelet coefficients. Then, rather than monitoring the original profile at n points, one instead monitors the reduced set of thresholded wavelet coefficients of the profile.

6.4.1 Monitoring User-Selected Coefficients

An example of such data reduction is in Jin and Shi (1999), where thresholding is used to remove or retain certain features within a profile that are useful for monitoring purposes. Jin and Shi apply thresholding to the wavelet coefficients for three purposes. First, they threshold for noise removal. This leaves a noise-free estimate of the profile for further analysis. Second, they threshold to retain only selected frequency bands of the profile that are related to particular process faults of interest. This step sets entire resolution levels of coefficients to zero. Finally, the remaining nonzero coefficients are thresholded iteratively to obtain an estimated profile at a desired preset accuracy limit P. The first and third threshold passes use the threshold rule in Equation (6.33). The value of λ is set as in Donoho and Johnstone (1998) for noise removal. For the accuracy limit, the value of λ is a function of P and the coefficients available at each successive iteration.

Jin and Shi segment the profile based on expert knowledge. The number of segments and the endpoints are chosen in order to identify areas where particular known types of faults may occur. As an example, they analyze stamp tonnage data on a sheet-metal press. A typical signal is shown in Figure 6.10. Note that this signal displays different types of behavior over different ranges of the signal. This profile displays both smooth and high-frequency behavior over different portions of the profile's support. An expert-provided segmentation is shown in Figure 6.11. Eight segments are

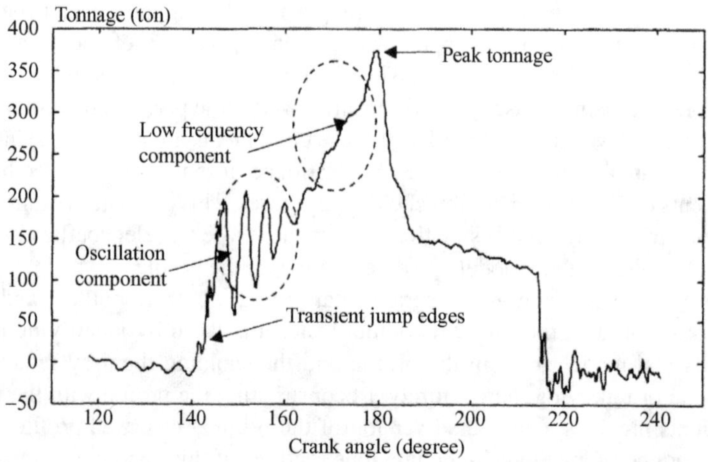

Figure 6.10 Tonnage signal as a function of angle. (Adapted from Jin and Shi 1999.)

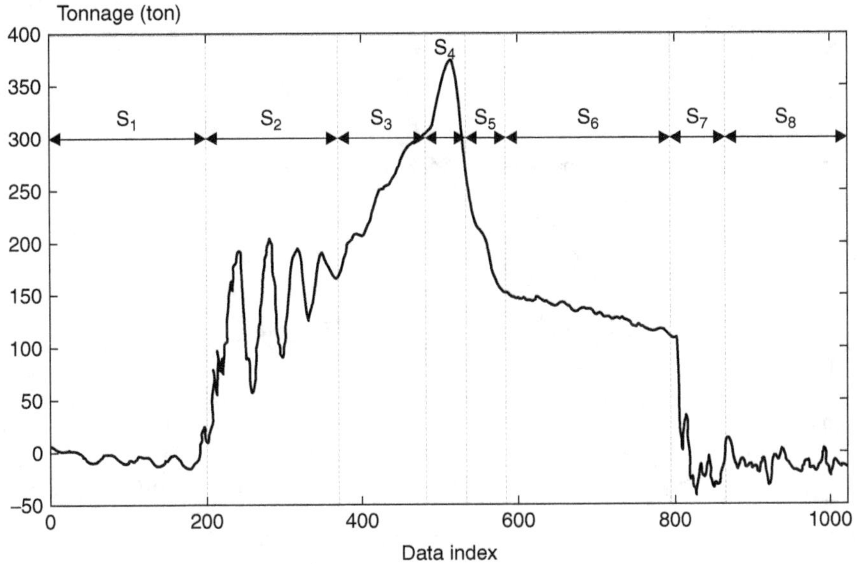

Figure 6.11 Tonnage signal segmented by expert input. (Adapted from Jin and Shi 1999.)

identified that partition the profile into regimes, where the profile displays markedly different characteristics. The accuracy limit P is specified for each segment.

There are many wavelets to choose from when performing wavelet analysis. Jin and Shi analyze the tonnage signal with the D3, a Daublet wavelet. This wavelet basis was chosen since it can accurately represent second-order polynomials with just the coarse portion of the wavelet transform. For this example, it is determined that second-order polynomials will suffice for the coarse approximation.

Example 6.7 The three passes of thresholding are applied to the tonnage signal profiles. Two tonnage profiles are considered. Each profile is sampled at $n = 1024$ equispaced points. One profile is a case of a run through the press that results in a sheet that is 10% thicker than usual, the other is 10% thinner. The first pass is for denoising. It is applied to the entire profile, not to individual segments. After this step of thresholding, the first profile has been reduced to 621 coefficients (of the original 1024) and the second to 640.

The next thresholding pass is done to remove unwanted frequency bands. This thresholding pass is applied to each segment separately since different segments have differing frequency bands of interest. In this pass, entire resolution levels j corresponding to the targeted frequency bands have all their coefficients set to 0 within the segment. The resolution levels are chosen by expert input. For the profiles above, this results in a reduction to 82 and 95 nonzero coefficients, respectively.

Finally, the third thresholding pass is applied to each profile, resulting is a final data set size of 66 and 73 nonzero coefficients, respectively. This is a significant data

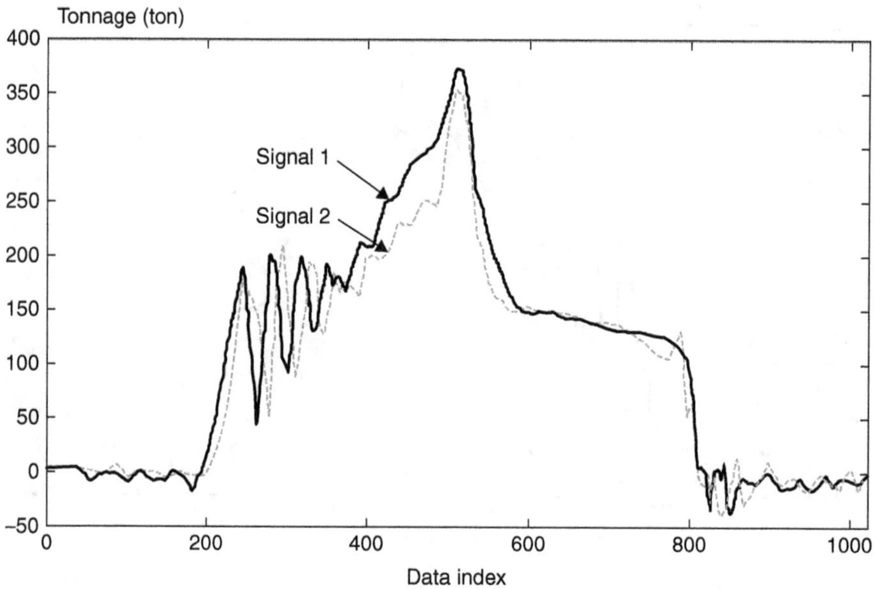

Figure 6.12 Wavelet thresholded estimates of two tonnage profiles. (Adapted from Jin and Shi 1999.)

reduction considering the original signals each consisted of 1024 wavelet coefficients. Figure 6.12 shows that despite over a 90% data reduction, the features of the original profiles are well modeled.

The wavelet coefficients are obtained for the entire signal without consideration of expert-imposed segmentation. After segmenting, some of the wavelet coefficients will be associated with wavelet functions whose support covers more than one segment. Care must be taken on how to determine which coefficients will belong in a particular segment when performing the second and third thresholding passes.

With expert knowledge of where faults may occur within a profile and how such faults correspond to a particular set of wavelet coefficients, the authors suggest use of an X-chart to monitor the appropriate wavelet coefficients. For example, to monitor for a change in the thickness of the sheet metal in this application it is suggested to monitor the 25th coefficient in the sixth resolution level of segment 3.

6.4.2 Monitoring Automatically Selected Coefficients

The above method requires that experts provide the statistic to be monitored, the statistic in this case being which particular set of wavelet coefficients to monitor. In Jin and Shi (2001), this method is modified to use wavelets to determine automatically which subset of coefficients should be monitored to detect faults with little to no information about the faults. In Phase I, The wavelet coefficients are obtained from a set of initial profiles, then thresholded to remove noise as discussed previously.

Hotelling's T^2 statistic is applied iteratively to the sets of thresholded wavelet coefficients to determine an in-control baseline set of wavelet coefficients. From this set, which after thresholding is greatly reduced in size from the initial length of the profile n, a subset of coefficients is selected to represent a fault.

In Phase II, an observed profile is decomposed into wavelet coefficients by the DWT and the coefficients specified in Phase I are kept. Faults are detected using control limits from a Hotelling T^2 statistic. When a fault is detected, it is classified by comparing the wavelet coefficients from the current profile to those in previously encountered faulty profiles. If the current fault is not sufficiently close to any existing class, then a new fault class is created. For each fault class, a subset of wavelet coefficients is selected automatically by choosing the coefficients that maximize the distinction between the fault classes.

As more profiles are observed, one of three events occurs. A new fault cluster is created with a defining subset of wavelet coefficients. Or, a profile matches an existing fault condition, in which case the wavelet coefficient subset may be updated with the new information from this faulty profile. Or, the profile may be in-control, in which case the description of the in-control profiles may be modified.

Another automatic selection method for wavelet coefficients that does not depend on expert opinion is in Lada et al. (2002). Similar to Jin and Shi (1999) and Jin and Shi (2001), the proposed method uses wavelet thresholding to reduce the dimensionality of the problem and remove nuisance noise.

A new type of thresholding is introduced, relative reconstruction error (RRE). RRE is a trade-off between fit to the profile and reduction in the number of nonzero wavelet coefficients. This method provides more reduction than many existing methods of thresholding. RRE takes the C largest coefficients that minimizes

$$RRE\,(C) = \frac{\sum_{i=1}^{n} (y_i - \hat{y}_{iC})^2}{\sum_{i=1}^{n} y_i^2} + \frac{C}{n}, \tag{6.35}$$

where $\hat{y}_{iC}$ is the fitted value using only the C largest wavelet coefficients.

For each of the Phase I profiles identified as in-control, RRE may select a different number of coefficients C. In this case, the union of coefficients selected by RRE for each profile is used. This set of specified coefficients will be used to describe the in-control profile, and only the corresponding coefficients are monitored in other profiles. Newly observed profiles are compared on this reduced size coefficient set and classified as either in-or out-of-control. The method assumes that a new profile has a possibly different mean than the in-control profiles, but the same covariance.

The authors randomly select a fixed number of in-control profiles multiple times. Sampling is done without replacement to ensure non-singularity of the covariance matrix estimate. From these samples, an in-control mean and covariance are calculated. Under normality assumptions, a new profile's coefficients are compared via a Hotelling T^2 statistic. The UCL in this case are from an F-distribution. If the profiles are not normal, a separate resampling method is used to derive a nonparametric distribution for the in-control profiles.

In Lada et al., the coefficient subset to be monitored is selected from the entire available set of wavelet coefficients. Zhou et al. (2006) provide a different way of selection of these coefficients. They determine which coefficients are to be monitored by choosing a maximum resolution level of the wavelet transform. They monitor all the coefficients existing in the lowest resolution level up to the specified maximum resolution level. This maximum level is determined by the user through an accuracy requirement. As the maximum level is increased, the fit between the profile and its wavelet estimate increases. By putting a limit on the accuracy, the user effectively determines the upper bound for the resolution levels of interest. Recall that the low-resolution levels have few coefficients, so this acts as a data-reduction step.

Once the resolution levels and their corresponding wavelet coefficients are selected in Zhou et al., a T^2 chart of the coefficients is used to monitor the profiles. In Phase I, it is used iteratively to determine the in-control condition. A T^2 chart based on the estimates from Phase I is used for Phase II. If an out-of-control profile is detected by this chart, then individual charts for each monitored wavelet coefficient are examined with a Bonferroni correction to the control limits. Using the locality property of wavelet estimation, these individual charts for specific coefficients can be associated with particular regions of the profile. Hence, the location of the fault within the profile may be estimated.

Example 6.8 As an example, profiles from a rapid thermal chemical vapor deposition (RTCVP) process are analyzed. Figure 6.13 shows four examples of in-control profiles, while Figure 6.14 contains graphs of four out-of-control profiles. Each profile is of length $n = 2^7 = 128$, but on the basis of the RRE coefficient selection rule, only 19 coefficients are identified as necessary to classify profiles. This identification, as in Jin and Shi (2001), is done without expert input. There were a total of 21 in-control profiles to begin with. Samples of size 5 were selected from these 21 profiles. The sample mean and covariance were estimated for in-control profiles by performing this sampling 40 times. The distribution was calculated nonparametrically using resampling, as well. For each of the out-of-control profiles, the p-values are all 0, successfully classifying all cases shown in Figure 6.14 as out-of-control profiles.

6.4.3 Thresholding on Prior Information

Jeong et al. (2006), like Lada et al. (2002), also provide a Phase II method for SPC based on wavelets. They shrink the wavelet coefficients of a newly observed profile by incorporating information from the known in-control profiles in the threshold rule. If θ^0 is the set of wavelet coefficients for an in-control profile, and $\tilde{\theta}$ is the set of coefficients of an observed profile, then their threshold is the hard threshold rule of

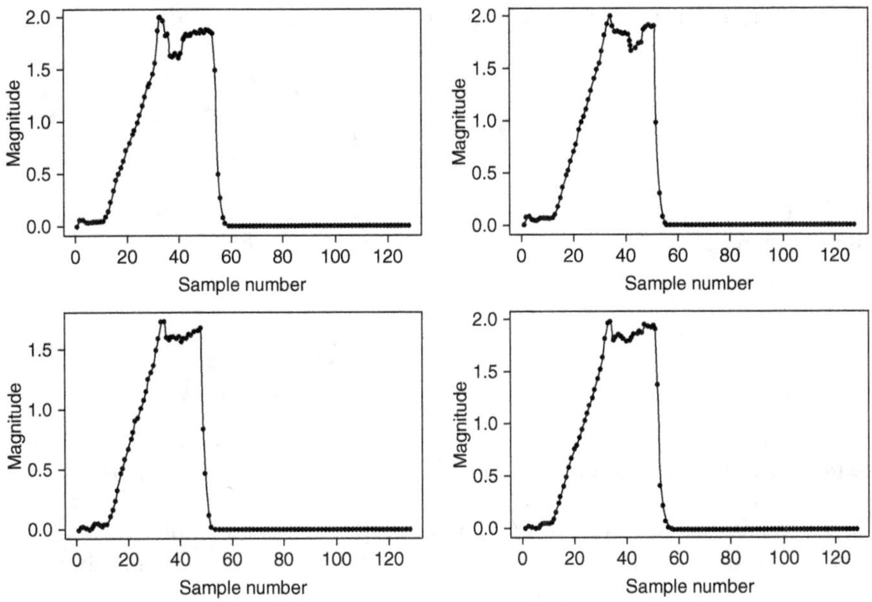

Figure 6.13 In-control profiles from the RTCVP process. (Adapted from Lada et al. 2002.)

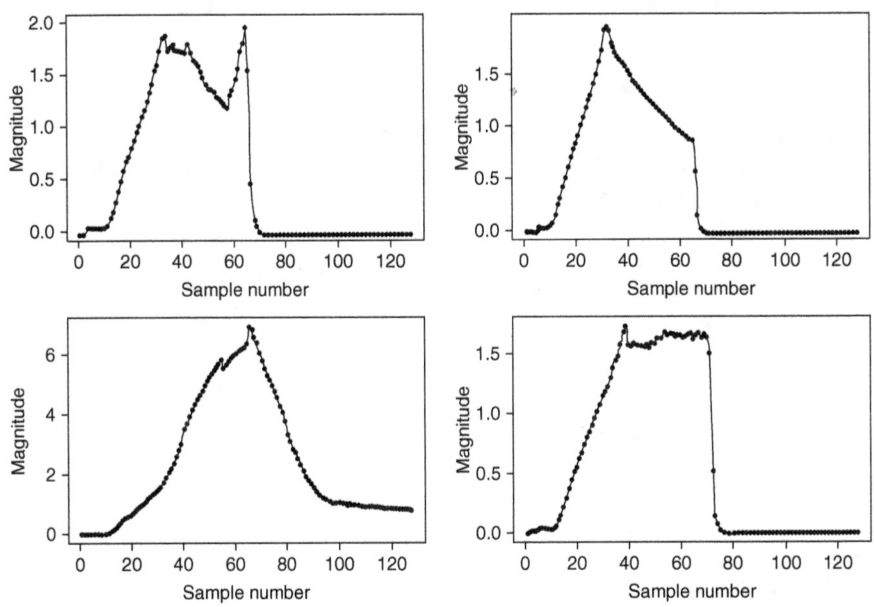

Figure 6.14 Out-of-control profiles from the RTCVP process. (Adapted from Lada et al. 2002.)

Donoho and Johnstone (1998) in Equation (6.33) applied to the difference:

$$\left(\tilde{\theta}_{jk} - \theta_{jk}^0\right)_* = \begin{cases} \tilde{\theta}_{jk} - \theta_{jk}^0, & \left|\tilde{\theta}_{jk} - \theta_{jk}^0\right| > \delta\sigma, \\ 0, & \left|\tilde{\theta}_{jk} - \theta_{jk}^0\right| > \delta\sigma. \end{cases} \tag{6.36}$$

By thresholding the difference, only those wavelet coefficients that deviate from the in-control coefficients are monitored, not the entire set of n coefficients. The test statistic used is

$$T^2 = \frac{1}{\sigma^2} \sum_{j,k} \left(\tilde{\theta}_{jk} - \theta_{jk}^0\right)_*^2, \tag{6.37}$$

with $\delta = \sqrt{2 \ln (n a_n)}$ and $a_n = \min \left\{ 4 \left[\sigma^{-1} \max_{j,k} \left(\tilde{\theta}_{jk} - \theta_{jk}^0 \right) \right]^{-4}, [\ln (n)]^{-2} \right\}$. It is assumed here that the variance σ^2 is known from the in-control profiles and does not change for out-of-control profiles. The distribution of T^2 is approximately normal and the UCL values can be found appropriately.

Based on simulations, their method is superior in terms of ARL to monitoring all coefficients. When compared to a method that monitors the subset of coefficients selected via thresholding as in Equation (6.33), the proposed method is generally more efficient, especially when the out-of-control profiles display very localized faults.

6.4.4 Wavelet Change-Point Method

Chicken et al. (2009) also apply thresholding to the difference of the Phase I known wavelet coefficients and the coefficients of a newly observed profile. Additionally, their method incorporates information from all previously observed, unclassified profiles to estimate the noise level σ, the magnitude of the fault detected, and the time at which the fault occurred. Thus, they analyze the profiles as a change-point problem. After observing T profiles, the hypotheses are

$$H_0 : \|f_t - f_0\|_2^2 = \int (f_t (x) - f_0 (x))^2 \, dx = 0, \quad t = 1, 2, \ldots, T \tag{6.38}$$

and

$$H_a : \|f_t - f_0\|_2^2 > 0, \quad t = \tau + 1, \tau + 2, \ldots, T. \tag{6.39}$$

Using a likelihood ratio approach, the test statistic is

$$h (\tau) = \frac{\hat{\gamma}}{2} \sum_{t=\tau+1}^{T} \left(\frac{w_t}{n} - 1 \right), \tag{6.40}$$

where w_t is the squared error between the tth observed profile's wavelet coefficients and the in-control standard's wavelet coefficients, and $\hat{\gamma}$ is

$$\hat{\gamma} = \frac{1}{T-\tau}\sum_{t=\tau+1}^{T}\hat{w}_t - \frac{1}{\tau}\sum_{t=1}^{\tau}\hat{w}_t. \tag{6.41}$$

Here, $\hat{w}_t$ is the thresholded difference of the wavelet coefficients from the observed profile and the in-control profile. The statistic h is used in a control chart by maximizing it over all τ and comparing it to the UCL. The UCL is found via simulation due to the complex form of the statistic.

Example 6.9 This change-point method is compared against three other wavelet methods through simulation. These other methods are the same as those in the simulation study of Jeong et al. (2006). Estimator M^1 monitors all the wavelet coefficients and hence possesses no data reduction property. M^2 monitors only those wavelet coefficients for a profile that are nonzero after VisuShrink thresholding Equation (6.34) is applied. M^3 is the estimate from Jeong et al. discussed above. M^* is the change-point estimator of Chicken et al. The nominal profile is from Mallat (2009) and is shown in Figure 6.15. Various fault conditions were added to this function, and normal errors with $\sigma = 1$ were used. Since the first three estimators are designed with σ known, all four are compared with this restriction in place. Three fault conditions are shown in Figure 6.16. In addition, two other faults were considered: one is a mean shift of the entire profile and the other is a localized shift from Jeong et al. (2006). The parameter a describes the integrated squared error between the in-control and out-of-control profiles. The appropriate signal-to-noise ratio in this application is the ratio of a to σ^2. For the simulated examples, this is

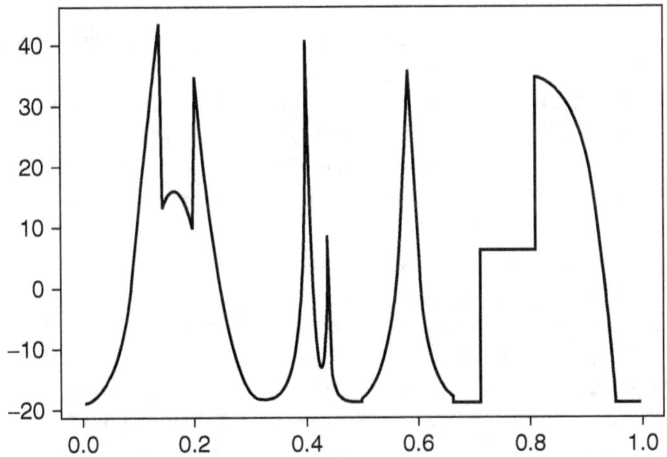

Figure 6.15 Piecewise smooth function. (Adapted from Mallat 2009.)

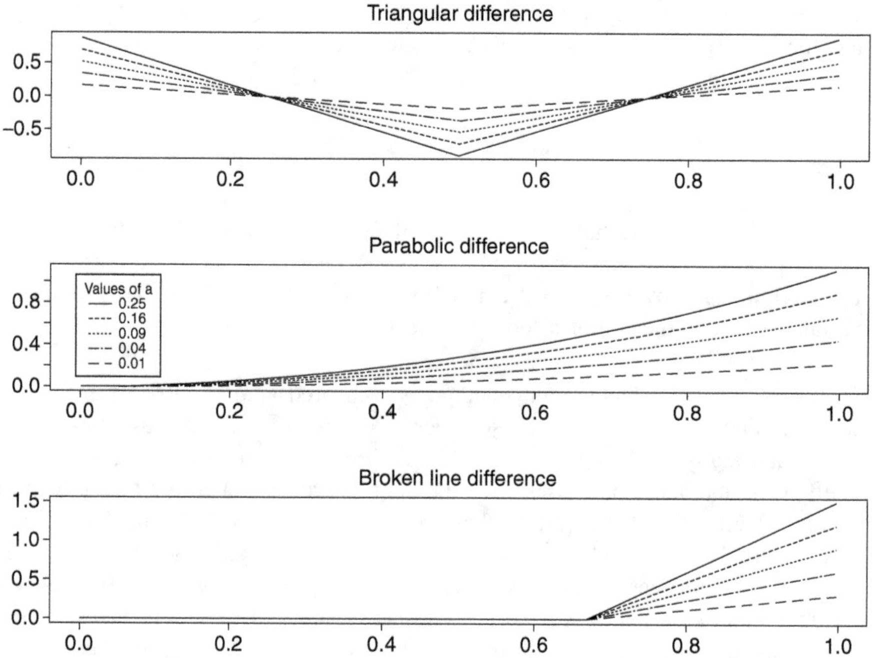

Figure 6.16 Out-of-control faults. (Adapted from Chicken et al. 2009.)

between $(4/9)\left(a/\sigma^2\right)$ and $\left(a/\sigma^2\right)$. Since $\sigma = 1$ and $a \leq 1/4$, these represent very small SNRs.

Table 6.2 shows the simulation results. The entries are ARLs with the standard deviation of the ARL in parentheses. Note that M^* is generally the most efficient at detecting any of the five types of faults simulated. It typically has both the smallest ARL and smallest standard deviation of the ARL.

Additionally, estimates for τ, σ, and a are available using the wavelet change-point estimate M^*. The other estimation methods do not provide these values. The estimate of τ is the value that maximizes h. Once this is determined, the estimate for a is $\hat{\gamma}$. The estimate of σ is estimated from the detail coefficients from the highest resolution level of all observed profiles.

Despite their different approaches, the wavelet methods in this section have much in common. Thresholding is used to significantly reduce the dimension of the data and provides a way of denoising the observed profile, removing unwanted nuisance variability. The resolution levels provided by wavelets allow the modeling of profiles whose features vary over their support (i.e., profiles that are smooth in some regions and irregular in others). Finally, wavelets provide a way of pinpointing locally the areas within a profile where a fault may possibly have occurred.

Table 6.2 ARL Comparisons of the Four Methods (Adapted from Chicken et al. 2009.)

Out-of-Control Profile	a 0.01	0.04	0.09	0.16	0.25
Horizontal Line					
M^1	122.94 (128.82)	34.76 (34.32)	7.97 (7.03)	2.27 (1.69)	1.18 (0.47)
M^2	91.36 (94.68)	14.48 (13.47)	2.62 (2.05)	1.14 (0.41)	1.00 (0.04)
M^3	71.11 (69.57)	20.76 (19.43)	5.90 (5.01)	2.05 (1.46)	1.17 (0.47)
M^*	42.45 (38.36)	2.50 (1.79)	1.14 (0.42)	1.01 (0.09)	1.00 (0.00)
Triangular					
M^1	131.50 (129.22)	36.26 (37.11)	8.32 (7.69)	2.28 (1.69)	1.19 (0.47)
M^2	90.19 (92.98)	14.33 (13.43)	2.64 (2.08)	1.11 (0.36)	1.01 (0.08)
M^3	72.87 (73.10)	21.01 (20.00)	5.88 (5.65)	2.09 (1.51)	1.16 (0.44)
M^*	97.94 (84.98)	7.72 (5.94)	1.52 (0.82)	1.03 (0.17)	1.00 (0.03)
Parabolic					
M^1	124.55 (123.86)	34.09 (33.72)	7.84 (7.11)	2.23 (1.62)	1.16 (0.44)
M^2	84.05 (83.28)	14.57 (13.92)	2.63 (2.05)	1.15 (0.40)	1.00 (0.04)
M^3	75.72 (73.49)	21.33 (20.57)	5.63 (5.45)	2.18 (1.61)	1.61 (0.43)
M^*	84.62 (72.88)	4.81 (3.80)	1.33 (0.63)	1.03 (0.18)	1.00 (0.00)
Broken Line					
M^1	128.24 (130.22)	37.21 (36.05)	7.93 (7.62)	2.27 (1.65)	1.22 (0.50)
M^2	91.49 (86.02)	15.52 (14.61)	2.49 (2.07)	1.15 (0.45)	1.00 (0.05)
M^3	75.16 (73.82)	20.25 (19.47)	5.52 (4.88)	2.08 (1.48)	1.16 (0.43)
M^*	96.96 (81.09)	8.98 (6.96)	1.60 (0.89)	1.05 (0.21)	1.00 (0.03)
Local Jumps					
M^1	122.22 (121.19)	35.37 (34.31)	7.70 (7.04)	2.24 (1.66)	1.16 (0.44)
M^2	90.17 (92.81)	14.16 (12.91)	2.58 (2.01)	1.12 (0.37)	1.00 (0.04)
M^3	72.28 (67.81)	20.72 (21.79)	5.90 (5.20)	2.03 (1.41)	1.16 (0.43)
M^*	111.73 (91.68)	11.54 (9.18)	2.09 (1.32)	1.07 (0.26)	1.00 (0.04)

REFERENCES

Apley, D. W. and Shi, J. (2001) A factor analysis method for diagnosing variability in multivariate manufacturing processes. *Technometrics*, **43**, 84–95.

Chang, S. I. and Yadama, S. (2010) Statistical process control for monitoring nonlinear profiles using wavelet filtering and B-spline approximation. *International Journal of Production Research*, **48**, 1049–1068.

Chicken, E., Pignatiello, J., and Simpson, J. (2009) Statistical process monitoring of nonlinear profiles using wavelets. *Journal of Quality Technology*, **41**, 198–212.

Daubechies, I. (1992) *Ten Lectures on Wavelets*. SIAM, Philadelphia.

Ding, Y., Zeng, L., and Zhou, S. (2006) Phase I analysis for monitoring nonlinear profiles in manufacturing processes. *Journal of Quality Technology*, **38**, 199–216.

Donoho, D. and Johnstone, I. (1998) Minimax estimation via wavelet shrinkage. *Annals of Statistics*, **26**, 879–921.

Gardner, M., Lu, J.-C., Gyurcsik, R., Wortman, J., Hornung, B., Heinisch, H., Rying, E., Rao, S., Davis, J., and Mozumder, P. (1997) Equipment fault detection using spatial signatures. components, packaging, and manufacturing technology, Part C, *IEEE Transactions on*, **20**, 295–304.

Jeong, M. K., Lu, J.-C., and Wang, N. (2006) Wavelet-based SPC procedure for complicated functional data. *International Journal of Production Research*, **44**, 729–744.

Jin, J. and Shi, J. (1999) Feature-preserving data compression of stamping tonnage information using wavelets. *Technometrics*, **41**, 327–339.

Jin, J. and Shi, J. (2001) Automatic Feature Extraction of Waveform Signals for In-Process Diagnostic Performance Improvement. *Journal of Intelligent Manufacturing*, **12**, 257–268.

King, E., Hart, J. D., and Wehrly, T. E. (1991) Testing the equality of two regression curves using linear smoothers. *Statistics and Probability Letters*, **12**, 239–247.

Lada, E., Lu, J.-C., and Wilson, J. (2002) A wavelet-based procedure for process fault detection. Semiconductor Manufacturing, *IEEE Transactions on*, **15**, 79–90.

Mallat, S. (2009) *A Wavelet Tour of Signal Processing*. Academic Press, Boston, 3rd ed.

Mason, R.L., T. N., and Young, J. (1995) Decomposition of T^2 for Multivariate Control Chart Interpretation. *Journal of Quality Technology*, **27**, 98–108.

Mason, R.L., T. N., and Young, J. (2001) A practical approach for interpreting multivariate T^2 control chart signals. *Journal of Quality Technology*, **29**, 396–406.

Shiau, J., Huang, H., Lin, S., and Tsai, M. (2009) Monitoring Nonlinear Profiles with Random Effects by Nonparametric Regression. *Communications in Statistics – Theory and Methods*, **38**, 1664–1679.

Sullivan, J. H. (2002) Detection of multiple change points from clustering individual observations. *Journal of Quality Technology*, **34**, 371–383.

Sullivan, J. H. and Woodall, W. H. (1995) Comparison of multivariate quality control charts for individual observations. *Journal of Quality Technology*, **28**, 398–408.

Tracy, N. D., Young, J. C., and Mason, R. L. (1992) Multivariate control charts for individual observations. *Journal of Quality Technology*, **24**, 88–95.

Walker, E. and Wright, S. P. (2002) Comparing curves using additive models. *Journal of Quality Technology*, **34**, 118–129.

Williams, J. D., Woodall, W. H., and Birch, J. B. (2007) Statistical monitoring of nonlinear product and process quality profiles. *Quality and Reliability Engineering International*, **23**, 925–941.

Winistorfer, P., Young, T., and Walker, E. (1996) Modeling and comparing vertical density profiles. *Wood and Fiber Science*, **28**, 133–141.

Zhang, H. and Albin, S. (2009) Detecting outliers in complex profiles using a control chart method. *IIE Transactions*, **41**, 335–345.

Zhou, S., Sun, B., and Shi, J. (2006). An SPC Monitoring System for Cycle-Based Waveform Signals Using Haar Transform. Automation Science and Engineering, *IEEE Transactions on*, **3**, 60–72.

CHAPTER 7

Multivariate Linear Profiles Monitoring

Rassoul Noorossana
Industrial Engineering Department, Iran University of Science and Technology, Tehran, Iran

Majid Eyvazian
Industrial Engineering Department, Iran University of Science and Technology, Tehran, Iran
and
Industrial Engineering Department, Khajeh Nasir Toosi University of Technology, Tehran, Iran

INTRODUCTION

In certain cases, quality of a process or a product can be effectively characterized by two or more linear profiles where response variables of interest are correlated. The statistical structure among the response variables and an explanatory variable considered in this chapter is referred to as multivariate linear profiles. As mentioned in Chapters 2 and 3, Parker et al. (2001) introduced a case study where the relationships between six response variables and six explanatory variables are investigated in a calibration process at NASA Langley Research Center. Mahmoud et al. (2007) studied the relationship between one response variable and one explanatory variable. Mahmoud (2008) also studied the relationship between one response variable and six explanatory variables. In this case study, each response variable is a linear function of all six explanatory variables where response variables are correlated. This well-known problem structure is referred to as multivariate multiple linear profiles. It is obvious that when linear profiles are monitored separately then correlation between response variables is ignored and misleading results should be expected. To overcome

Statistical Analysis of Profile Monitoring, First Edition. Edited by Rassoul Noorossana, Abbas Saghaei and Amirhossein Amiri.

this problem, the use of methods that consider the multivariate structure between response variables is inevitable. Noorossana et al. (2010b) studied a real case study at the body shop of an automotive plant where the functional relationship between a set of responses and one explanatory variable are linear and response variables are correlated. In this case study, the purpose is to investigate calibration between desired force and the real force exerted by a 1600-ton hydraulic press machine. They discussed this case as multivariate simple linear profiles.

In this chapter, multivariate simple linear profiles monitoring and multivariate multiple linear profiles monitoring will be discussed. In Section 7.1, we discuss Phase I and Phase II methods on monitoring multivariate simple linear profiles and multivariate simple linear profiles. In Section 7.2, multivariate multiple linear profiles and methods for monitoring multivariate multiple linear profiles, both in Phase I and Phase II are discussed. In addition, a diagnostic method for Phase I analysis and a change point approach in Phase II for monitoring multivariate multiple linear profiles are presented.

7.1 MONITORING MULTIVARIATE SIMPLE LINEAR PROFILES

In this section, we assume that there are several response variables and one explanatory variable. Each response variable is a linear function of the explanatory variable and response variables are correlated. This well-known structure is referred to as multivariate simple linear profiles, where simple refers to one explanatory variable and multivariate refers to several response variables. In Section 7.1.1, we discuss Phase I monitoring methods and Section 7.1.2 presents Phase II monitoring methods for multivariate simple linear profiles.

7.1.1 Phase I Multivariate Simple Linear Profiles Monitoring

Let us assume that there are m samples available from a historical data set. In each sample, we have n_k values for explanatory variable (sample size) and for each value of explanatory variable there are p corresponding response values. Hence, for the kth random sample we have n observations as $x_i, y_{i1k}, y_{i2k}, \ldots, y_{ipk}$, where $i = 1, 2, \ldots, n_k$, $k = 1, 2, \ldots, m$, and x_i is the ith value of the explanatory variable and $y_{i1k}, y_{i2k}, \ldots,$ and y_{ipk} are the corresponding response values. In each sample, the relationship for relating the explanatory variable x to the response variables (y) is given by

$$\mathbf{Y}_k = \mathbf{X}_k\mathbf{B}_k + \mathbf{E}_k, \quad k = 1, 2, \ldots, m, \tag{7.1}$$

or equivalently,

$$\begin{bmatrix} y_{11k} & y_{12k} & \cdots & y_{1pk} \\ y_{21k} & y_{22k} & \cdots & y_{2pk} \\ \vdots & \vdots & \ddots & \vdots \\ y_{n_k1k} & y_{n_k2k} & \cdots & y_{n_kpk} \end{bmatrix} = \begin{bmatrix} 1 & x_1 \\ 1 & x_2 \\ \vdots & \vdots \\ 1 & x_{n_k} \end{bmatrix} \begin{bmatrix} \beta_{01k} & \beta_{02k} & \cdots & \beta_{0pk} \\ \beta_{11k} & \beta_{12k} & \cdots & \beta_{1pk} \end{bmatrix}.$$

$$+\begin{bmatrix} \varepsilon_{11k} & \varepsilon_{12k} & \cdots & \varepsilon_{1pk} \\ \varepsilon_{21k} & \varepsilon_{22k} & \cdots & \varepsilon_{2pk} \\ \vdots & \vdots & \ddots & \vdots \\ \varepsilon_{n_k 1k} & \varepsilon_{n_k 2k} & \cdots & \varepsilon_{n_k pk} \end{bmatrix}, \tag{7.2}$$

where $\mathbf{Y}_k = (\mathbf{y}_{1k}, \mathbf{y}_{2k}, \ldots, \mathbf{y}_{n_k k})^{\mathrm{T}}$ is an $n_k \times p$ matrix of response variables for the kth sample. Each row of $\mathbf{Y}_k$ (or $\mathbf{y}_{ik}$) contains the values of the p response variables corresponding to a specific value of the explanatory variable (x_i). The matrix $\mathbf{X}_k = [\mathbf{1} \quad \mathbf{x}_k]$ is an $n_k \times 2$ matrix of explanatory variable with $\mathbf{x}_k$ representing the vector consisting of the values of the explanatory variable. We assume that the matrix $\mathbf{X}_k$ is constant from sample to sample.

Figure 7.1 illustrates the framework for the multivariate simple linear profiles in Phase I, where the relationships between four response variables ($p = 4$) and one explanatory variable are linear. The notations explained earlier and used in Equations (7.1) and (7.2) can be easily followed from in this figure.

As shown in Figure 7.1, n_k shows the number of x-values which is equal to 11 and $i = 1, 2, \ldots, n_k$ shows the observation number in each sample. The value of the parameter p represents the number of response values observed at each x-value. In Figure 7.1, the value of p is equal to 4 corresponding to the four profiles and $j = 1, 2, \ldots, p$ shows the jth response corresponding to each x-value. The parameter m represents the number of samples, which is equal to 15 and $k = 1, 2, \ldots, m$ shows the sample number.

The matrix $\mathbf{E}_k = (\varepsilon_{1k}, \varepsilon_{2k}, \ldots, \varepsilon_{n_k k})^{\mathrm{T}}$ is an $n_k \times p$ matrix of random errors for the kth sample. Each row of $\mathbf{E}_k$ (or ε_{ik}) contains p random error terms corresponding to the p values of the response variables. According to multivariate linear regression literature, the vector of error terms for each observation (ε_{ik}) has a p-variate normal distribution with p-variate mean vector of $\underline{\mathbf{0}}$ and $p \times p$ variance–covariance matrix of $\boldsymbol{\Sigma}_k$. In addition, it is typically assumed that the vectors of error terms (rows of $\mathbf{E}_k$ or ε_{ik}) are uncorrelated with each other indicating no within profiles autocorrelation. The matrix $\mathbf{B}_k = (\beta_{0k}, \beta_{1k})^{\mathrm{T}}$ is a $2 \times p$ matrix consisting of simple linear regression parameters. The ordinary least squares estimator for $\mathbf{B}_k = (\beta_{0k}, \beta_{1k})^{\mathrm{T}}$, which minimizes the value of $\left|(\mathbf{Y}_k - \mathbf{X}_k\hat{\mathbf{B}}_k)^{\mathrm{T}}(\mathbf{Y}_k - \mathbf{X}_k\hat{\mathbf{B}}_k)\right|$ is given by

$$\hat{\mathbf{B}}_k = (\mathbf{X}_k^{\mathrm{T}}\mathbf{X}_k)^{-1}\mathbf{X}_k^{\mathrm{T}}\mathbf{Y}_k = (\hat{\beta}_{0k}, \hat{\beta}_{1k})^{\mathrm{T}} = \begin{bmatrix} \hat{\beta}_{01k} & \cdots & \hat{\beta}_{0pk} \\ \hat{\beta}_{11k} & \cdots & \hat{\beta}_{1pk} \end{bmatrix}$$

$$= \begin{bmatrix} \bar{y}_{.1k} - \hat{\beta}_{1jk}\bar{x}_k & \cdots & \bar{y}_{.pk} - \hat{\beta}_{1jk}\bar{x}_k \\ \dfrac{S_{xy(1)}}{S_{xx}} & \cdots & \dfrac{S_{xy(p)}}{S_{xx}} \end{bmatrix}, \tag{7.3}$$

where $\bar{y}_{.jk} = \frac{1}{n_k}\sum_{i=1}^{n_k} y_{ijk}$, $\bar{x}_k = \frac{1}{n_k}\sum_{i=1}^{n_k} x_{ik}$, $S_{xy(j)} = \sum_{i=1}^{n_k} (x_{ik} - \bar{x}_k)y_{ijk}$, and $S_{xx} = \sum_{i=1}^{n_k} (x_i - \bar{x}_k)^2$ (see Rencher 2002, p. 337).

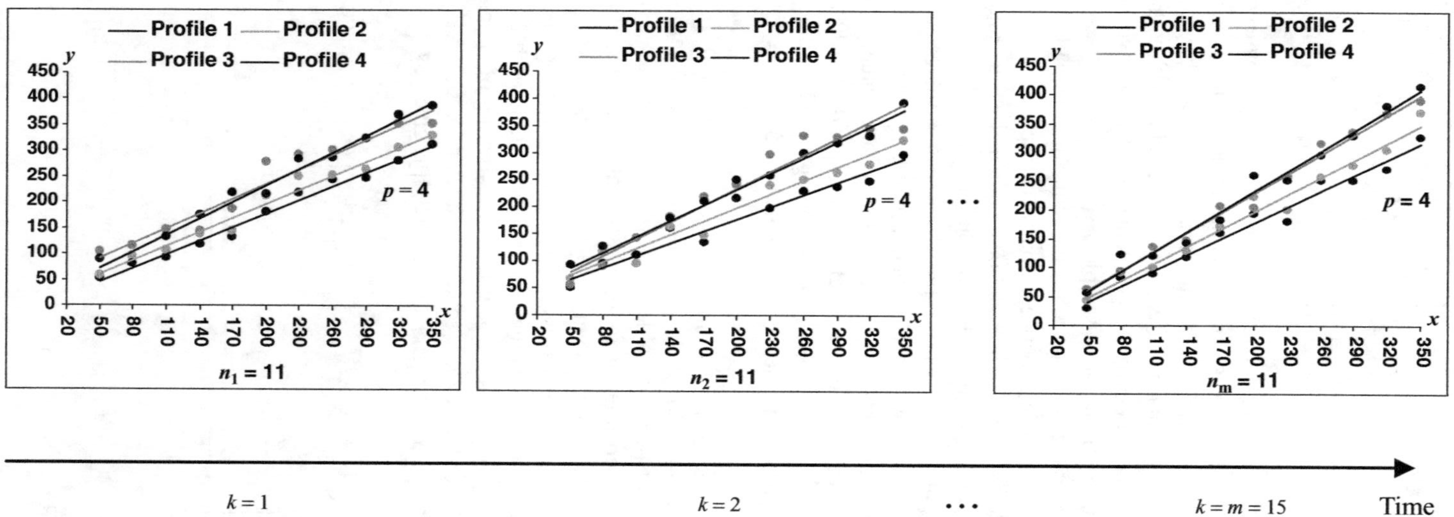

Figure 7.1 The framework for a multivariate simple linear profile.

7.1.1.1 T^2 Control Chart Approaches

On the basis of successive differences, Noorossana et al. (2008) proposed two T^2 methods for monitoring multivariate simple linear profiles in Phase I. In the first method, they first rewrote matrix $\hat{\mathbf{B}}_k$ as a $1 \times 2p$ vector denoted by $\hat{\boldsymbol{\beta}}_k^{\mathrm{T}}$, where

$$\hat{\boldsymbol{\beta}}_k^{\mathrm{T}} = (\hat{\beta}_{01k}, \hat{\beta}_{02k}, \ldots, \hat{\beta}_{0pk}, \hat{\beta}_{11k}, \hat{\beta}_{12k}, \ldots, \hat{\beta}_{1pk}) \tag{7.4}$$

is a multivariate normal random vector with mean vector

$$\boldsymbol{\beta}_k^{\mathrm{T}} = (\beta_{01k}, \beta_{02k}, \ldots, \beta_{0pk}, \beta_{11k}, \beta_{12k}, \ldots, \beta_{1pk}) \tag{7.5}$$

and $2p \times 2p$ covariance matrix $\Sigma_{\hat{\beta}}$.

The average vector $\bar{\hat{\boldsymbol{\beta}}}^{\mathrm{T}}$ is computed using $\bar{\hat{\boldsymbol{\beta}}}^{\mathrm{T}} = m^{-1} \sum_{k=1}^{m} \hat{\boldsymbol{\beta}}_k^{\mathrm{T}}$. The covariance matrix $\Sigma_{\hat{\beta}}$ is estimated using successive differences of the regression parameters estimators. Williams et al. (2007) defined $\hat{v}_i = \hat{\boldsymbol{\beta}}_{i+1}^{\mathrm{T}} - \hat{\boldsymbol{\beta}}_i^{\mathrm{T}}$, $i = 1, 2, \ldots, m-1$ and stacked these $m-1$ difference vectors into matrix $\hat{\mathrm{V}}$ as

$$\hat{\mathrm{V}} = \begin{bmatrix} \hat{v}_1 \\ \hat{v}_2 \\ \vdots \\ \hat{v}_{m-1} \end{bmatrix}. \tag{7.6}$$

The estimator for the covariance matrix is provided by

$$\mathrm{S}_{\hat{\beta}} = \frac{\hat{\mathrm{V}}^{\mathrm{T}} \times \hat{\mathrm{V}}}{2 \times (m-1)}. \tag{7.7}$$

The resulting T_k^2 is given by

$$T_k^2 = (\hat{\boldsymbol{\beta}}_{\mathrm{k}} - \bar{\hat{\boldsymbol{\beta}}})^{\mathrm{T}} \mathrm{S}_{\hat{\beta}}^{-1} (\hat{\boldsymbol{\beta}}_{\mathrm{k}} - \bar{\hat{\boldsymbol{\beta}}}). \tag{7.8}$$

The upper control limit (UCL) for the T^2 statistic in Equation (7.8) is chosen to give a specified probability of Type I error. Sullivan and Woodall (1996) showed that a T^2 control chart based on the values of T_k^2 is effective in detecting step and ramp shifts in the mean vector in Phase I.

In the second method, they coded the x-values to change the average to zero. Hence, Equation (7.1) can be written as:

$$\mathbf{Y}_k = \mathbf{U}\mathbf{B}_k' + \mathbf{E}_k, k = 1, 2, \ldots, m \tag{7.9}$$

where $\mathbf{U} = [\mathbf{1} \quad \mathbf{u}]$ is an $n \times 2$ matrix of coded explanatory variable and $\mathbf{u}$ is equal to $(x_1 - \bar{x}, x_2 - \bar{x}, \ldots, x_n - \bar{x})^{\mathrm{T}}$. The matrix $\mathbf{B}_k' = (\boldsymbol{\beta}_{0k}', \boldsymbol{\beta}_{1k}')^{\mathrm{T}}$ is a $2 \times p$ matrix of the simple linear regression parameters with $\beta'_{1k} = \beta_{1k}$ and $\beta'_{0k} = \beta_{0k} + \bar{x}\beta_{1k}$. For the kth sample, the least squares estimator for β'_{0k} is $\hat{\beta}'_{0k} = (\bar{y}_{.1k}, \bar{y}_{.2k}, \ldots, \bar{y}_{.pk})$, while the least squares estimator for β'_{1k} is the same as that for β_{1k}, i.e., $(\frac{S_{xy_{1k}}}{S_{xx}}, \frac{S_{xy_{2k}}}{S_{xx}}, \ldots, \frac{S_{xy_{pk}}}{S_{xx}})$ (see Kim et al. (2003) for details). It can be shown that $\hat{\beta}'_{0k}$ and $\hat{\beta}'_{1k}$ are $1 \times p$ multivariate normal random vectors, which are independent of each other. Thus, they applied separate multivariate control charts to monitor the Y-intercepts and slopes. They proposed the use of two separate T^2 control charts for monitoring Y-intercepts and slopes vectors. The use of separate control charts facilitates interpretation of changes in the process parameters.

The vectors $\bar{\hat{\beta}}'_0$ and $\bar{\hat{\beta}}'_1$ are computed using $\bar{\hat{\beta}}'_0 = m^{-1} \sum_{k=1}^{m} \hat{\beta}'_{0k}$ and $\bar{\hat{\beta}}'_1 = m^{-1} \sum_{k=1}^{m} \hat{\beta}'_{1k}$, respectively. The covariance matrices $\sum_{\hat{\beta}'_0}$ and $\sum_{\hat{\beta}'_1}$ can be estimated on the basis of successive differences of the regression parameters estimators as follows:

$$S_{\hat{\beta}'_0} = \frac{\hat{\mathbf{V}}_0^{\mathrm{T}} \times \hat{\mathbf{V}}_0}{2 \times (m-1)} \quad \text{and} \quad S_{\hat{\beta}'_1} = \frac{\hat{\mathbf{V}}_1^{\mathrm{T}} \times \hat{\mathbf{V}}_1}{2 \times (m-1)}, \tag{7.10}$$

where

$$\hat{\mathbf{V}}_0 = \begin{bmatrix} \hat{v}_{01} \\ \hat{v}_{02} \\ \vdots \\ \hat{v}_{0(m-1)} \end{bmatrix} = \begin{bmatrix} \hat{\beta}'_{02} - \hat{\beta}'_{01} \\ \hat{\beta}'_{03} - \hat{\beta}'_{02} \\ \vdots \\ \hat{\beta}'_{0m} - \hat{\beta}'_{0(m-1)} \end{bmatrix} \text{ and } \hat{\mathbf{V}}_1 = \begin{bmatrix} \hat{v}_{11} \\ \hat{v}_{12} \\ \vdots \\ \hat{v}_{1(m-1)} \end{bmatrix} = \begin{bmatrix} \hat{\beta}'_{12} - \hat{\beta}'_{11} \\ \hat{\beta}'_{13} - \hat{\beta}'_{12} \\ \vdots \\ \hat{\beta}'_{1m} - \hat{\beta}'_{1(m-1)} \end{bmatrix}. \tag{7.11}$$

Thus, the resulting T^2 statistics for monitoring the Y-intercepts and slopes, denoted by T_{Ik}^2 and T_{Sk}^2, respectively, are given by

$$T_{Ik}^2 = (\hat{\beta}'_{0k} - \bar{\hat{\beta}}'_0)\mathbf{S}_{\hat{\beta}'_0}^{-1}(\hat{\beta}'_{0k} - \bar{\hat{\beta}}'_0)^{\mathrm{T}} \text{ and } T_{Sk}^2 = (\hat{\beta}'_{1k} - \bar{\hat{\beta}}'_1)\mathbf{S}_{\hat{\beta}'_1}^{-1}(\hat{\beta}'_{1k} - \bar{\hat{\beta}}'_1)^{\mathrm{T}}. \tag{7.12}$$

The values for upper control limits are determined through simulation such that a specified overall probability of Type I error is achieved.

7.1.1.2 Principal Components Analysis Approaches

Noorossana et al. (2008) suggested an approach based on principal components analysis for monitoring multivariate simple linear profiles in Phase I. When there are several correlated response variables (*y*s) corresponding to each value of the explanatory variable, then principal components analysis can be used to reduce the dimensionality of the response variables as well as to make the response variables uncorrelated.

In their proposed method, they first pooled all the m samples into one sample of size mn. Then, for each observation, they calculated the scores for the principal components of the p response variables and chose the principal components, which explained most of the variation. They considered these few principal components as new response variable and modeled the relationship between the new response variables and explanatory variable. Since the new response variables (first few principal components) are uncorrelated, the relationship between these new response variables and explanatory variable can be monitored using separate control charts. To do this, they proposed the use of the likelihood ratio method proposed by Mahmoud et al. (2007), which is much more effective in detecting shift(s) than other competing methods. In other words, each new profile is monitored by a separate likelihood ratio control chart. The upper control limit for each control chart is chosen to give an overall false alarm probability of size α. Note that each principal component follows a normal distribution because it is a linear combination of the response variables. Hence, normality assumption of the response variables in the likelihood ratio method by Mahmoud et al. (2007) is satisfied.

It should be noted that the principal components approach might not detect shifts in the directions of other principal components not considered in the analysis. To protect against this shortcoming, one can supplement a principal components approach with other control charts to monitor the profiles.

7.1.1.3 Wilks' Lambda Approach

Noorossana et al. (2008) extended the F method proposed by Mahmoud and Woodall (2004) for monitoring multivariate simple linear profiles in Phase I. Similar to Mahmoud and Woodall (2004), they recommended the use of indicator variables (dummy variables) in a multivariate simple linear regression model to compare the multivariate regression lines associated with each sample.

Let us assume there are m samples of size n, where each sample can be represented as $(x_{ik}, y_{i1k}, y_{i2k}, \ldots, y_{ipk})$, for $i = 1, 2, \ldots, n$ and $k = 1, 2, \ldots, m$. This information will be used to fit a multivariate simple linear regression model to the process data. The similarity of the multivariate regression lines for all samples is an issue that needs to be investigated. The first step is to pool all the m samples into one sample of size mn. Then, $m - 1$ indicator variables are defined as

$$\begin{cases} Z_{ij} = 1 & \text{if observation } i \text{ is from sample } j \\ Z_{ij} = 0 & \text{otherwise,} \quad i = 1, 2, \ldots, mn \text{ and } j = 1, 2, \ldots, m - 1. \end{cases}$$

The mth sample is called the reference sample. Finally, we fit the following regression model to the pooled data:

$$\begin{aligned} \mathbf{y}_i = {} & \boldsymbol{\beta}_0 + \boldsymbol{\beta}_1 X_i + \boldsymbol{\beta}_{01} Z_{1i} + \boldsymbol{\beta}_{02} Z_{2i} + \cdots + \boldsymbol{\beta}_{0m'} Z_{m'i} \\ & + \boldsymbol{\beta}_{11} Z_{1i} X_i + \boldsymbol{\beta}_{12} Z_{2i} X_i + \cdots + \boldsymbol{\beta}_{1m'} Z_{m'i} X_i + \boldsymbol{\varepsilon}_i \end{aligned} \tag{7.13}$$

for $i = 1, 2, \ldots, mn$, where $m' = m - 1$, $\mathbf{y}_i$ is a $1 \times p$ vector of response variables for the ith observation of the pooled data, $\boldsymbol{\varepsilon}_i$ is a $1 \times p$ vector of error terms, and $\boldsymbol{\beta}_0$, $\boldsymbol{\beta}_1$, $\boldsymbol{\beta}_{0j}$ and $\boldsymbol{\beta}_{1j}$ for $j = 1, 2, \ldots, m - 1$, are $1 \times p$ vectors of regression parameters.

To test for the equality of m regression lines, the following hypotheses are tested:

$$H_0 : \beta_{01} = \beta_{02} = \cdots = \beta_{0m'} = \beta_{11} = \beta_{12} = \cdots = \beta_{1m'} = 0,$$
$$H_1 : H_0 \text{ is not true}$$

where $\mathbf{0}$ is a $1 \times p$ vector of zeros.

Under the null hypothesis, the reduced model is given as

$$\mathbf{y}_i = \beta_0 + \beta_1 x_i + \varepsilon_i, \quad i = 1, 2, \ldots, mn. \tag{7.14}$$

The usual test statistic for testing H_0 is

$$\Lambda = \frac{\left|\mathbf{Y}'\mathbf{Y} - \hat{\mathbf{B}}_\mathbf{f}'\mathbf{X}_\mathbf{f}'\mathbf{Y}\right|}{\left|\mathbf{Y}'\mathbf{Y} - \hat{\mathbf{B}}_\mathbf{r}'\mathbf{X}_\mathbf{r}'\mathbf{Y}\right|}, \tag{7.15}$$

where $|\cdot|$ refers to the determinant of the matrix and $\mathbf{Y} = (\mathbf{y}_1, \mathbf{y}_2, \ldots, \mathbf{y}_{mn})^{\mathrm{T}}$ is an $mn \times p$ matrix of the response variables for the pooled data. The matrix $\mathbf{X}_\mathbf{f}$ is an $mn \times 2m$ matrix of explanatory variables and $\hat{\mathbf{B}}_\mathbf{f}$ is a $2m \times p$ matrix of regression parameters both associated with the full model. The matrix $\mathbf{X}_\mathbf{r}$ is an $mn \times 2$ matrix of explanatory variables and $\hat{\mathbf{B}}_\mathbf{r}$ is a $2 \times p$ matrix of regression parameters both associated with the reduced model.

This test statistic Λ follows Wilks' lambda distribution with p, $2(m-1)$, and $m(n-2)$ degrees of freedom under the null hypothesis. The null hypothesis H_0 is rejected if $\Lambda < \Lambda_{\alpha,p,2(m-1),m(n-2)}$, where $\Lambda_{\alpha,p,2(m-1),m(n-2)}$ is the critical value for the Wilks' lambda statistic, which is available in most multivariate references such as Rencher (2002).

In order to test the equality of covariance matrices, they used the statistic proposed by Lee et al. (1977). For the kth sample, an unbiased estimator of Σ_k is given by

$$\mathbf{S}_k = \frac{\mathbf{Y}_k'\mathbf{Y}_k - \hat{\mathbf{B}}_k'\mathbf{X}'\mathbf{Y}_k}{n-2}, \tag{7.16}$$

where $\mathbf{Y}_k$ is an $n \times p$ matrix of response variables for the kth sample. The matrix $\mathbf{X}_k = [\mathbf{1} \quad \mathbf{x}]$ is an $n \times 2$ matrix of explanatory variables and $\hat{\mathbf{B}}_k$ is the least squares estimator for $\mathbf{B}_k$ for the kth sample. To construct the test, we calculate M as

$$M = \frac{|\boldsymbol{S}_1|^{(n-2)/2}\,|\boldsymbol{S}_2|^{(n-2)/2}\cdots|\boldsymbol{S}_m|^{(n-2)/2}}{\left|\boldsymbol{S}_{pl}\right|^{m(n-2)/2}} = \frac{(|\boldsymbol{S}_1| \times |\boldsymbol{S}_2| \times \cdots \times |\boldsymbol{S}_m|)^{(n-2)/2}}{\left|\boldsymbol{S}_{pl}\right|^{m(n-2)/2}}, \tag{7.17}$$

where $\boldsymbol{S}_{pl}$ is the pooled sample covariance matrix defined by

$$\boldsymbol{S}_{pl} = \frac{\sum_{k=1}^{m}(n-2)\boldsymbol{S}_k}{\sum_{k=1}^{m}(n-2)} = \frac{\sum_{k=1}^{m}\boldsymbol{S}_k}{m}. \tag{7.18}$$

The statistic M is a modification of the likelihood ratio and varies between 0 and 1, with values near 1 favoring equality of covariance matrices and values near 0 leading to the rejection of equality of the covariance matrices. The exact upper percentage points of $-2\ln \mathrm{M}$ are given in Lee et al. (1977).

Noorossana et al. (2008) compared the performance of their proposed method for Phase I monitoring of multivariate simple linear profiles in terms of the overall probabilities of an out-of-control signal. Their simulation showed that Wilks' lambda approach performs better than other methods in detecting shifts in regression parameters. In addition, principal components approach works well when the number of principal components used increases.

Example 7.1 In order to illustrate the applicability of the proposed methods in real world, Noorossana et al. (2008) employed the suggested control schemes in a calibration application in a 1600-ton hydraulic press at the body shop of an automotive manufacturing company. In this application, the purpose was to investigate calibration of the real force applied by a hydraulic press machine. An important input value in this press machine is the nominal force exerted by cylinders on the metal plates in order to produce the desired parts. For each value of the nominal force (explanatory variable) there are four real forces measured in the four cylinders (response variables) of the machine. The problem was to investigate the deviation between the nominal force (x) and the real forces (y_1, y_2, y_3, and y_4) exerted on each cylinder. It should be noted that significant differences between the desired force and the real forces, which could be due to oscillation in oil temperature, variation in oil volume, daily set-up changes, etc., might lead to defective or low-quality parts. To avoid this problem, the calibration process of the real force values should be monitored over time to decide whether the recalibration is needed.

Their studies showed that a simple linear regression model characterizes the relationship between each response variable and the explanatory variable. Since response variables are correlated, the relationship among the response variables and the single explanatory variable should be modeled in the context of multivariate simple linear profiles. It is well known that if the correlation structure between response variables is overlooked or ignored by monitoring each profile separately, then misleading results should be expected.

To perform the analysis, 15 samples of size 11 corresponding to nominal force values of 50, 80, 110, 140, 170, 200, 230, 260, 290, 320, and 350 are collected and the corresponding output force associated with each cylinder is measured (The data set is available upon request from the authors.). It should be pointed out that prior to applying Phase I approaches, diagnostic checking on the assumptions of linearity of the regression lines, normality, independency of error vectors, and the correlation between response variables should be performed. Preliminary analysis indicates that there is no violation of the model assumptions.

Table 7.1 shows the least squares estimates for the regression coefficients of four simple linear profiles in each sample.

Table 7.1 The Least Squares Estimates for the Regression Coefficients of Four Simple Linear Profiles

	y_1(Cylinder 1)		y_2(Cylinder 2)		y_3(Cylinder 3)		y_4(Cylinder 4)	
Sample	Intercept	Slope	Intercept	Slope	Intercept	Slope	Intercept	Slope
1	1.02	0.88	13.18	0.91	44.00	0.96	19.19	1.07
2	27.50	0.75	33.42	0.83	27.31	1.04	37.52	0.98
3	−4.89	0.92	5.23	0.96	20.18	1.03	20.01	1.07
4	−5.66	0.90	−3.62	0.98	23.59	1.07	19.82	1.06
5	−13.27	0.91	−6.59	1.00	−0.81	1.05	15.75	1.13
6	3.09	0.86	4.74	0.98	5.13	1.11	12.65	1.10
7	18.35	0.81	31.04	0.87	38.60	1.00	21.88	1.06
8	5.93	0.85	15.21	0.92	37.75	1.00	−3.36	1.15
9	−8.50	0.91	−9.91	1.02	15.57	1.05	0.80	1.13
10	0.25	0.82	6.71	0.90	1.97	1.10	11.61	1.06
11	−18.04	0.90	−8.45	0.98	17.22	1.04	16.71	1.05
12	15.55	0.78	13.69	0.89	26.96	1.02	13.28	1.05
13	4.60	0.86	0.36	0.98	6.39	1.09	4.65	1.10
14	−18.48	0.97	−6.11	1.03	4.43	1.18	8.98	1.05
15	−6.24	0.92	−3.50	1.00	3.06	1.13	−0.21	1.16

Using the data set, the pooled covariance matrix of the response variables is estimated as

$$\mathbf{S}_{pl} = \left(\sum_{k=1}^{15} (n-2)\mathbf{S}_k\right) \Big/ \left(\sum_{k=1}^{m} (n-2)\right) = (1/m)\sum_{k=1}^{15} \mathbf{S}_k$$

$$= \begin{bmatrix} 330.18 & 327.38 & 152.87 & 96.86 \\ 327.38 & 427.00 & 237.83 & 105.11 \\ 152.87 & 237.83 & 704.88 & -60.98 \\ 96.86 & 105.11 & -60.98 & 371.78 \end{bmatrix}.$$

Figure 7.2 shows the control charts based on T^2 methods. The T^2 statistics were calculated using Equations (7.8) and (7.12), respectively. First graph shows the T^2 statistics corresponding to the first T^2 method and the second and third graphs show the T^2 statistics for intercepts and slops associated with the second T^2 method, respectively. The upper control limits are determined to obtain a probability of 0.04 for Type I error. As shown in Figure 7.2, all multivariate simple linear profiles are in-control based on T^2 methods.

For principle components analysis approach, they considered the first and second principal components as new response variables. The likelihood ratio statistics proposed by Mahmoud et al. (2007) are calculated and the results are summarized in Table 7.2. In this table, $lrtc_1$ and $lrtc_2$ are the corresponding likelihood ratio statistics

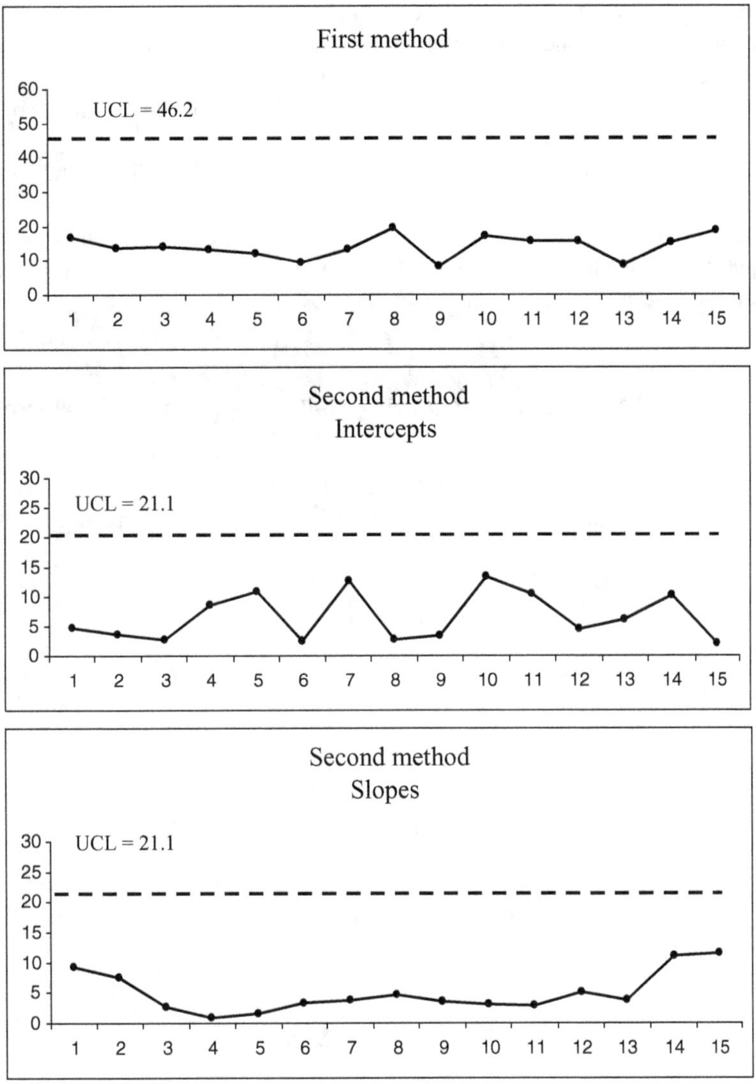

Figure 7.2 Control charts for T^2 methods.

Table 7.2 The Values of Likelihood Ratio Statistics for the First and Second Principal Components Denoted by $lrtc_1$ and $lrtc_2$

m_1	1	2	3	4	5	6	7	8	9	10	11	12	13	14
$lrtc_1$	0.8	2.55	2.08	1.95	1.11	0.7	2.45	3.05	2.05	1.58	1.01	2.22	3.26	1.36
$lrtc_2$	0.4	0.05	0.18	0.31	1.34	1.6	1.28	0.91	2.27	1.66	1.36	0.95	1.21	0.25

proposed by Mahmoud et al. (2007) for the first and the second principal components, respectively. The threshold limits of the likelihood ratio statistics are set as equal to 4.85 to achieve a probability of Type I error of 0.04. In this example, the maximum of the statistics values does not exceed the threshold limits. Hence, it can be concluded that the process is in-control.

For Method D, the Wilks' lambda statistic used for the equality of the regression parameters is equal to 0.43, which leads to a p-value of 0.22 showing that the process is statistically in-control. In addition, the value of the statistic $(-2\log(M))$ is equal to 215 leading to a p-value of 0.48 indicating that the hypothesis of equal variance–covariance matrices cannot be rejected.

7.1.2 Phase II Methods for Monitoring Multivariate Simple Linear Profiles

Assume that for the kth random sample collected over time, we have n observations $(x_{ik}, y_{i1k}, y_{i2k}, \ldots, y_{ipk})$ where $i = 1, 2, \ldots, n$ and $k = 1, 2, \ldots$. The relationship between the observations when the process is in-control can be represented as

$$\mathbf{y}_{ik} = \boldsymbol{\beta}_0 + x_i\boldsymbol{\beta}_1 + \boldsymbol{\varepsilon}_{ik}, \quad i = 1, 2, \ldots, n, \tag{7.19}$$

or equivalently

$$\mathbf{Y}_k = \mathbf{XB} + \mathbf{E}_k. \tag{7.20}$$

It should be noted that in Phase II, it is assumed that parameters $\mathbf{B} = (\boldsymbol{\beta}_0, \boldsymbol{\beta}_1)^{\mathrm{T}}$ and $\sum$ are known. As mentioned in the previous section, for the kth sample, the least squares estimator of $\mathbf{B} = (\boldsymbol{\beta}_0, \boldsymbol{\beta}_1)^{\mathrm{T}}$ is given by

$$\hat{\mathbf{B}}_k = (\hat{\boldsymbol{\beta}}_{0k}, \hat{\boldsymbol{\beta}}_{1k})^{\mathrm{T}} = (\mathbf{X}^{\mathrm{T}}\mathbf{X})^{-1}\mathbf{X}^{\mathrm{T}}\mathbf{Y}_k. \tag{7.21}$$

Matrix $\hat{\mathbf{B}}_k$ can be rewritten as a $1 \times 2p$ vector denoted by $\hat{\boldsymbol{\beta}}_k^{\mathrm{T}}$, where

$$\hat{\boldsymbol{\beta}}_k^{\mathrm{T}} = (\hat{\beta}_{01k}, \hat{\beta}_{02k}, \ldots, \hat{\beta}_{0pk}, \hat{\beta}_{11k}, \hat{\beta}_{12k}, \ldots, \hat{\beta}_{1pk}) \tag{7.22}$$

is a multivariate normal random vector with mean vector

$$\boldsymbol{\beta} = (\beta_{01}, \beta_{02}, \ldots, \beta_{0p}, \beta_{11}, \beta_{12}, \ldots \beta_{1p}) \tag{7.23}$$

and a $2p \times 2p$ covariance matrix $\sum_{\beta}$. The elements of $\sum_{\beta}$ are given by the following expressions:

$$\begin{aligned}
\mathrm{Cov}(\hat{\beta}_{0uk}, \hat{\beta}_{0vk}) &= \sigma_{uv}\left(\frac{1}{n} + \frac{\bar{x}^2}{S_{xx}}\right) = \rho_{uv}\sigma_u\sigma_v\left(\frac{1}{n} + \frac{\bar{x}^2}{S_{xx}}\right), \\
\mathrm{Cov}(\hat{\beta}_{1uk}, \hat{\beta}_{1vk}) &= \frac{\sigma_{uv}}{S_{xx}} = \frac{\rho_{uv}\sigma_u\sigma_v}{S_{xx}}, \quad \text{and} \\
\mathrm{Cov}(\hat{\beta}_{0uk}, \hat{\beta}_{1vk}) &= -\sigma_{uv}\frac{\bar{x}}{S_{xx}} = -\rho_{uv}\sigma_u\sigma_v\frac{\bar{x}}{S_{xx}},
\end{aligned} \tag{7.24}$$

where σ_{uv} is the uth row of the vth column of the covariance matrix $\sum$ and $\rho_{uv} = \sigma_{uv}/(\sigma_u.\sigma_v)$ is the uth row of the vth column of the correlation matrix. The proofs are provided in Appendix A.

7.1.2.1 T^2 Control Chart Approaches

Three methods for monitoring multivariate simple linear profiles in Phase II were proposed by Noorossana et al. (2010a). In the first method, they first rewrote matrix $\hat{\mathbf{B}}_k$ as a $1 \times 2p$ vector denoted by $\hat{\boldsymbol{\beta}}_k^{\mathrm{T}}$, where

$$\hat{\boldsymbol{\beta}}_k^{\mathrm{T}} = (\hat{\beta}_{01k}, \hat{\beta}_{02k}, \ldots, \hat{\beta}_{0pk}, \hat{\beta}_{11k}, \hat{\beta}_{12k}, \ldots, \hat{\beta}_{1pk}) \tag{7.25}$$

is a multivariate normal random vector with mean vector

$$\boldsymbol{\beta}^{\mathrm{T}} = (\beta_{01}, \beta_{02}, \ldots, \beta_{0p}, \beta_{11}, \beta_{12}, \ldots \beta_{1p}) \tag{7.26}$$

and a $2p \times 2p$ covariance matrix $\sum_{\boldsymbol{\beta}}$. They showed that the elements of $\sum_{\boldsymbol{\beta}}$ are given by the following expressions:

$$\begin{aligned}
\mathrm{Cov}(\hat{\beta}_{0uk}, \hat{\beta}_{0vk}) &= \sigma_{uv}\left(\frac{1}{n} + \frac{\bar{x}^2}{S_{xx}}\right) = \rho_{uv}\sigma_u\sigma_v\left(\tfrac{1}{n} + \tfrac{\bar{x}^2}{S_{xx}}\right), \\
\mathrm{Cov}(\hat{\beta}_{1uk}, \hat{\beta}_{1vk}) &= \frac{\sigma_{uv}}{S_{xx}} = \frac{\rho_{uv}\sigma_u\sigma_v}{S_{xx}}, \quad \text{and} \\
\mathrm{Cov}(\hat{\beta}_{0uk}, \hat{\beta}_{1vk}) &= -\sigma_{uv}\frac{\bar{x}}{S_{xx}} = -\rho_{uv}\sigma_u\sigma_v\frac{\bar{x}}{S_{xx}},
\end{aligned} \tag{7.27}$$

where σ_{uv} is the uvth element of the covariance matrix $\sum$ and $\rho_{uv} = \sigma_{uv}/(\sigma_u.\sigma_v)$ is the uvth element of the correlation matrix.

They proposed multivariate exponentially weighted moving average (MEWMA) approach to monitor multivariate linear profiles parameters. For the kth sample, if the process is in-control, $\hat{\boldsymbol{\beta}}_k$ is a $2p \times 1$ multivariate normal random vector with mean vector $\boldsymbol{\beta}^{\mathrm{T}}$ and known covariance matrix $\sum_{\beta}$. They used the MEWMA control chart proposed by Lowry et al. (1992) to monitor the $2p \times 1$ vector of $\hat{\boldsymbol{\beta}}_k$ as follows:

$$\mathbf{z}_{k,\boldsymbol{\beta}} = \theta(\hat{\boldsymbol{\beta}}_k - \boldsymbol{\beta})^{\mathrm{T}} + (1-\theta)\mathbf{z}_{k-1,\boldsymbol{\beta}}, \quad k = 1, 2 \ldots. \tag{7.28}$$

The parameter $\theta(0 < \theta \leq 1)$ is the smoothing parameter and $\mathbf{z}_{0,\beta}$ is $1 \times 2p$ vector of zeros. When the process is in control, $\mathbf{z}_{k,\beta}$ is a multivariate normal random vector with mean vector of zeros and known covariance matrix of $\sum_{z,\beta} = \{\theta/(2-\theta)\} \sum_{\beta}$. Therefore, they proposed the chart statistic as

$$T^2_{\mathbf{Z}(k,\boldsymbol{\beta})} = \mathbf{z}_{k,\boldsymbol{\beta}} \sum_{z,\beta}^{-1} \mathbf{z}^{\mathrm{T}}_{k,\boldsymbol{\beta}}. \tag{7.29}$$

The upper control limit for the above statistic is determined by simulation to give a specified in-control average run length (ARL).

7.1.2.2 *The Combined MEWMA and Chi-square Control Chart Approach*

The second method is the combined MEWMA and chi-square control charts scheme proposed by Noorossana et al. (2010a). For the kth random sample collected over time, $\bar{\boldsymbol{\varepsilon}}_k = (\bar{\varepsilon}_{1k}, \bar{\varepsilon}_{2k}, \ldots, \bar{\varepsilon}_{pk})$ is a $1 \times p$ vector of average error where $\bar{\varepsilon}_{jk} = n^{-1}\sum_{i=1}^{n} \varepsilon_{ijk}$, $j = 1, 2, \ldots, p$. The average error vector $\bar{\boldsymbol{\varepsilon}}_k$ follows a multivariate normal distribution with mean vector zero and known covariance matrix $\boldsymbol{\Sigma}_{\bar{\varepsilon}} = n^{-1}\boldsymbol{\Sigma}$. For the kth sample, they used the following MEWMA statistic:

$$\mathbf{z}_{k,\boldsymbol{\varepsilon}} = \theta\bar{\boldsymbol{\varepsilon}}_k + (1-\theta)\mathbf{z}_{k-1,\varepsilon}, \quad k = 1, 2 \ldots, \tag{7.30}$$

where $\mathbf{z}_{0,\varepsilon}$ is the $1 \times p$ vector of zeros. When process is in control, $\mathbf{z}_{k,\varepsilon}$ is a multivariate normal random vector with mean vector zero and known covariance matrix $\sum_{z,\varepsilon} = \{\theta/(2-\theta)\} \sum_{\bar{\varepsilon}} = \theta/[n(2-\theta)]\} \sum$. Therefore, they proposed the kth chart statistic for the MEWMA chart as

$$T^2_{Z(k,\varepsilon)} = \mathbf{z}_{k,\varepsilon} \sum_{\mathbf{z},\varepsilon}^{-1} \mathbf{z}^{\mathrm{T}}_{k,\varepsilon}. \tag{7.31}$$

The upper control limit for the MEWMA control chart is chosen to give a specified in-control ARL.

In order to monitor the process variability, they proposed a chi-square statistic in combination to a MEWMA control chart as

$$\chi^2_k = \sum_{i=1}^{n} \left(\boldsymbol{\varepsilon}_{ik} \sum^{-1} \boldsymbol{\varepsilon}^{\mathrm{T}}_{ik} \right), \tag{7.32}$$

where $\boldsymbol{\varepsilon}_{ik}$ is the ith row of the error term matrix. It should be noted that $\boldsymbol{\varepsilon}_{ik}$ is a multivariate normal random vector with mean vector zero and known covariance matrix $\boldsymbol{\Sigma}$. When process is in-control, χ^2_k is a chi-square random variable with np

degrees of freedom. Hence, a $100(1-\alpha)$ percentile of the chi-square distribution with np degrees of freedom is used to construct an upper control limit as $UCL = \chi^2_{np,\alpha}$.

7.1.2.3 *The Three MEWMA Control Charts*

The third method proposed by Noorossana et al. (2010a) is based on the method proposed by Kim et al. (2003). Following Kim et al. (2003), Noorossana et al. (2010a) coded the x-values to change the average to zero. If the x-values are coded then the $1 \times p$ vector of the y-intercepts and the $1 \times p$ vector of the slopes will be independent from each other. Hence, $\mathbf{y}_{ik}$ can be rewritten as

$$\mathbf{y}_{ik} = \boldsymbol{\beta}_0' + x_i'\boldsymbol{\beta}_1' + \boldsymbol{\varepsilon}_{ik}, \quad i = 1, 2, \ldots, n, \tag{7.33}$$

where $x_i' = (x_i - \bar{x})$, $\boldsymbol{\beta}_1' = \boldsymbol{\beta}_1$, $\boldsymbol{\beta}_0' = \boldsymbol{\beta}_0 + \bar{x}\mathbf{J}$, and $\mathbf{J}$ is a vector of 1's. For the kth sample, the least squares estimator of $\boldsymbol{\beta}_0'$ is $\hat{\boldsymbol{\beta}}_{0k}' = (\bar{y}_{.1k}, \bar{y}_{.2k}, \ldots, \bar{y}_{.pk})$ and the least squares estimator of $\boldsymbol{\beta}_1'$ is the same as that for $\boldsymbol{\beta}_1$. When process is in-control, $\hat{\boldsymbol{\beta}}_{0k}'$ and $\hat{\boldsymbol{\beta}}_{1k}'$ will be independent normal random vectors. Thus, they applied separate MEWMA control charts for $\boldsymbol{\beta}_0'$ and $\boldsymbol{\beta}_1'$. In order to detect shifts in the process variability, they used another MEWMA control chart for monitoring the covariance matrix of the error terms. Using three separate control charts for the y-intercepts, slopes, and process variability leads to an easier interpretation of any process change.

For the chart designed for monitoring y-intercepts, they used the MEWMA statistic proposed by Lowry et al. (1992) defined as

$$\mathbf{z}_{Ik} = \theta(\hat{\boldsymbol{\beta}}_{0k}' - \boldsymbol{\beta}_0') + (1-\theta)\mathbf{z}_{I(k-1)}, \quad k = 1, 2, \ldots, \tag{7.34}$$

where $\mathbf{z}_{I0}$ is a $1 \times p$ vector of zeros and $\theta (0 < \theta < 1)$ is the smoothing parameter. They showed that when the process is in-control, $\mathbf{z}_{Ik}$ is a multivariate normal random vector with mean vector zero and known covariance matrix $\sum_{\mathbf{z}_I} = \{\theta/[n(2-\theta)]\}\sum$. Therefore, they used the kth chart statistic as

$$T_{Ik}^2 = \mathbf{z}_{Ik} \sum_{z_I}^{-1} \mathbf{z}_{Ik}^{\mathrm{T}}. \tag{7.35}$$

The upper control limit for the above statistic is determined by simulation to give a specified in-control ARL.

For the MEWMA chart for monitoring the slopes, they used the MEWMA statistic defined as

$$\mathbf{z}_{Sk} = \theta(\hat{\boldsymbol{\beta}}_{1k}' - \boldsymbol{\beta}_1') + (1-\theta)\mathbf{z}_{S(k-1)}, \quad k = 1, 2, \ldots, \tag{7.36}$$

where $\mathbf{z}_{S0}$ is a $1 \times p$ vector of zeros. They showed that when process is in-control, $\mathbf{z}_{Sk}$ is a multivariate normal random vector with mean vector zero and known covariance matrix $\sum_{\mathbf{z}_S} = \{\theta/[S_{xx}(2-\theta)]\}\sum$. Therefore, they used the kth chart statistic as

$$T_{Sk}^2 = \mathbf{z}_{Sk} \sum_{z_S}^{-1} \mathbf{z}_{Sk}^{\mathrm{T}}. \tag{7.37}$$

The MEWMA control chart gives an out-of-control signal when $T^2_{Sk} > h_S$, where $h_S(>0)$ is determined by simulation to achieve a specified in-control ARL.

For monitoring the process variability, they applied the logarithmic transformation of $\chi^2_k = \sum_{i=1}^{n} \varepsilon_{ik} \sum^{-1} \varepsilon^{\mathrm{T}}_{ik}$ and used a MEWMA chart statistic defined by

$$z_{Ek} = \max\{\theta \ln(\chi^2_k) + (1-\theta) z_{E(k-1)}, np\}, \quad k = 1, 2, \ldots, \tag{7.38}$$

where np is the mean of chi-square random variable with np degrees of freedom and z_{E0} is set equal to np. The MEWMA control chart gives an out-of-control signal when $z_{Ek} > h_E$, where $h_E(>0)$ is determined to give a specified in-control ARL.

7.1.3 Performance Comparison of the Methods

Noorossana et al. (2010a) compared the performance of proposed methods for Phase II monitoring of multivariate simple linear profiles in terms of ARL. Their simulation showed that the single MEWMA control chart approach and combined MEWMA and chi-square control chart approach perform slightly better than the three MEWMA Control Charts for detecting sustained shifts in the profiles parameters. However, the three MEWMA Control Charts chart is superior to the combined MEWMA and chi-square control chart in detecting small shifts and is superior to the single MEWMA method in detecting different shifts of covariance matrix elements.

7.2 MONITORING MULTIVARIATE MULTIPLE LINEAR PROFILES

In this section, we assume that there are some dependent quality characteristics as response variables that are modeled as functions of some explanatory variables. Each response variable is linear function of the explanatory variables and again response variables are correlated. This structure is referred to as multivariate multiple linear profiles where multiple refers to several explanatory variables and multivariate refers to several response variables. In Section 7.2.1, we present Phase I methods for monitoring multivariate multiple linear profiles and then in Section 7.2.2, discuss about Phase II methods for monitoring multivariate multiple linear profiles.

7.2.1 Phase I Methods for Monitoring Multivariate Multiple Linear Profiles

Let us assume that there are m samples available from a historical data set. For the kth sample, we have observations $(x_{1ik}, x_{2ik}, \ldots, x_{qik}, y_{1ik}, y_{2ik}, \ldots, y_{pik})$, $i = 1, 2, \ldots, n_k$, $k = 1, 2, \ldots, m$, where n_k is the size of sample k, m is the number of samples, and p and q are the number of response and explanatory variables, respectively. The multivariate multiple linear model, which relates the response variables with explanatory variables is given by

$$\mathbf{Y}_k = \mathbf{X}_k \mathbf{B}_k + \mathbf{E}_k \tag{7.39}$$

or equivalently,

$$\begin{bmatrix} y_{11k} & y_{12k} & \cdots & y_{1pk} \\ y_{21k} & y_{22k} & \cdots & y_{2pk} \\ \vdots & \vdots & \ddots & \vdots \\ y_{n_k1k} & y_{n_k2k} & \cdots & y_{n_kpk} \end{bmatrix} = \begin{bmatrix} 1 & x_{11} & \cdots & x_{1q} \\ 1 & x_{21} & \cdots & x_{2q} \\ \vdots & \vdots & \ddots & \vdots \\ 1 & x_{n_k1} & \cdots & x_{n_kq} \end{bmatrix} \begin{bmatrix} \beta_{01} & \beta_{02} & \cdots & \beta_{0p} \\ \beta_{11} & \beta_{12} & \cdots & \beta_{1p} \\ \vdots & \vdots & \ddots & \vdots \\ \beta_{q1} & \beta_{q2} & \cdots & \beta_{qp} \end{bmatrix}$$

$$+ \begin{bmatrix} \varepsilon_{11k} & \varepsilon_{12k} & \cdots & \varepsilon_{1pk} \\ \varepsilon_{21k} & \varepsilon_{22k} & \cdots & \varepsilon_{2pk} \\ \vdots & \vdots & \ddots & \vdots \\ \varepsilon_{n_k1k} & \varepsilon_{n_k2k} & \cdots & \varepsilon_{n_kpk} \end{bmatrix}, \tag{7.40}$$

where $\mathbf{Y}_k$ is an $n_k \times p$ matrix of response variables for the kth sample, $\mathbf{X}_k$ is an $n_k \times (q+1)$ matrix of explanatory variables, $\mathbf{B}_k$ is a $(q+1) \times p$ matrix of the regression parameters, and $\mathbf{E}_k$ is an $n_k \times p$ matrix of the error terms. It is assumed that the error terms has a p-variate multivariate normal distribution with p-variate mean vector of $\underline{\mathbf{0}}$ and $p \times p$ variance–covariance matrix of $\mathbf{\Sigma}$. The ordinary least square (OLS) estimator of the matrix $\mathbf{B}_k$ is as follows:

$$\hat{\mathrm{B}}_k = (\mathrm{X}_k^{\mathrm{T}}\mathrm{X}_k)^{-1}\mathrm{X}_k^{\mathrm{T}}\mathrm{Y}_k. \tag{7.41}$$

Noorossana et al. (2010b) extended four methods for Phase I monitoring of multivariate multiple linear profiles. They extend the T^2 method, Wilks' lambda approach, and principal components approach of Noorossana et al. (2008) and proposed an LRT method for monitoring of multivariate multiple linear regression profiles.

7.2.1.1 T^2 Control Chart Approach

Noorossana et al. (2010b) extend the T^2 method of Noorossana et al. (2008) for monitoring of multivariate multiple linear profiles. Similar to Noorossana et al. (2008), they first rewrote the matrix $\hat{\mathbf{B}}_k$ as a $1 \times (q+1)p$ multivariate normal random vector and denote it by $\hat{\boldsymbol{\beta}}_k$ as follows:

$$\hat{\boldsymbol{\beta}}_k = (\hat{\beta}_{01k}, \hat{\beta}_{02k}, \ldots, \hat{\beta}_{0pk}, \hat{\beta}_{11k}, \hat{\beta}_{12k}, \ldots \hat{\beta}_{1pk}, \ldots, \hat{\beta}_{q1k}, \hat{\beta}_{q2k}, \ldots \hat{\beta}_{qpk})^{\mathrm{T}}. \tag{7.42}$$

Then, they construct the T^2 statistic for the kth sample as following:

$$T_k^2 = (\hat{\boldsymbol{\beta}}_k - \bar{\hat{\boldsymbol{\beta}}})^{\mathrm{T}} \mathrm{S}_{\beta}^{-1} (\hat{\boldsymbol{\beta}}_k - \bar{\hat{\boldsymbol{\beta}}}), \tag{7.43}$$

where similar to Noorossana et al. (2008), $\bar{\hat{\boldsymbol{\beta}}}$ is computed by using $\bar{\hat{\boldsymbol{\beta}}} = \frac{1}{m}\sum_{k=1}^{m} \hat{\boldsymbol{\beta}}_k$ and $\mathbf{S}_{\beta}$ is the estimator of the covariance matrix of $\hat{\boldsymbol{\beta}}_k$ based on the successive differences of the regression parameters vector estimators. The upper control limit for this control chart is chosen to give a specified probability of Type I error.

7.2.1.2 Likelihood Ratio Approach

The second method of Noorossana et al. (2010b) was the extension of the method proposed by Mahmoud et al. (2007) to the multivariate multiple linear regression profiles. With a step shift in one or more regression parameters after sample m_1, they assumed the following models:

$$\begin{aligned} \mathrm{Y}_k &= \mathrm{X}_k\mathrm{B}_{k1} + \mathrm{E}_{k1}, \quad k = 1, 2, \ldots, m_1 \\ \mathrm{Y}_k &= \mathrm{X}_k\mathrm{B}_{k2} + \mathrm{E}_{k2}, \quad k = m_1 + 1, \ldots, m, \end{aligned} \tag{7.44}$$

where $\mathbf{E}_{k1} \sim MN(\underline{\mathbf{0}}, \boldsymbol{\Sigma}_1)$ and $\mathbf{E}_{k2} \sim MN(\underline{\mathbf{0}}, \boldsymbol{\Sigma}_2)$.

To test the hypotheses

$$\begin{aligned} &H_0 : \mathrm{B}_{k1} = \mathrm{B}_{k2} = \mathrm{B}, \text{ and } \Sigma_1 = \Sigma_2 = \Sigma \\ &H_1 : H_0 \text{ is not true,} \end{aligned} \tag{7.45}$$

they showed that the likelihood ratio statistic is given by the following equation:

$$lrt_{m1} = N \log\left|\hat{\Sigma}\right| - N_1 \log\left|\hat{\Sigma}_1\right| - N_2 \log\left|\hat{\Sigma}_2\right|, \tag{7.46}$$

where N, N_1, and N_2 are the total number of observations, the total number of observations prior to the change point, and the total number of observations following the change point, respectively. $\left|\hat{\boldsymbol{\Sigma}}\right|$ is the determinant of the error terms variance–covariance matrix estimator for the multivariate regression model fitted for all of samples pooled into one sample of size N, and $\left|\hat{\boldsymbol{\Sigma}}_1\right|$ and $\left|\hat{\boldsymbol{\Sigma}}_2\right|$ are the determinants of the error terms variance–covariance matrix estimator for the multivariate regression model fitted for all of samples prior to m_1 pooled into one sample of size N_1, and for all of samples following m_1 pooled into one sample of size N_2, respectively. $\left|\hat{\boldsymbol{\Sigma}}\right|$, $\left|\hat{\boldsymbol{\Sigma}}_1\right|$, and $\left|\hat{\boldsymbol{\Sigma}}_2\right|$ are computed as follows:

$$\hat{\boldsymbol{\Sigma}} = \frac{(\mathbf{Y} - \mathbf{X}\hat{\mathbf{B}})^{\mathrm{T}}(\mathbf{Y} - \mathbf{X}\hat{\mathbf{B}})}{N}, \tag{7.47}$$

$$\hat{\boldsymbol{\Sigma}}_1 = \frac{(\mathbf{Y}_1 - \mathbf{X}_1\hat{\mathbf{B}}_1)^{\mathrm{T}}(\mathbf{Y}_1 - \mathbf{X}_1\hat{\mathbf{B}}_1)}{N_1}, \tag{7.48}$$

and

$$\hat{\boldsymbol{\Sigma}}_2 = \frac{(\mathbf{Y}_2 - \mathbf{X}_2\hat{\mathbf{B}}_2)^{\mathrm{T}}(\mathbf{Y}_2 - \mathbf{X}_2\hat{\mathbf{B}}_2)}{N_2}, \tag{7.49}$$

where $\hat{\mathbf{B}}$, $\hat{\mathbf{B}}_1$, and $\hat{\mathbf{B}}_2$ are the multivariate regression parameters' estimator matrices calculated on the basis of all samples, samples prior to the change point, and samples following the change point, respectively.

Based on Mahmoud et al. (2007), Noorossana et al. (2010b)proposed the corrected likelihood ratio statistic as follows:

$$lrtc_{m1} = \frac{lrt_{m1}}{E(lrt_{m1})}, \tag{7.50}$$

where $E(lrt_{m1})$ is the expected value of lrt_{m1} statistic under H_0 and can be determined by simulation. The threshold for this control chart is chosen to give a specified probability of Type I error.

7.2.1.3 Wilks' Lambda Approach

The third method of Noorossana et al. (2010b) was the extension of the Wilks' lambda approach proposed by Noorossana et al. (2008) to the multivariate multiple linear regression profiles. Let us assume there are m samples each with size of n_k. Similar to Noorossana et al. (2008), the first step is to pool all the m samples into one sample of size $N = \sum_{k=1}^{m} n_k$. Then, $m - 1$ indicator variables are created as

$$\begin{cases} Z_{ki} = 1 & \text{if observation } i \text{ is from sample } k \\ Z_{ki} = 0 & \text{otherwise,} \quad i = 1, 2, \ldots, N \quad k = 1, 2, \ldots, m-1. \end{cases} \tag{7.51}$$

The mth sample is called the reference sample. Finally, the following regression model is fitted to the pooled data:

$$\begin{aligned} \boldsymbol{y}_i = {} & \hat{\boldsymbol{\beta}}_0 + \hat{\boldsymbol{\beta}}_1 x_{1i} + \cdots + \hat{\boldsymbol{\beta}}_q x_{qi} + \hat{\boldsymbol{\beta}}_{01} Z_{1i} + \hat{\boldsymbol{\beta}}_{11} Z_{1i} x_{1i} + \cdots + \hat{\boldsymbol{\beta}}_{q1} Z_{1i} x_{qi} + \cdots \\ & + \hat{\boldsymbol{\beta}}_{0m'} Z_{m'i} + \hat{\boldsymbol{\beta}}_{1m'} Z_{m'i} x_{1i} + \cdots + \hat{\boldsymbol{\beta}}_{qm'} Z_{m'i} x_{qi} + \varepsilon_i, \quad i = 1, 2, \ldots, N, \end{aligned} \tag{7.52}$$

where $m' = m - 1$, $\boldsymbol{y}_i$ is a $1 \times p$ response variables for the ith observation in the pooled sample and $\boldsymbol{\beta}_0, \boldsymbol{\beta}_1, \ldots, \boldsymbol{\beta}_q, \boldsymbol{\beta}_{01}, \ldots, \boldsymbol{\beta}_{qm'}$ are $1 \times p$ vector of regression parameters.

To check for the equality of the m multivariate regression lines the following hypotheses are tested:

$$\begin{aligned} & H_0 : \boldsymbol{\beta}_{01} = \boldsymbol{\beta}_{11} = \cdots = \boldsymbol{\beta}_{q1} = \cdots = \boldsymbol{\beta}_{0m'} = \boldsymbol{\beta}_{1m'} = \cdots = \boldsymbol{\beta}_{qm'} = \mathbf{0} \\ & H_1 : H_0 \text{ is not true.} \end{aligned} \tag{7.53}$$

Under the null hypothesis, the reduced model is as follows:

$$\boldsymbol{y}_i = \hat{\boldsymbol{\beta}}_0 + \hat{\boldsymbol{\beta}}_1 x_{1i} + \cdots + \hat{\boldsymbol{\beta}}_q x_{qi} + \varepsilon_i, \quad i = 1, 2, \ldots, N. \tag{7.54}$$

The test statistic for testing H_0 is

$$\Lambda = \frac{\left|\mathbf{Y}^{\mathrm{T}}\mathbf{Y} - \hat{\mathbf{B}}^{\mathrm{T}}\mathbf{X}^{\mathrm{T}}\mathbf{Y}\right|}{\left|\mathbf{Y}^{\mathrm{T}}\mathbf{Y} - \hat{\mathbf{B}}_r^{\mathrm{T}}\mathbf{X}_r^{\mathrm{T}}\mathbf{Y}\right|}, \tag{7.55}$$

where $\mathbf{Y} = (\mathbf{y}_1, \mathbf{y}_2, \ldots, \mathbf{y}_N)^{\mathrm{T}}$ is an $N \times p$ matrix of response variables for the pooled data, $\mathbf{X}$ is an $N \times [m(q+1)]$ matrix of explanatory variables, $\hat{\mathbf{B}}$ is an $[m(q+1)] \times p$ matrix of the regression parameters for full model in Equation (7.52). $\mathbf{X}_r$ is an $N \times (q+1)$ matrix of explanatory variables and $\hat{\mathbf{B}}_r$ is an $(q+1) \times p$ matrix of the regression parameters for the reduced model in Equation (7.54).

This test statistic follows Wilks' lambda distribution with p, $(q+1)(m-1)$, and $N - m(q+1)$ degrees of freedom under the null hypothesis. The null hypothesis H_0 is rejected if $\Lambda < \Lambda_{\alpha,\, p,\, (q+1)(m-1),\, N-m(q+1)}$.

In order to test the equality of covariance matrices, Noorossana et al. (2008) proposed the statistic of Lee et al. (1977), $-2\log(M)$, where M statistic is given by

$$M = \frac{|\boldsymbol{S}_1|^{(n_1-q-1)/2}\,|\boldsymbol{S}_2|^{(n_2-q-1)/2}\cdots|\boldsymbol{S}_m|^{(n_m-q-1)/2}}{\left|\boldsymbol{S}_{pl}\right|^{(N-m(q+1))/2}}, \tag{7.56}$$

in which $\boldsymbol{S}_k$ $(k = 1, 2, \ldots, m)$ and $\boldsymbol{S}_{pl}$ are computed by

$$\boldsymbol{S}_k = \frac{\mathbf{Y}_k^{\mathrm{T}}\mathbf{Y}_k - \hat{\mathbf{B}}_k^{\mathrm{T}}\mathbf{X}_k^{\mathrm{T}}\mathbf{Y}_k}{n_k - (q+1)} \tag{7.57}$$

and

$$\boldsymbol{S}_{pl} = \frac{\sum_{k=1}^{m}[n_k - (q+1)]\boldsymbol{S}_k}{\sum_{k=1}^{m}[n_k - (q+1)]} \tag{7.58}$$

The M statistic varies between 0 and 1, with values near 1 favoring equality of covariance matrices.

7.2.1.4 Principal Components Analysis Approach

The fourth method is the Principal components analysis method proposed by Noorossana et al. (2008). In this method, they used principal components analysis to reduce the number of response variables as well as to make the correlated response variables independent. They first pooled all the m samples into one sample of size N and used the principal components of the response variables, which explain the most variation in the response variables. Then, they modeled the relationships between scores of the principal components and the explanatory variables. Since the new response variables are independent, the relationship between new response variables and explanatory variables were monitored separately. Finally, they used the likelihood ratio method by Mahmoud et al. (2007) to monitor the relationships between the new response variables and the explanatory variable in Phase I.

7.2.2 Performance Comparison of the Methods

Noorossana et al. (2010b) compared the performance of proposed methods for Phase I monitoring of multivariate multiple linear profiles in terms of the overall probabilities of an out-of-control signal. The results showed that the likelihood ratio approach and Wilks' lambda approach work well in detecting shifts in regression parameters. Likelihood ratio approach is superior to Wilks' lambda statistics in detecting step shifts in the process parameters. However, the Wilks' lambda approach is better than the likelihood ratio approach in detecting outliers.

7.2.3 Diagnostic Aid

Noorossana et al. (2010b) proposed a diagnostic procedure to find the out-of-control samples based on Wilks' lambda approach. In their proposed approach, if the equality of regression parameters or equality of covariance matrices is rejected, each sample is compared with the *m*th sample to find the out-of-control sample. In other words, the equality of the regression parameters and the equality of their covariance matrices ($\boldsymbol{\Sigma}_j = \boldsymbol{\Sigma}_m$) based on *j*th and *m*th samples are tested. If one of the mentioned hypotheses is rejected, then its corresponding diagnostic test is performed. If the hypothesis of equality of covariance matrices is rejected, the statistic $-2\log(M)$ is used in which M is calculated as follows:

$$M = \frac{\left|\boldsymbol{S}_j\right|^{(n_j-(q+1))/2}\left|\boldsymbol{S}_m\right|^{(n_m-(q+1))/2}}{\left|\boldsymbol{S}_{pl}\right|^{(n_j+n_m-2(q+1))/2}}, \tag{7.59}$$

where $\boldsymbol{S}_k$ and $\boldsymbol{S}_{pl}$ are calculated using the following equations:

$$\boldsymbol{S}_k = \frac{\mathbf{Y}_k^{\mathrm{T}}\mathbf{Y}_k - \hat{\mathbf{B}}_k^{\mathrm{T}}\mathbf{X}_k^{\mathrm{T}}\mathbf{Y}_k}{n_k - (q+1)} \tag{7.60}$$

and

$$\boldsymbol{S}_{pl} = \frac{[n_j - (q+1)]\boldsymbol{S}_j + [n_m - (q+1)]\boldsymbol{S}_m}{[n_j - (q+1)] + [n_m - (q+1)]}. \tag{7.61}$$

If the equality of regression parameters is rejected, a method similar to the approach by Mahmoud (2008) is used. Similar to Mahmoud (2008), they first combined the *j*th and *m*th samples. Then they decomposed the Wilks' lambda statistic to $m-1$ Wilks' statistics with p, $q+1$, and $n_j + n_m - 2(q+1)$ degrees of freedom, each obtained by comparing the *j*th sample with the last sample.

Example 7.2 As mentioned in Chapter 2, one of the important profile examples is a calibration application presented in Parker et al. (2001), in which, the purpose is to investigate replicated calibrations of a force balance used in wind tunnel experiments at NASA Langley Research Center. In this case, as stated in Chapter 2 there are six forces and moments (explanatory variables) and six electrical measurements of forces and moments (response variables) that are modeled as functions of forces

Table 7.3 In-Control Regression Parameters

	Intercept	Normal	Axial	Pitch	Roll	Yaw	Side
Y_1	−0.05	10	−0.01	−0.03	0.26	0	0.03
Y_2	0.48	0.24	21.01	−0.09	0.03	−0.12	0.01
Y_3	0.37	0.09	0.01	6.81	0.04	0.02	−0.03
Y_4	0.04	0	0	0	10.53	−0.47	0.21
Y_5	0.09	−0.02	0	0.01	0.02	7	−0.37
Y_6	0.01	0.04	0	−0.01	0.18	−0.34	11.46

and moments. Mahmoud et al. (2007) studied the relationship between one response variable and one explanatory variable and Mahmoud (2008) studied the relationship between one response variable and six explanatory variable.

In this example, the relationships between response variables and explanatory variables can be modeled as multivariate multiple liner profiles. Thus, Noorossana et al. (2010b) applied their proposed Phase I multivariate multiple linear profile methods to this case. Since their proposed methods are sensitive to the departure from normality and linearity, the values of the response variables were simulated. They generated multivariate multiple linear regression data set by using the real x-values used in NASA. The parameters of the in-control regression parameters are presented in Table 7.3.

The errors were simulated from a multivariate normal distribution with mean vector of zeros and covariance matrix with unit variances and covariance of 0.2 for each pair of error variables. The data set used in this example, consists of 11 samples ($m = 11$) each with sample sizes equal to 64, 73, or 74.

To evaluate the performance of methods, two out-of-control situations were considered. In the first out-of-control situations, the intercepts of the six profiles in sample 6 were changed to 1, 1, 2, 2, 1, and 2, respectively. In the second out-of-control situation, the slope of the independent variable "Normal" for the first profile in Sample 9 was changed from 10 to 0.26.

They applied only Wilks' lambda and the likelihood ratio approaches to the data set, because of the superior performance of these methods. The values of the *lrtc* statistics that are computed by using Equation (7.50) are presented in Table 7.4.

The upper control limit for the *lrtc* statistics is set equal to 1.52 to give a probability of Type I error of 0.04. The maximum of the $lrtc_{m1}$ statistics is greater than the upper control limit. This result shows that the process is out-of-control.

The value of the Wilks' lambda statistic is equal to 0.0000047 and the corresponding p-value is equal to zero. Then the equality of the regression parameters is rejected. The value of the statistic $(-2\log(M))$ is equal to 204.07 with a p-value

Table 7.4 The Values of *lrtc* Statistics for the 11 Samples

m_1	1	2	3	4	5	6	7	8	9	10
$lrtc_{m1}$	36.081	47.133	59.702	70.999	81.646	92.211	103.5	114.21	45.749	34.764

Table 7.5 The Values of the Wilks' Statistics and Corresponding *p*-Values

Sample	Wilks' Statistic	*p*-Value
1	0.683	0.267
2	0.752	0.854
3	0.691	0.331
4	0.731	0.742
5	0.745	0.821
6	0.233	**0.000**
7	0.637	0.152
8	0.674	0.349
9	0.000	**0.000**
10	0.734	0.760

of 0.848, which shows that the hypothesis of equal variance–covariance matrices is not rejected.

Since the equality of the regression parameters is rejected, they decompose the Wilks' statistic to determine the out-of-control sample. The values of Wilks' statistics and the corresponding p-values are given in Table 7.5, which shows that the sixth and ninth samples are out-of-control.

7.2.4 Phase II Methods for Monitoring Multivariate Multiple Linear Profiles

Eyvazian et al. (2010) proposed some methods for monitoring multivariate multiple linear profiles in Phase II. Assume that for the kth random sample collected over time we have n observations as $(x_{1i}, x_{2i}, \ldots, x_{qi}, y_{1ik}, y_{2ik}, \ldots, y_{pik})$, $i = 1, 2, \ldots, n$, where p and q are the number of response and explanatory variables, respectively. When the process is in statistical control, the model that relates the response variables with explanatory variables can be represented as following:

$$\mathbf{Y}_k = \mathbf{XB} + \mathbf{E}_k. \tag{7.62}$$

It should be noted that in Phase II it is assumed that parameters $\mathbf{B}$ and Σ are known. As mentioned in the previous section, for the kth random sample, the OLS estimator of the matrix $\mathbf{B}_k$ is as follows:

$$\hat{\mathbf{B}}_k = (\mathbf{X}_k^{\mathrm{T}}\mathbf{X}_k)^{-1}\mathbf{X}_k^{\mathrm{T}}\mathbf{Y}_k. \tag{7.63}$$

Eyvazian et al. (2010) proposed four methods for monitoring multivariate multiple linear profiles in Phase II. The first and second methods were MEWMA methods. The third method was an extension of the likelihood ratio method proposed by Zhang et al. (2009). Finally, the fourth method was the same as the combined MEWMA and chi-square control charts procedure proposed by Noorossana et al. (2010a).

7.2.4.1 The MEWMA Method

In the first method, Eyvazian et al. (2010) rewrote the matrix $\hat{\mathbf{B}}_k$ as a $((q+1)p) \times 1$ vector and denote it by $\hat{\boldsymbol{\beta}}_k$, where:

$$\hat{\boldsymbol{\beta}}_k = (\hat{\beta}_{01k}, \hat{\beta}_{11k}, \ldots, \hat{\beta}_{q1k}, \hat{\beta}_{02k}, \hat{\beta}_{12k}, \ldots, \hat{\beta}_{q2k}, \ldots, \hat{\beta}_{0pk}, \hat{\beta}_{1pk}, \ldots, \hat{\beta}_{qpk})^{\mathrm{T}}. \tag{7.64}$$

When the process is in-control, $\hat{\boldsymbol{\beta}}_k$ is a multivariate normal random vector with mean vector and covariance matrix equal to

$$E(\hat{\boldsymbol{\beta}}_k) = \boldsymbol{\beta} = (\beta_{01}, \beta_{11}, \ldots, \beta_{q1}, \beta_{02}, \beta_{12}, \ldots, \beta_{q2}, \ldots, \beta_{0p}, \beta_{1p}, \ldots, \beta_{qp})^{\mathrm{T}} \tag{7.65}$$

$$\boldsymbol{\Sigma}_{\hat{\boldsymbol{\beta}}_k} = \begin{pmatrix} \boldsymbol{\Sigma}_{11} & \boldsymbol{\Sigma}_{12} & \cdots & \boldsymbol{\Sigma}_{1p} \\ \boldsymbol{\Sigma}_{21} & \boldsymbol{\Sigma}_{22} & \cdots & \boldsymbol{\Sigma}_{2p} \\ \vdots & \vdots & \ddots & \vdots \\ \boldsymbol{\Sigma}_{p1} & \boldsymbol{\Sigma}_{p2} & \cdots & \boldsymbol{\Sigma}_{pp} \end{pmatrix}, \tag{7.66}$$

where $\boldsymbol{\Sigma}_{hj} = [\mathbf{X}^{\mathrm{T}}\mathbf{X}]^{-1}\sigma_{hj}$ is a $(q+1) \times (q+1)$ and σ_{hj} is the hjth element of the covariance matrix $\boldsymbol{\Sigma}$. They used the MEWMA control chart proposed by Lowry et al. (1992) to monitor the regression parameters and define the $\mathbf{z}_k$ as follows:

$$\mathbf{z}_k = \lambda(\hat{\boldsymbol{\beta}}_k - \boldsymbol{\beta}) + (1-\lambda)\mathbf{z}_{(k-1)}, \tag{7.67}$$

where, $\mathbf{z}_k$ is a multivariate normal random vector and $\mathbf{z}_0$ is $((q+1)p) \times 1$ vector of zeros.

Therefore, they proposed the chart statistic as

$$T^2_{Z_k} = \mathbf{z}_k^{\mathrm{T}} \sum_z^{-1} \mathbf{z}_k, \tag{7.68}$$

in which Σ_z is equal to $\frac{\lambda}{2-\lambda}\Sigma_{\hat{\beta}_k}$.The upper control limit for the above statistic is determined by simulation to give a specified in-control ARL.

7.2.4.2 MEWMA Method in Low-Dimensional Space

The performance of T^2 methods gets worse when the dimension of coefficients vector increases. As discussed in Chapter 4, Mahmoud (2008) proposed a method to reduce the number of profile parameters in Phase I monitoring of multiple linear profile. In this method, one monitors the parameters of a simple linear regression including intercept, slope, and standard deviation instead of parameters of the multiple linear regression model. Eyvazian et al. (2010) used this idea to decrease the dimension of previous method from $(q+1)p$ to $2p$. In this method, first the following model is fitted to each profile data set at sample k:

$$\begin{bmatrix} y_{1jk} \\ y_{2jk} \\ \vdots \\ y_{njk} \end{bmatrix} = \begin{bmatrix} 1 & u_{1j} \\ 1 & u_{2j} \\ 1 & \vdots \\ 1 & u_{nj} \end{bmatrix} \begin{bmatrix} A_{0jk} \\ A_{1jk} \end{bmatrix} + \begin{bmatrix} \varepsilon_{1jk} \\ \varepsilon_{2jk} \\ \vdots \\ \varepsilon_{njk} \end{bmatrix}, \tag{7.69}$$

where u_{ij} is equal to $E(y_{ij}) = \beta_{0j} + \beta_{1j}x_{1i} + \cdots + \beta_{qj}x_{qi}$. Since the response variables are correlated, the OLS estimators for A_{0jk} and A_{1jk} in different profiles are correlated and should not be monitored separately. Therefore, the estimated parameters for each sample are written as

$$\hat{\mathbf{A}}_k = (\hat{A}_{01k}, \hat{A}_{11k}, \hat{A}_{02k}, \hat{A}_{12k}, \ldots, \hat{A}_{0pk}, \hat{A}_{1pk})^{\mathrm{T}}. \tag{7.70}$$

When the process is in-control, the expected value and variance–covariance matrix for $\hat{\mathbf{A}}_k$ are as follows, respectively:

$$E(\hat{\mathbf{A}}_k) = \mathbf{A} = (A_{01}, A_{11}, A_{02}, A_{12}, \ldots, A_{0p}, A_{1p})^{\mathrm{T}} = (0, 1, 0, 1, \ldots, 0, 1)^{\mathrm{T}} \tag{7.71}$$

$$\mathbf{\Sigma}_{\hat{\mathbf{A}}_k} = \begin{pmatrix} \mathbf{\Sigma}'_{11} & \mathbf{\Sigma}'_{12} & \cdots & \mathbf{\Sigma}'_{1p} \\ \mathbf{\Sigma}'_{21} & \mathbf{\Sigma}'_{22} & \cdots & \mathbf{\Sigma}'_{2p} \\ \vdots & \vdots & \ddots & \vdots \\ \mathbf{\Sigma}'_{p1} & \mathbf{\Sigma}'_{p2} & \cdots & \mathbf{\Sigma}'_{pp} \end{pmatrix}, \tag{7.72}$$

in which $\mathbf{\Sigma}'_{hj}$ is a 2×2 matrix and equals to

$$\mathbf{\Sigma}'_{hj} = \begin{bmatrix} \mathrm{cov}(\hat{A}_{0hk}, \hat{A}_{0jk}) & \mathrm{cov}(\hat{A}_{0hk}, \hat{A}_{1jk}) \\ \mathrm{cov}(\hat{A}_{1hk}, \hat{A}_{0jk}) & \mathrm{cov}(\hat{A}_{1hk}, \hat{A}_{1jk}) \end{bmatrix}. \tag{7.73}$$

The elements of $\mathbf{\Sigma}'_{hj}$ are given by

$$\mathrm{cov}(\hat{A}_{0hk}, \hat{A}_{0jk}) = \sigma_{hj} \sum_{i=1}^{n} \left(\frac{1}{n} + \frac{(u_{ih} - \bar{u}_h)\bar{u}_h}{S_{u_h u_h}} \right) \left(\frac{1}{n} + \frac{(u_{ij} - \bar{u}_j)\bar{u}_j}{S_{u_j u_j}} \right) \tag{7.74}$$

$$\mathrm{cov}(\hat{A}_{1hk}, \hat{A}_{1jk}) = \frac{S_{u_h u_j}\sigma_{hj}}{S_{u_h u_h} S_{u_j u_j}} \tag{7.75}$$

$$\mathrm{cov}(\hat{A}_{0hk}, \hat{A}_{1jk}) = -\frac{S_{u_h u_j}\sigma_{hj}}{S_{u_h u_h} S_{u_j u_j}} \bar{u}_h \tag{7.76}$$

and equivalently,

$$\operatorname{cov}(\hat{A}_{1hk}, \hat{A}_{0jk}) = -\frac{S_{u_h u_j}\sigma_{hj}}{S_{u_h u_h} S_{u_j u_j}} \bar{u}_j, \tag{7.77}$$

in which $S_{u_v u_w} = \sum_{i=1}^{n} (u_{iv} - \bar{u}_v)(u_{iw} - \bar{u}_w)$.

Finally, similar to the previous method, they used the MEWMA procedure proposed by Lowry et al. (1992) to monitor the new regression parameters vector over time.

7.2.4.3 *Likelihood Ratio Method*

As discussed in Chapter 2, Zhang et al. (2009) proposed a control chart based on the likelihood ratio for monitoring simple linear profiles. Eyvazian et al. (2010) extended this method to the multivariate multiple linear regression profile case. They showed that for the kth sample collected over time, the likelihood ratio statistic is given as:

$$LRT_k = n \log |\boldsymbol{\Sigma}| - n \log \left|\hat{\boldsymbol{\Sigma}}_k\right| + C_k - np, \tag{7.78}$$

where C_k is equal to $\sum_{\mathbf{i=1}}^{\mathbf{n}} (\boldsymbol{y}_{ik} - \boldsymbol{x}_i\mathbf{B})\boldsymbol{\Sigma}^{-1}(\boldsymbol{y}_{ik} - \boldsymbol{x}_i\mathbf{B})^{\mathrm{T}}$ in which $(\boldsymbol{y}_{ik} - \boldsymbol{x}_i\mathbf{B})$ is the ith row of matrix $(\mathbf{Y}_k - \mathbf{XB})$. In addition, $\hat{\boldsymbol{\Sigma}}_k$ is the maximum likelihood estimator of $\boldsymbol{\Sigma}$ for the kth sample and equals to:

$$\hat{\boldsymbol{\Sigma}}_k = \frac{(\mathbf{Y}_k - \mathbf{X}\hat{\mathbf{B}}_k)^{\mathrm{T}}(\mathbf{Y}_k - \mathbf{X}\hat{\mathbf{B}}_k)}{n}, \tag{7.79}$$

where $\hat{\mathbf{B}}_k$ is the maximum likelihood estimator of $\mathbf{B}$ for the kth sample which is the same as OLS estimator of $\mathbf{B}$.

They substituted $\hat{\mathbf{B}}_k$ in Equation (7.79) with its EWMA statistic given by

$$\mathrm{E}\hat{\mathbf{B}}_k = \lambda\hat{\mathbf{B}}_k + (1-\lambda)\mathrm{E}\hat{\mathbf{B}}_{(k-1)} \tag{7.80}$$

and defined the statistic S_k as

$$\mathrm{S}_k = \frac{(\mathbf{Y}_k - \mathbf{X}(\mathrm{E}\hat{\mathbf{B}}_k))^{\mathrm{T}}(\mathbf{Y}_k - \mathbf{X}(\mathrm{E}\hat{\mathbf{B}}_k))}{n}, \tag{7.81}$$

where $\mathrm{E}\hat{\mathbf{B}}_0 = \mathbf{B}$.

In addition, they replaced the statistics ES_k and $\mathrm{E}C_k$ by $\hat{\boldsymbol{\Sigma}}_k$ and C_k in Equation (7.78), as follows:

$$\mathrm{ES}_k = \lambda\mathrm{S}_k + (1-\lambda)\mathrm{ES}_{k-1} \tag{7.82}$$

$$\mathrm{E}C_k = \lambda C_k + (1-\lambda)\mathrm{E}C_{k-1}, \tag{7.83}$$

where $\mathrm{E}\hat{\mathrm{S}}_0 = \boldsymbol{\Sigma}$ and $\mathrm{E}C_0 = np$.

Therefore, Equation (7.78), can be rewritten as given below:

$$\mathrm{E}LRT_k = n \log |\mathbf{\Sigma}| - n \log |\mathrm{E}\mathbf{S}_k| + \mathrm{E}C_k - np. \tag{7.84}$$

The upper control limit for the $\mathrm{E}LRT_k$ statistic is obtained by simulation to give a specified in-control ARL.

7.2.4.4 Combined MEWMA and Chi-Square Control Chart Method

The fourth method used by Eyvazian et al. (2010) was the same as the combined MEWMA and chi-square control charts scheme proposed by Noorossana et al. (2010a) to compare the performance of this method with other proposed methods.

7.2.5 Performance Comparison of the Methods

Eyvazian et al. (2010) compared the performance of proposed methods for Phase II monitoring of multivariate multiple linear profiles in terms of ARL. Their result showed that the MEWMA and chi-square control chart control approach and the EWMA method in low-dimensional space perform better than the other methods in detecting shifts in the regression coefficients. However, in detecting shift in the elements of variance–covariance matrix, likelihood ratio method, and the MEWMA and chi-square control chart approach perform better than other methods.

7.2.6 Change Point Estimator

Eyvazian et al. (2010) proposed a change point method to estimate the location of a shift in Phase II monitoring of multivariate multiple linear profiles. They used the maximum likelihood estimator (MLE) of the change point to estimate the change point. When an approach signals at subgroup k, the value of t that maximizes the likelihood function is the MLE for the change point. The estimator is computed by

$$\hat{t} = \underset{1 \le t < k}{\arg\ \max}(lr_{t,k}), \tag{7.85}$$

in which $lr_{t,k}$ is generalized likelihood ratio statistic and is equal to:

$$lr_{t,k} = (k - t)n\left[\log|\Sigma| - \log\left|\hat{\Sigma}_{t,k}\right| - p\right] + \sum_{j=t+1}^{k}\sum_{i=1}^{n}(\boldsymbol{y}_{ik} - \boldsymbol{x}_{ik}\mathbf{B})\Sigma^{-1}(\boldsymbol{y}_{ik} - \boldsymbol{x}_{ik}\mathbf{B})^{\mathrm{T}} \tag{7.86}$$

$\hat{\Sigma}_{t,k}$ is calculated using Equation (7.79) after pooling samples $t+1, t+2, \ldots,$ and k as below:

$$\hat{\mathbf{\Sigma}}_{t,k} = \frac{(\mathbf{Y}_{t,k} - \mathbf{X}_{t,k}\hat{\mathbf{B}}_{t,k})^{\mathrm{T}}(\mathbf{Y}_{t,k} - \mathbf{X}_{t,k}\hat{\mathbf{B}}_{t,k})}{n(k-t)}, \tag{7.87}$$

where $\mathbf{X}_{t,k}$ and $\mathbf{Y}_{t,k}$ are the $n(k-t) \times (q+1)$ matrix of explanatory variables and $n(k-t) \times p$ matrix of response variables, respectively and $\hat{\mathbf{B}}_{t,k}$ is the OLS estimator of the matrix $\mathbf{B}$.

REFERENCES

Eyvazian, M., Noorossana, R., Saghaie, A., and Amiri, A. (2010) Phase II Monitoring of Multivariate Multiple Linear Regression Profiles. Published online in Quality and Reliability Engineering International, DOI:10.1002/qre.1119.

Kim, K., Mahmoud, M. A., and Woodall, W. H. (2003) On the monitoring of linear profiles. *Journal of Quality Technology*, **35**, 317–328.

Lee, J. C., Chang, T. C., and Krishnaiah, P. R. (1977) Approximations to the distributions of the likelihood ratio statistics for testing certain structures on the covariance matrices of real multivariate normal populations. *Multivariate Analysis*, **4**, 105–118.

Lowry, C. A., Woodall, W. H., Champ, C. W., and Rigdon, S. E. (1992) A multivariate exponentially weighted moving average control chart. *Technometrics*, **34**, 46–53.

Mahmoud, M. A. (2008) Phase I analysis of multiple linear regression profiles. *Communications in Statistics, Simulation and Computation*, **37**, 2106–2130.

Mahmoud, M. A., Parker, P. A., Woodall, W. H., and Hawkins, D. M. (2007) A change point method for linear profile data. *Quality and Reliability Engineering International*, **23**, 247–268.

Mahmoud, M. A. and Woodall, W. H. (2004) Phase I monitoring of linear profiles with calibration application. *Technometrics*, **46**, 380–391.

Noorossana, R., Eyvazian, M., Amiri, A., and Vaghefi, A. (2008) Phase I Monitoring of Multivariate Simple Linear Profiles, Technical Report, Industrial Engineering Department, Iran University of Science and Technology (in Farsi).

Noorossana, R., Eyvazian, M., and Vaghefi, A. (2010a) Phase II monitoring of multivariate simple linear profiles. *Computers and Industrial Engineering*, **58**(4), 563–570.

Noorossana, R., Eyvazian, M., Amiri, A., and Mahmoud, M. A. (2010b) Statistical monitoring of multivariate multiple linear regression profiles in phase I with calibration application. *Quality and Reliability Engineering International*, **26**(3), 291–303.

Parker, P. A., Morton, M., Draper, N. R., and Line, W. P. (2001), "A Single-Vector Force Calibration Method Featuring the Modern Design of Experiments,". Proceedings of the American Institute of Aeronautics and Astronautics 39th Aerospace Sciences Meeting & Exhibit, Reno, Nevada.

Rencher, A. C. (2002) Methods of Mulrivariate Analysis, 2nd Edition, John Wiley & Sons, Canada.

Sullivan, J. H. and Woodall, W. H. (1996) A comparison of multivariate control charts for individual observations. *Journal of Quality Technology*, **28**, 398–408.

Williams, J. D., Woodall, W. H., and Birch, J. B. (2007) Statistical monitoring of nonlinear product and process quality profiles. *Quality and Reliability Engineering*, **23**(8), 925–941.

Zhang, J., Li, Z., and Wang, Z. (2009) Control chart based on likelihood ratio for monitoring linear profiles. *Computational Statistics and Data Analysis*, **53**, 1440–1448.

CHAPTER 8

Statistical Process Control for Geometric Specifications

Bianca Maria Colosimo and Quirico Semeraro
Department of Mechanical Engineering, Politecnico di Milano, Milano, Italy

Massimo Pacella
Dipartimento di Ingegneria dell'Innovazione, University of Salento, Lecce, Italy

INTRODUCTION

In the area of statistical process control (SPC), approaches for *profile monitoring* proposed in the literature (Woodall et al. 2004; Woodall 2007) can be classified with reference to the modeling technique considered (linear or nonlinear regression, nonparametric regression or approaches for multivariate data reduction as the principal/independent component analysis) as well as to the type of application faced (i.e., calibration study, process signal, or geometric specification monitoring).

With reference to the type of application faced, most of the studies on profile monitoring deal with calibration studies (Stover and Brill 1998; Kang and Albin 2000; Kim et al. 2003; Mahmoud and Woodall 2004; Chang and Gan 2006; Gupta et al. 2006; Zou et al. 2006; Mahmoud et al. 2007) where the profiles that have to be monitored are straight lines. A second stream of application concerns monitoring of signals coming out from sensored machines. For example, Jin and Shi (1999, 2001) and Ding et al. (2006) referred to profiles representing force and torque signals collected from online sensors on a press in a stamping process. The third stream of applications concerns the use of profile monitoring for quality control of *geometric specifications* (Colosimo and Pacella 2007, 2010; Colosimo et al. 2008, 2010). The

Statistical Analysis of Profile Monitoring, First Edition. Edited by Rassoul Noorossana, Abbas Saghaei and Amirhossein Amiri.

present chapter specifically deals with this kind of applications of profile monitoring, which is significant in the manufacturing field.

In manufacturing of mechanical components, technical drawings show an increasing number of geometric specifications (aka geometric tolerances). This is mainly due to the deeper interest in relating the characteristics of a geometric feature (i.e., a machined profile or surface) to the functional properties of the produced item. A geometric specification is a product specification concerning a geometric feature, i.e., a shape in a two-dimensional (2D) or in the three-dimensional (3D) space. This specification can concern simple (e.g., roundness and cylindricity) or even complex shapes (free-forms) and is designed for constraining the deviation of the actual feature from the ideal one, since this deviation can affect the functional properties of the produced item.

A manufacturing process may experience changes due to the material machined, improper setup, errors of the operator, wear or sudden changes of the machine conditions, etc. Usually, these changes may cause deteriorated process performance, i.e., the process may produce an increased number of nonconforming geometric features, i.e., profiles and surfaces that deviate excessively from the ideal or nominal shape. In this framework, a statistical process monitoring procedure should quickly detect, by issuing an alarm, any deviation of the machined shape from the usual pattern obtained when the process is in its in-control state.

The widespread use of modern measurement systems in manufacturing such as coordinate measuring machines (CMMs), which decisively reduce the complexity and the time required for checking geometric specifications, also contribute to the increasing important role that profiles and surfaces play in modern manufacturing of mechanical components. These machines, which can be controlled either manually by an operator or automatically by a computer, are the most general-purpose devices nowadays available in the industry for measuring tasks. Measurements are defined by a probe attached to a moving axis of the machine. This probe (which can be mechanical, optical, or laser, among others) assesses the position of a target in a given reference system. By precisely recording the coordinates of the targets, points are generated which can then be analyzed for describing the geometric feature. Machined profiles and surfaces can be thought as functional data if a subset of the coordinates, which are used to describe the position of each sampled point, can be represented as function of one or more independent variables. When the independent variables are fixed, they may act as a counter of the sampled point.

Consider a 2D circular profile described in a polar reference system. The radius can be modeled as a function of the angular position. For example, Figure 8.1 depicts a circular profile, which consists of a set of $P = 36$ points located on equally spaced angular positions. The independent variable is the angular position $\theta_p = (p-1)(2\pi/P)$, where $p = 1, \ldots, P$ and $\theta_p \in [0, 2\pi]$ radiants (alternatively, θ_p may be expressed in degrees as it is in Figure 8.1), while the dependent one is the radius $y(\theta_p)$. The circular profile in the polar diagram of Figure 8.1 can be equivalently represented as in Figure 8.2.

Similarly, consider a 3D cylindrical surface described in a cylindrical reference system, the surface can be modeled by representing the radius as a function of both the

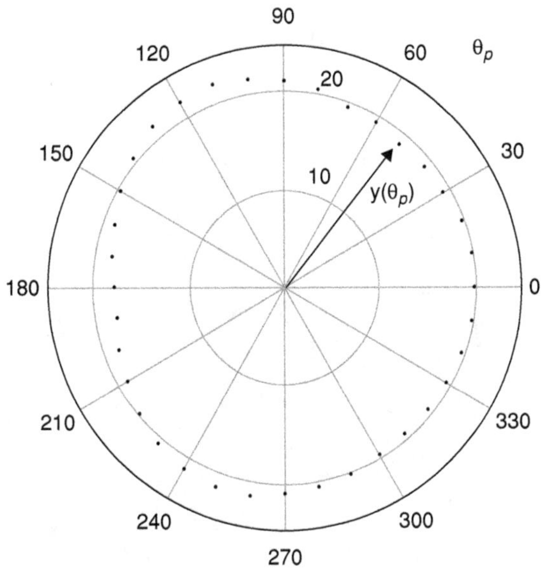

Figure 8.1 An example of circular profile in a polar reference system (angular positions in degrees); the arrow shows the measure at a given angular position (6th point, angular position 50°).

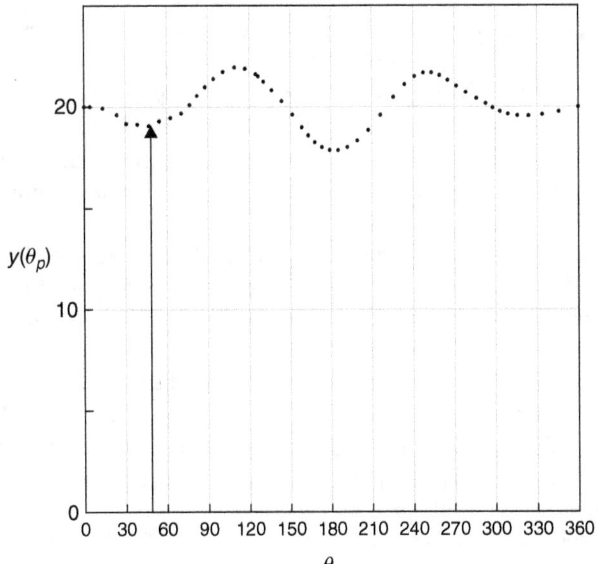

Figure 8.2 An alternative representation of the circular profile depicted in Figure 8.1; the arrow shows the measure at a given angular position (6th point, angular position 50°).

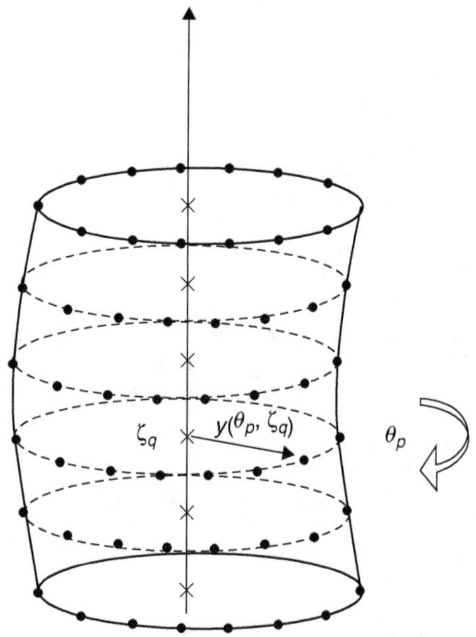

Figure 8.3 An example of cylindrical surface in a cylindrical reference system (function of the angular and vertical position); the arrow shows the measure at a given angular and vertical position.

angular and vertical position. Figure 8.3 depicts a cylindrical surface, which consists of a discrete set of points sampled on fixed and deterministic angular positions, $\theta_p = (p-1)(2\pi/P)$ where $p = 1, \ldots, P$ and $\theta_p \in [0, 2\pi]$ (θ_p may be expressed in degrees), as well as at fixed and deterministic vertical positions $\zeta_q = 2\frac{q-1}{Q-1} - 1$ where $q = 1, \ldots, Q$ and $\zeta_q \in [-1, 1]$. ζ_q is a normalized variable with reference to the vertical position of the point. In Figure 8.3 the independent variables are θ_p and ζ_q, while the dependent one is $y\left(\theta_p, \zeta_q\right)$.

Colosimo and Pacella (2007) and Colosimo et al. (2008) dealt with 2D roundness profiles obtained by lathe-turning, showing that both principal component analysis (PCA) and spatial autoregressive regression (SARX) models can be used for modeling and then monitoring the geometric profile. By combining these models with control charting, the authors showed how out-of-controls of the manufactured profile can be easily and quickly detected. Subsequently, Colosimo and Pacella (2010) compared performance of different approaches (as the previous two model-based approaches as well as a different simpler approach) to outline scenarios in which a specific approach outperforms the others in geometric error monitoring. Eventually, Colosimo et al. (2010), who used as case study the cylindricity of lathe-turned items, showed that 3D surfaces can be modeled as functional data as well and surface monitoring can be considered as a generalization of profile monitoring. The SARX model for 2D roundness profiles in (Colosimo et al. 2008) was exploited as starting reference to identify the parametric models of the large-scale pattern characterizing all the 3D cylindrical surfaces. The model was further extended to represent the spatial correlation characterizing adjacent points on each machined surface, i.e., a parametric model of cylindrical surfaces able to represent both the large-scale and the small-scale (spatial correlation) characteristics of the machined surface.

The present chapter summarizes results obtained in the aforementioned papers in order to show how profile monitoring can be effectively used for monitoring geometric features. A manufacturing reference case study, which is presented in Subsections 8.1.1 and 8.1.2, is used for this aim. The remainder of the chapter is structured as follows. In Section 8.1, examples of geometric feature concerning circularity are introduced. Section 8.2 presents two analytical methods, i.e., regression and PCA, for modeling the functional relationship representing the manufactured shape and for checking whether or not the estimated model is stable over time. Section 8.3 shows how traditional approaches for SPC can be used for monitoring geometric form errors. Given their easiness of use, these approaches are viable solutions for the industrial practice. In Section 8.4, a comparison study is presented in order to allow one to select a specific approach in a given production scenario. Section 8.5 discusses a possible generalization of the approaches for modeling 3D functional data and how profile monitoring can be generalized to surface monitoring. Conclusions and final remarks are given in Section 8.6.

8.1 EXAMPLES OF GEOMETRIC FEATURE CONCERNING CIRCULARITY

8.1.1 An Example of 2D Geometric Feature Concerning Circularity

Assume we collect a group of N 2D profiles, where each profile consists of P measurements observed at a fixed set of locations. Let $y_n(p)$ denote the dependent variable measured at a specific location of index p on the nth profile, ($p = 1, \ldots, P$ and $n = 1, \ldots, N$). In the case of circular profiles, each consisting of P points sampled on a fixed set of equally spaced positions, the dependent variable $y_n(p)$ is the radial deviation from the nominal radius measured at the angular positions $\theta_p = (p-1)\frac{2\pi}{P}$.

The real case study presented by Colosimo et al. (2008) is used as reference throughout this chapter for 2D profiles. In particular, the case study refers to circular profiles measured on $N = 100$ items produced by lathe-turning starting from C20 carbon steel cylinders (original diameter 30 mm, final diameter 26 mm). One circular profile was measured for each item, at a given distance from the spindle, by using a CMM. Each circular profile consisted of $P = 748$ equally spaced measurements of the radius, where each measurement corresponds to a given angular position. The original measurements were scaled by subtracting the least-squares estimate of the radius and centered at the least-squares estimate of the centre. Hence, each point in a circular profile can be seen as the radial deviation from the mean radius observed at a given angle and this is why observations on a profile can be either positive or negative. A graphical representation of the whole experimental data set, in which the independent variable is the location index on the part and the dependent one is the

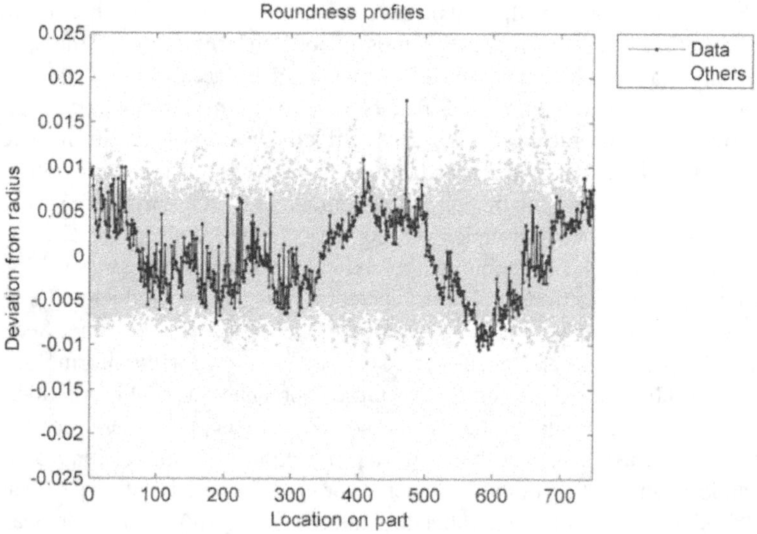

Figure 8.4 100 roundness profiles of 748 points each (one out of the 100 profiles in bold line); vertical axis scale (mm); the independent variable is the location index on the part (from 1 to 748).

deviation from the nominal radius in mm, is shown in Figure 8.4 (where 1 out of the 100 profiles is in bold line).

From Figure 8.4 it seems that no systematic pattern is characterizing the roundness profiles of the reference case study. This appearance hides a common problem of shape analysis, which consists in feature registration or alignment. In fact, the profiles shown in Figure 8.4 are actually misaligned because of the random contact angle of the turning process. Figure 8.5 shows the aligned roundness profiles after that a proper registration procedure is implemented on the data. The goal of the registration procedure implemented is to minimize the phase delay caused by the random contact angle (Colosimo and Pacella 2007). From a visual inspection of Figure 8.5, it can be easily observed that the roundness profiles share a common shape (pattern), i.e., the turning process leaves a specific *manufacturing signature* on the machined components (Colosimo et al. 2008).

For a circular feature, the geometric form error of the actual profile from the ideal geometry is represented by a synthetic variable. In the case of a circular profile, such a value is referred to as *out-of-roundness* (OOR). The OOR is computed as the difference between the maximum and the minimum radial distances of the actual geometric feature with respect to a predetermined center, which is the center of the so-called *substitute geometry*. Geometrically, this corresponds to find two concentric circles, one circumscribing and one inscribing the profile sampled on the manufactured feature. The OOR is then estimated by the width of the annulus determined by these two concentric circles (see Figure 8.6).

Several algorithms and methods can be implemented in order to determine the common center of the circumscribed circle and of the inscribed one. Previous studies

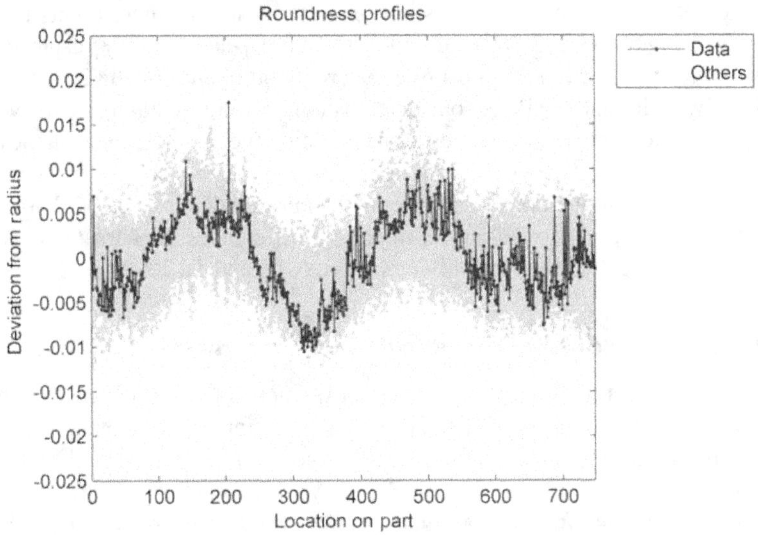

Figure 8.5 Hundred roundness profiles of 748 points each; aligned data (one out of the 100 profiles in bold); vertical axis scale (mm); the independent variable is the location index on the part (from 1 to 748).

(Colosimo et al. 2008) showed that no difference in profile monitoring results can be found if the minimum zone (MZ) rather than the least-squares (LS) approach is used for computing the form error. This is why the LS algorithm is considered as reference from now on.

The LS algorithm (i.e., the widely used procedure for form tolerance evaluation) minimizes the sum of squared deviations of measured points from the fitted feature.

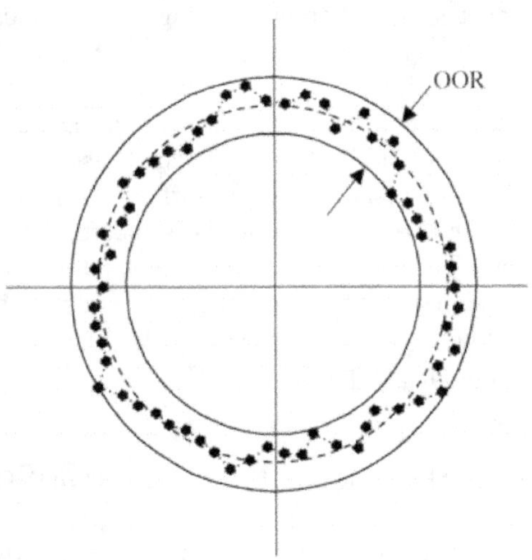

Figure 8.6 A schematic representation of the OOR value for a circular profile; the dashed line represents the substitute geometry (Adapted from Colosimo et al. 2008.)

The LS method associates one substitute feature to measurement points (e.g., one circle for roundness) and calculates the maximum peak-to-valley distance of the measurement points from the substitute feature (Moroni and Pacella 2008).

Usually, a circular profile is considered as conforming to requirements when the OOR value is lower than the geometric specification (i.e., the tolerance on the technical drawing). However, it should be noted that many different circular profiles, which can induce different functional properties of the machined items, can be characterized by the same OOR value resulting in important differences in characteristics of the profile, for example, in the assembly precision, (Cho and Tu 2002).

8.1.2 An Example of 3D Geometric Feature Concerning Cylindricity

Assume we collect a group of N 3D surfaces, where each surface consists of $P \times Q$ measurements observed at a fixed grid of locations. Let $y_n(p, q)$ denote the dependent variable measured at a specific location of indices p and q on the nth surface ($p = 1, \ldots, P, q = 1, \ldots, Q$ and $n = 1, \ldots, N$). In the case of cylindricity, $y_n(p, q)$ is the radial deviation from the nominal cylinder, which is expressed in a cylindrical reference system, i.e., at angular positions $\theta_p = (p - 1)(2\pi/P)$ and vertical positions $\zeta_q = 2\frac{q-1}{Q-1} - 1$ ($\theta_p \in [0, 2\pi]$ and $\zeta_q \in [-1, 1]$).

As previously described, the case study that consists of $N = 100$ items produced by lathe-turning starting from C20 carbon steel cylinders is also used as a reference for 3D surfaces. In particular, one cylindrical surface was measured for each item produced by lathe-turning using a CMM. A set of $P = 68$ generatrices was sampled on each surface and a set of $Q = 61$ points was measured on each of the 68 generatrices. Therefore, a set of $P \times Q = 4148$ points, equally distributed on the cylindrical surface, was eventually available for each produced item.

Among different geometric specifications, cylindricity plays an important role when functionality relates to rotation or coupling of mechanical components (e.g., shafts and holes). According to the standards, the cylindricity is the condition of a surface of revolution in which all the points of the surface are equidistant from a common axis. The deviation of the actual shape from the ideal one is specified by the cylindricity error, which is the minimum radial distance between two coaxial cylinders that contain among them the actual surface. This error is often referred to as *out-of-cylindricity* (OOC). According to the standard, the OOC error has to be computed in two steps. The first step consists in computing the reference cylinder, which is fitted on the actual data by using one of several possible criteria (e.g., LS, MZ). Then, the deviations of the actual shape from the reference cylinder is computed and possibly modeled in the 3D space.

8.2 CONTROL CHARTS FOR PROFILE MONITORING

Many approaches proposed in the literature for profile monitoring focus on the case of linear profiles only. However, when geometric features are of interest, approaches focusing on nonlinear profiles should be more appropriate (Jin and Shi 1999, 2001;

Young et al. 1999; Walker and Wright 2002; Ding et al. 2006; Williams et al. 2007; Jensen et al. 2008; Zou et al. 2008; Zhang and Albin 2009).

Another important issue related to profile monitoring when geometric features are of interest is the autocorrelation of data. Walker and Wright (2002) outlined that "autocorrelation is frequently present in data that are observed within small intervals of time or space. Indeed, the data analyzed exhibited a significant amount of positive autocorrelation" (Walker and Wright 2002, p. 124). Despite the specific mention to the autocorrelation problem, the approach proposed by Walker and Wright (2002) was based on independent data, as most of the other approaches focused on profiles monitoring. The only exception to this general rule is the papers by Jensen et al. (2008) and Colosimo et al. (2008, 2010). In particular, while the first paper dealt with autocorrelated profile data, the second ones focused on spatially autocorrelated data and are more suitable for modeling geometric form features (i.e., profiles and surfaces measured on machined items). As a matter of fact, when data refer to geometric form errors, measurements are often spatially correlated because they are obtained in similar conditions of the machining process and related to similar (local) properties of the machined material. Spatial autocorrelation, which allows one to represent contiguity in space, is different from temporal autocorrelation (time-series models). In the case of profiles, contiguity in space implies that the dependency among data on a profile is bidirectional (i.e., a given point may be correlated to points located on its left and on its right, regardless of the specific direction), while time-series models are suitable to represent just a one-direction dependency (i.e., past data influence future ones) (Whittle 1954). Second, spatial models allow one to represent a specific type of relationship among points observed in closed profiles (e.g., roundness profile). In fact, when data on a closed or circuit profile are sequentially numbered (by defining an arbitrary starting point), observations at the beginning and at the end of the profile are spatially correlated. Colosimo et al. (2008, 2010) presented an approach for modeling roundness profiles based on fitting a SARX model (Cressie 1993). The regression model presented in these papers is summarized in Subsection 8.2.1.

8.2.1 The Regression-Based Approach

With reference to the case study presented in Subsection 8.1.1, assume we organize data into column vectors $\mathbf{y}_n' = \left[\, y_n(1) \ldots y_n(p) \ldots y_n(P) \,\right]$, the general SARX model can be written in matrix notation as follows:

$$\begin{aligned} \mathbf{y}_n &= \mathbf{X}\mathbf{b}_n + \boldsymbol{\upsilon}_n, \\ (\mathbf{I} - \mathbf{R}_n)\,\boldsymbol{\upsilon}_n &= \boldsymbol{\varepsilon}_n, \\ \mathbf{R}_n &= \sum_{s=1}^{S} a_{sn}\mathbf{W}^{(s)}. \end{aligned} \tag{8.1}$$

The first expression in Equation (8.1) describes the $P \times 1$ vector of response for the nth profile $\mathbf{y}_n$ as formed by a *large-scale* and a *small-scale* component (Cressie 1993). The large-scale component is given by $\mathbf{X}\mathbf{b}_n$ where $\mathbf{X}$ is a $P \times R$ matrix of R regressor variables that are assumed to be known and constant and $\mathbf{b}'_n = \left[b_{1n} \cdots b_{rn} \cdots b_{Rn} \right]$ is the $R \times 1$ vector of regression parameters that are normally distributed with mean $\boldsymbol{\beta}' = \left[\beta_1 \cdots \beta_r \cdots \beta_R \right]$ and covariance matrix $\mathbf{B}$ ($\mathbf{b}_n \sim N(\boldsymbol{\beta}, \mathbf{B})$).

The small-scale component is the $P \times 1$ vector of error terms $\boldsymbol{v}_n$ in Equation (8.1). Error terms are assumed to be spatially correlated and are represented as a spatial autoregressive process (SAR) of order S. The SAR(S) model is given by the last two expressions of Equation (8.1), where $\mathbf{I}$ is the $P \times P$ identity matrix, $\boldsymbol{\varepsilon}_n$ is a $P \times 1$ vector of independently and normally distributed residuals $\left(\boldsymbol{\varepsilon}_n \sim N\left(\mathbf{0}, \sigma^2\mathbf{I}\right)\right)$, and $\mathbf{a}'_n = \left[a_{1n} \cdots a_{sn} \cdots a_{Sn} \right]$ is the vector of coefficients of the SAR(S) model for the nth profile, which is assumed to be normally distributed with mean $\boldsymbol{\alpha}' = \left[\alpha_1 \cdots \alpha_s \cdots \alpha_S \right]$ and covariance matrix $\mathbf{A}$ ($\mathbf{a}_n \sim N(\boldsymbol{\alpha}, \mathbf{A})$).

The $P \times P$ matrix $\mathbf{W}^{(s)}$ of elements $w^{(s)}(p_1, p_2)$ $(p_1, p_2 = 1, \ldots, P)$ represents the core of the small-scale model, since it is the sth order adjacent matrix $(s = 1, \ldots, S)$. $w^{(1)}(p_1, p_2)$ is set equal to 1 if the p_1th point is the neighbor of the p_2th, and 0 otherwise. Analogously, the element of a second-order adjacency matrix, $w^{(2)}(p_1, p_2)$, is set equal to 1 if the p_1th point is a neighbor of the original first-generation neighbors of the p_2th point, and so on. By definition, all the adjacency matrices are binary (i.e., they can have only 0 and 1 elements) and symmetric matrices whose diagonal elements are equal to 0 (Cressie 1993).

For each profile, two vectors of coefficients $\mathbf{b}_n$ and $\mathbf{a}_n$ are considered. In order to let the model have the most general form, we further assume that these two vectors could also be correlated, i.e., cov $(\mathbf{b}_n, \mathbf{a}_n) = \mathbf{D}$ (Colosimo and Pacella 2010). In other words, with reference to the parametric model structure given in Equation (8.1), we merge the two vectors characterizing the observed pattern into a C-length single coefficient vector related to the nth profile (where $C = R + S$):

$$\mathbf{c}'_n = \left[\mathbf{b}'_n \quad \mathbf{a}'_n \right] = \left[b_{1n} \cdots b_{rn} \cdots b_{Rn} \, a_{1n} \cdots a_{sn} \cdots a_{Sn} \right],$$

$$\mathbf{c}_n \sim N(\boldsymbol{\mu}, \boldsymbol{\Sigma}), \text{ where } \boldsymbol{\mu}' = \left[\boldsymbol{\beta}' \quad \boldsymbol{\alpha}' \right], \quad \boldsymbol{\Sigma} = \begin{bmatrix} \mathbf{B} & \mathbf{D} \\ \mathbf{D}' & \mathbf{A} \end{bmatrix}, \tag{8.2}$$

the model for the nth profile shown in Equation (8.1) is completely defined by a SARX model that requires one to estimate the $C = R + S$ parameters, which are the components of the vector $\mathbf{c}_n$, and the residuals' variance σ^2.

Let $\hat{\mathbf{c}}'_n = \left[\hat{\mathbf{b}}'_n \; \hat{\mathbf{a}}'_n \right]$ represent the vector of C parameters' estimates for the jth profile. A T^2 control chart can be designed with reference to the statistics

$$T_n^2 = (\hat{\mathbf{c}}_n - \boldsymbol{\mu})' \boldsymbol{\Sigma}^{-1} (\hat{\mathbf{c}}_n - \boldsymbol{\mu}), \tag{8.3}$$

where $\boldsymbol{\mu}$ and $\boldsymbol{\Sigma}$ are the mean vector and covariance matrix of the coefficients (estimated from a set of in-control profiles), respectively. Williams et al. (2006) studied the performance of different control limits to be used for the T_n^2. When the number of samples N is at least twice the number of parameters estimated $(C + [C(C+1)/2])$, the following upper control limit (UCL) can be used:

$$UCL = \chi^2_{\alpha,C}, \tag{8.4}$$

where $\chi^2_{\alpha,C}$ is the $100\,(1-\alpha)$ percentile of the chi-square distribution with C degrees of freedom. The $P \times 1$ vector of estimated residuals associated with the nth profile can be described as follows:

$$\mathbf{e}_n = (\mathbf{I} - \mathbf{R}_n)\left(\mathbf{y}_n - \mathbf{X}\hat{\mathbf{b}}_n\right), \tag{8.5}$$

where $\mathbf{R}_n = \sum_{s=1}^{S} \hat{a}_{sn}\mathbf{W}^{(s)}$. The estimated variance of residuals $\hat{\sigma}_n^2$ is given by

$$\hat{\sigma}_n^2 = \frac{\mathbf{e}'_n\mathbf{e}_n}{P-1}. \tag{8.6}$$

In order to monitor the residual variance, a traditional Shewhart-type control chart can be used. Let σ^2 denote the actual residuals' variance, the upper and lower control limits can be computed as follows:

$$\begin{aligned} UCL &= \frac{\sigma^2}{P-1}\chi^2_{\alpha/2,P-1}, \\ CL &= \sigma^2, \\ LCL &= \frac{\sigma^2}{P-1}\chi^2_{1-(\alpha/2),P-1}, \end{aligned} \tag{8.7}$$

where $\chi^2_{\alpha/2,P-1}$ and $\chi^2_{1-(\alpha/2),P-1}$ are respectively the upper and lower $\alpha/2$ percentage points of the chi-square distribution with $P-1$ degrees of freedom associated with the residuals (Montgomery 2004).

8.2.1.1 Application of the Regression-Based Approach

The SARX model in Equation (8.1) can be fitted to the data of the case study in previous subsection 8.1.1. In the case of circular profiles, the large-scale component in Equation (8.1) can be expressed as linear combination of *harmonics* (Cho and Tu 2001). Therefore, the element of the rth column and nth row of $\mathbf{X}$ in Equation (8.1), say $x_r(p)$, can be expressed either as $x_r(p) = \cos\left(h\theta_p\right)$ or $x_r(p) = \sin\left(h\theta_p\right)$, i.e., as a sinusoidal function of frequency equal to $h\,(2\pi/P)$ rad/sample. $h \in \{1, \ldots, P/2\}$ is an index, which represents the undulations per revolution (upr) of the sinusoidal function.

Two harmonics were selected for modeling the radial deviations in the actual case study, namely the second and the third one ($h = 2, 3$). Indeed, the process signature was mainly affected by ovality and triangularity. The oval contour was possibly due to a bi-lobe error motion affecting the spindle's lathe or to eccentricity caused by an improper setup, while the tri-lobe pattern was due to a similar error motion of the spindle. Hence, matrix $\mathbf{X}$ in Equation (8.1) has $R = 4$ columns. The pth row of matrix $\mathbf{X}$ is equal to $\left[\cos\left(2\theta_p\right)\ \sin\left(2\theta_p\right)\ \cos\left(3\theta_p\right)\ \sin\left(3\theta_p\right)\right]$.

As for the small-scale component of the actual roundness profiles, the vector error terms $\boldsymbol{v}_n$ in Equation (8.1) was modeled as a SAR model of order 2 ($S = 2$) using the algorithm implemented in the spatial econometrics toolbox (LeSage 1999).

Table 8.1 Parameters for the Distribution of SARX Coefficients $\mathbf{c}_n \sim N(\boldsymbol{\mu}, \boldsymbol{\Sigma})$ for the Actual Roundness Data of the Reference Case Study

$$\boldsymbol{\beta}' = \begin{bmatrix} -0.0341 & 0.0313 & 0.0080 & -0.0322 \end{bmatrix}$$

$$\boldsymbol{\alpha}' = \begin{bmatrix} 0.3021 & 0.2819 \end{bmatrix}$$

$$\mathbf{B} = \begin{bmatrix} 4.0646 & -2.0200 & 0.6540 & 0.2652 \\ -2.0200 & 3.8961 & 1.4851 & 0.0614 \\ 0.6540 & 1.4851 & 2.2346 & -0.1074 \\ 0.2652 & 0.0614 & -0.1074 & 3.1214 \end{bmatrix} \times 10^{-4}$$

$$\mathbf{D} = \begin{bmatrix} -0.8844 & -2.4101 \\ -1.2123 & 1.9568 \\ -1.1844 & 0.5958 \\ -1.4993 & -3.7224 \end{bmatrix} \times 10^{-4}$$

$$\mathbf{A} = \begin{bmatrix} 38.0199 & 15.8999 \\ 15.8999 & 43.2491 \end{bmatrix} \times 10^{-4}$$

Given the identified model, each profile of index n was associated to a vector of $C = 4 + 2$ parameters $\mathbf{c}_n = \begin{bmatrix} b_{1n} \ b_{2n} \ b_{3n} \ b_{4n} \ a_{1n} \ a_{2n} \end{bmatrix}$. We observed that the C-length parameter vector $\mathbf{c}_n$ changes from profile to profile according to a C-variate normal distribution of mean $\boldsymbol{\mu}$ and covariance matrix $\boldsymbol{\Sigma}$, i.e., $\mathbf{c}_n \sim N(\boldsymbol{\mu}, \boldsymbol{\Sigma})$ (the Shapiro–Wilk normality test for multivariate distribution was used, with a p-value equal to 0.03121). Table 8.1 summarizes the vector $\boldsymbol{\mu} = \begin{bmatrix} \boldsymbol{\beta}' & \boldsymbol{\alpha}' \end{bmatrix}$ and the matrix $\boldsymbol{\Sigma} = \begin{bmatrix} \mathbf{B} & \mathbf{D} \\ \mathbf{D}' & \mathbf{A} \end{bmatrix}$ estimated on the actual data.

Hence, by considering the values of $\boldsymbol{\mu}$ and $\boldsymbol{\Sigma}$ in Table 8.1, instances of the C-length parameter vector $\mathbf{c}_n$ can be simulated on a computer (the command "mvnrnd" of MATLAB was used in our work). By combining instances of the C-length parameter vector $\mathbf{c}_n$ to instances of the P-length vector of residuals ε_n the SARX model in Equation (8.1) can be used to simulate on a computer realistic roundness profiles (Colosimo and Pacella 2010). In particular, the residuals ε_n were simulated as independently and normally distributed, with zero mean and common standard deviation σ (the command "randn" of MATLAB was used in our work). The numerical value $\sigma = 9.2244 \cdot 10^{-4}$ mm was estimated from actual data and used for the simulations of roundness profiles in our work.

As for example, we considered a set of $N = 100$ roundness profiles obtained from computer simulation. Two control charts (i.e., a multivariate control chart on the estimated coefficients $\hat{\mathbf{c}}_n'$ and a univariate control chart on the estimated residuals' variance $\hat{\sigma}_n^2$) were designed on the simulated data. Assuming a nominal false-alarm probability $\alpha' = 1\%$, the Type I error probability of each chart was set equal to $\alpha = 1 - \sqrt{1 - \alpha'} = 0.5012\%$. Given that $C = 6$ and $N = 100$ in our case, the condition suggested in Williams et al. (2006), i.e., $N > 2(C + [C(C+1)/2])$, to apply the asymptotic control limit given in Equation (8.4) holds. Therefore, the control limit of the T^2 control chart is given by $UCL = \chi^2_{0.005012,6} = 18.5416$. Figure 8.7 depicts the T^2 control chart which indicates no out-of-control profiles.

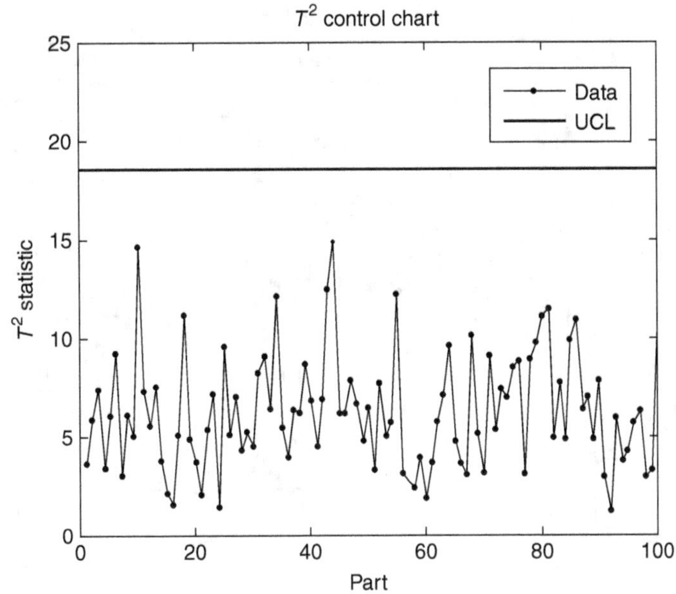

Figure 8.7 T^2 control chart for Phase I of 100 roundness profiles (simulated data).

Similarly, in order to monitor the residuals' variance, a control chart was implemented on the set of 100 simulated roundness profiles. The control limits in Equation (8.7) were used for this control chart. An estimate of the residuals' variance equal to 8.4717×10^{-7}, which was obtained as mean of the residuals' variance for each of the $N = 100$ simulated profile, was considered for the limits. The numerical values $UCL = 9.7539 \times 10^{-7}$ and $LCL = 7.2935 \times 10^{-7}$ were obtained. Figure 8.8 depicts the σ^2 control chart which indicates no out-of-control profiles.

8.2.2 The PCA-Based Approach

Ramsay and Silverman (2005) presented an extension of PCA to functional data, i.e., an approach that allows one to find a set of orthonormal functions (also called functional principal components—PCs), so that the original data can be approximated in terms of a linear combination of these basic functions. In particular, Ramsay and Silverman (2005) showed that, in the case of equally spaced observations, the easiest way to compute the PCs consists of modeling the curve data sampled at regular intervals as a multivariate vector, and performing a traditional PCA on the set of samples collected over different curves.

When the PCA outlines a set of significant PCs to be retained, the coefficients (or loadings) defining these significant PCs can be interpreted as eigenfunctions (also called empirical orthogonal functions). These eigenfunctions do not have a parametric expression and are empirical, since they are obtained by the data at hand. A rough sketch of how PCA works on profile data is discussed below.

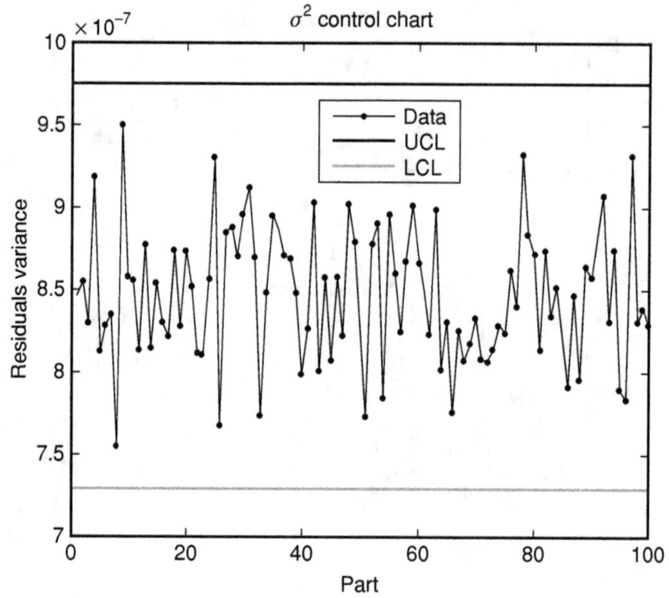

Figure 8.8 Shewhart control chart for monitoring the residual variance of Phase I roundness profiles (simulated data).

Assume we organize N vectors of $P \times 1$ profile data $\mathbf{y}_p$ into an $N \times P$ matrix $\mathbf{Y}$ whose nth row is the transpose of the nth data vector $\mathbf{y}_n$. PCA consists of performing a spectral decomposition of the covariance matrix of $\mathbf{Y}$. The covariance matrix describes the variability of the data observed at each location with respect to the mean value observed at the same location in all the profiles.

Let $\mathbf{S}$ be the covariance matrix, i.e., $\mathbf{S} = [1/(N-1)] \sum_{n=1}^{N} (\mathbf{y}_n - \bar{\mathbf{y}})(\mathbf{y}_n - \bar{\mathbf{y}})'$, where $\bar{\mathbf{y}} = (1/N) \sum_{n=1}^{N} \mathbf{y}_n$ is the sample mean profile, the spectral decomposition consists of finding the $P \times P$ matrices $\mathbf{U}$ and $\mathbf{L}$ that satisfy the following relationship:

$$\mathbf{U}'\mathbf{S}\mathbf{U} = \mathbf{L}, \tag{8.8}$$

where $\mathbf{L}$ is a diagonal matrix that contains the eigenvalues of $\mathbf{S}$ (say l_p), while $\mathbf{U}$ is an orthonormal matrix whose pth column $\mathbf{u}_p$ is the pth eigenvector of $\mathbf{S}$ (the so-called loadings).

With reference to the nth profile $\mathbf{y}_n$, denote by $\mathbf{z}_n$ the vector

$$\mathbf{z}_n = \mathbf{U}'(\mathbf{y}_n - \bar{\mathbf{y}}) = \begin{bmatrix} z_{n1} & \dots & z_{np} & \dots & z_{nP} \end{bmatrix}', \tag{8.9}$$

where z_{np} are the so-called scores. Each profile can then be expressed as a linear combination of loadings $\mathbf{u}_n$, where the weights of the linear combination are the scores z_{np}

$$\mathbf{y}_n = \bar{\mathbf{y}} + z_{n1}\mathbf{u}_1 + \cdots + z_{np}\mathbf{u}_p + \cdots + z_{nP}\mathbf{u}_P. \tag{8.10}$$

Since the PCs are statistically uncorrelated and each PC has variance equal to the corresponding eigenvalue (l_p), we can rank the PCs according to the associated eigenvalue and decide to retain just the most important ones (i.e., the ones which are associated with larger variances). Different approaches can be used to select the proper set of PCs (Jolliffe 2002; Jackson 2003). For instance, cross-validation can be effectively used to choose the number M of significant PCs (Colosimo and Pacella 2007). When a subset M of the whole number of P PCs is retained ($M < P$), the original profile can be estimated as follows:

$$\hat{\mathbf{y}}_{n(M)} = \bar{\mathbf{y}} + z_{n1}\mathbf{u}_1 + \cdots + z_{nm}\mathbf{u}_m + \cdots + z_{nM}\mathbf{u}_M. \tag{8.11}$$

Similar to the regression-based approach, also in the case of PCA a T^2 control chart can be used for monitoring the vector of the first M retained PCs. In this case, the T^2 statistic is given by (Jackson 2003)

$$T_n^2 = \frac{z_{n1}^2}{l_1} + \cdots + \frac{z_{n2}^2}{l_m} + \cdots + \frac{z_{nM}^2}{l_M}. \tag{8.12}$$

If the profile changes in a direction orthogonal to that of the first M PCs, previous T^2 control chart will not be able to issue an alarm. For this reason, another control chart based on the Q statistic (sometimes referred to as the squared prediction error or SPE control chart) also has to be used (Jackson 2003). Given the estimate in Equation (8.11), the Q statistic can be computed as the sum of the squared errors obtained by reconstructing each observation by the first M PCs:

$$Q_n = \left(\mathbf{y}_n - \hat{\mathbf{y}}_{n(M)}\right)' \left(\mathbf{y}_n - \hat{\mathbf{y}}_{n(M)}\right). \tag{8.13}$$

The upper control limit of the T^2 statistics in Equation (8.12) can be computed as follows (Williams et al. 2006).

$$UCL = \chi^2_{\alpha,M}, \tag{8.14}$$

where $\chi^2_{\alpha,M}$ is the $100\,(1-\alpha)$ percentile of the chi-square distribution with M degrees of freedom. With reference to the Q statistic, according to Nomikos and MacGregor (1995) the upper control limit can be computed as

$$UCL = g\chi^2_{\alpha,h}, \tag{8.15}$$

where g and h can be estimated as $\hat{g} = \hat{\sigma}_Q^2/\left(2\bar{Q}\right)$, $\hat{h} = 2\bar{Q}^2/\hat{\sigma}_Q^2$ while $\bar{Q}$ and $\hat{\sigma}_Q^2$ are the sample mean and the sample variance obtained by computing the Q statistics via Equation (8.13) for the set of the N profiles.

8.2.2.1 Application of the PCA-Based Approach

With reference to the aforementioned data set of $N = 100$ simulated roundness profiles, we used a cross-validation approach in order to determine the number of

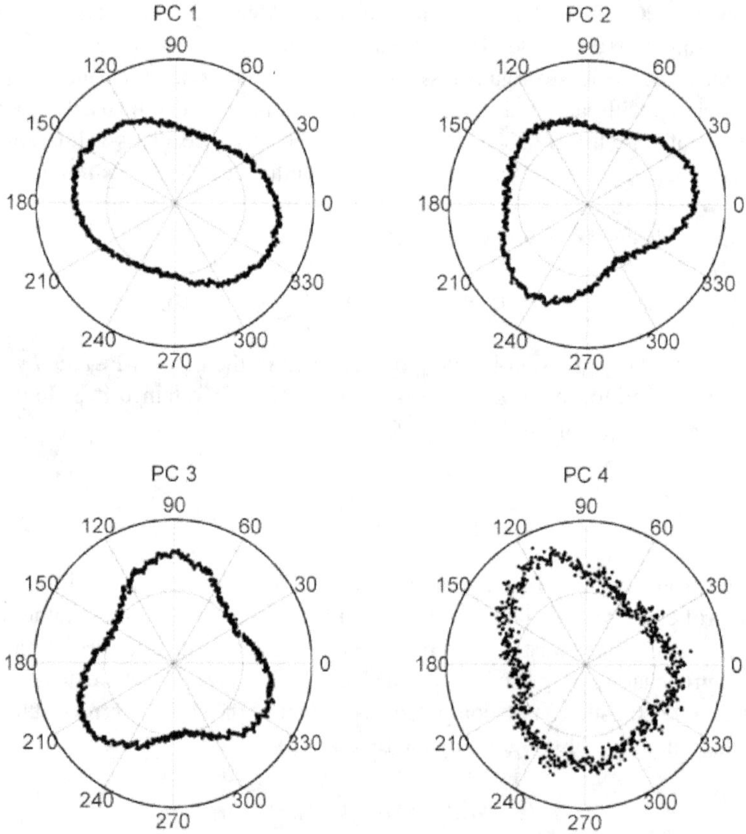

Figure 8.9 First four eigenfunctions related to the four retained PCs (computed on simulated data).

significant PCs to be retained (Colosimo and Pacella 2007). With reference to the simulated roundness profiles the number of significant PCs is equal to $M = 4$.

Once the PCA has been performed, the retained PCs should be interpreted to gain more insight into the systematic pattern characterizing the machined profiles. To this aim, each eigenfunction $\mathbf{u}_p$ (i.e., the coefficients of each eigenvector also known as loadings) can be graphically presented as a function of the location. Following this practice, Figure 8.9 depicts the polar diagrams of the first four eigenfunctions $(\mathbf{u}_1, \mathbf{u}_2, \mathbf{u}_3, \mathbf{u}_4)$, which are related to the four retained PCs. The first PC, which describes the most important component of variability, presents a bi-lobe form. This qualitative observation indicates that the main variability around the mean profile is due to a periodic function characterized by a frequency of 2 upr. The second PC is a mixture of a bi-lobe and a tri-lobe contour. This mixture is obtained by combining two periodic functions, namely, a 2 and a 3 upr harmonic. The third PC presents a tri-lobe contour, while the fourth PC is again a mixture of a bi-lobe and a tri-lobe contour.

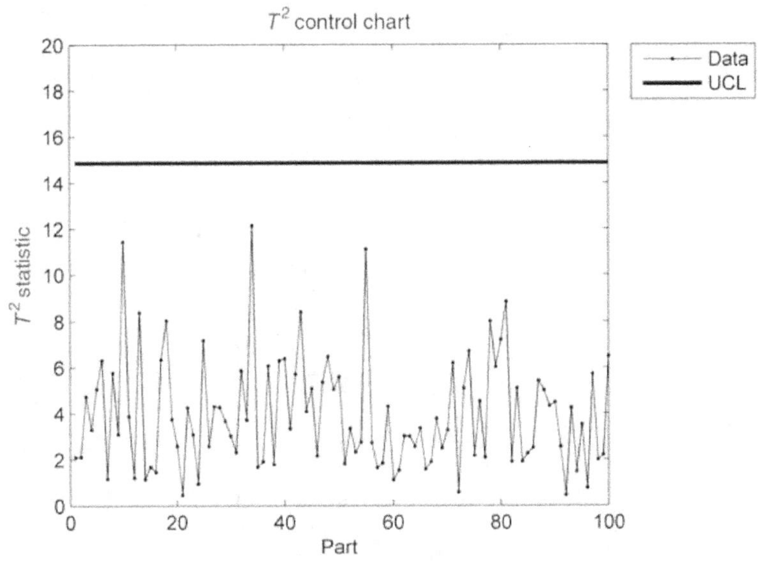

Figure 8.10 T^2 control chart for Phase I of 100 roundness profiles (simulated data).

Given the PCA model, a multivariate control chart on the scores $(z_{n1}, z_{n2}, z_{n3}, z_{n4})$ and a univariate control chart on the Q statistic have to be designed. Assuming a nominal false-alarm probability $\alpha' = 1\%$, the Type I error probability of each chart is set equal to $\alpha = 1 - \sqrt{1-\alpha'} = 0.5012\%$. The control limit of the T^2 control chart in Equation (8.14) is $UCL = \chi^2_{0.005012,4} = 14.8546$. The UCL of the Q control chart is in Equation (8.15) and is about equal to 1.325×10^{-3}.

Figure 8.10 depicts the PCA-based T^2 control chart, which indicates no out-of-control conditions for the simulated profiles. Similarly, Figure 8.11 depicts the Q control chart, which indicates two out-of-control conditions for the 52nd and 54th profile, where the Q statistic resulted slightly above the UCL. As we dealt with simulated in-control profiles no special causes existed to explain these out-of-controls. Therefore, they can be considered false alarms released by the Q control chart and probably due to the different structure of the error terms' correlation, which indeed changes from profile to profile.

8.3 SIMPLE APPROACHES FOR MONITORING MANUFACTURED PROFILES: THE INDUSTRIAL PRACTICE

8.3.1 Control Chart on Geometric Errors

When quality of a manufactured product is related to a geometric specification (represented by a tolerance value), the Shewhart's control chart can still be used for process monitoring. In fact, the information related to a geometric feature must be summarized in just one synthetic variable to decide whether the machined item

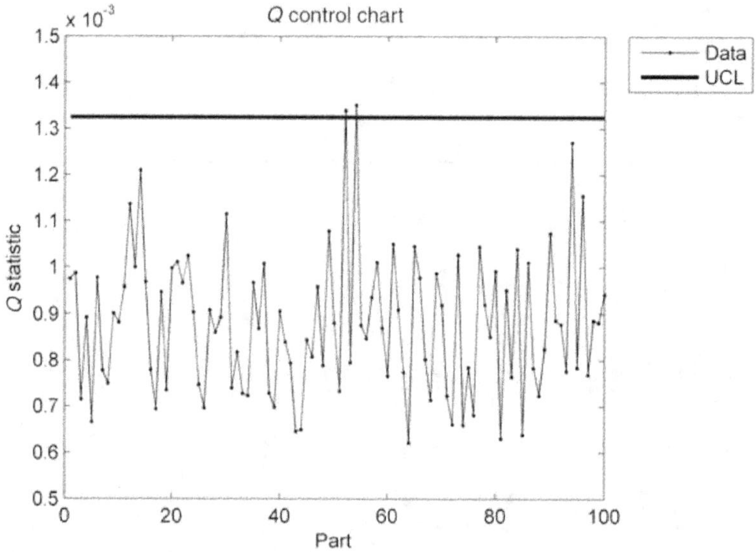

Figure 8.11 Q control chart for Phase I of 100 roundness profiles (simulated data).

has to be scrapped/reworked or can be considered conforming to the requirement. In the case of a circular profile, the OOR value is computed. Then, the circular feature is considered conforming to the requirement if the corresponding OOR is not greater than the tolerance value.

Since the OOR values have to be computed for quality inspection of manufactured items, these values can also be exploited for process monitoring. Indeed, the simplest approach for process monitoring consists of designing a Shewhart's control chart of the OOR values. In order to exemplify this approach, the set of 100 roundness profiles previously obtained from simulation is considered as reference. Table 8.1 reports the OOR values (LS method) obtained for each item. The implementation developed by NPL (the UK's National Measurement Laboratory) and based on the Least Squares Geometric Elements (LSGE) library for MATLAB was considered (Forbes 1994).

With reference to the 100 samples in Phase I, summarized in Table 8.2, an individual's control charts on the OOR values was designed. The Anderson–Darling test was implemented in order to detect any departures from normality for the distribution function of the 100 OOR values. From numerical computation on the data reported in Table 8.1, the set of 100 OORs resulted to be normally distributed (p-value equal to 0.559). Therefore, a control chart for individuals was designed for the sequence of sample OOR values (with a nominal false-alarm probability $\alpha = 1\%$). This control chart, in which no out-of-control alarms are released, is showed in Figure 8.12.

8.3.2 The Location Control Chart

A different approach, aimed at combining simplicity with the need of keeping all the information of data observed at each location of the machined feature is

Table 8.2 OOR Values (100 Simulated Profiles) on the basis of the LS Algorithm; Scale OOR (mm)

	OOR		OOR		OOR		OOR		OOR
1	0.0143	21	0.0125	41	0.0150	61	0.0154	81	0.0137
2	0.0164	22	0.0099	42	0.0127	62	0.0163	82	0.0128
3	0.0137	23	0.0144	43	0.0169	63	0.0142	83	0.0074
4	0.0155	24	0.0135	44	0.0097	64	0.0129	84	0.0126
5	0.0127	25	0.0180	45	0.0126	65	0.0146	85	0.0087
6	0.0139	26	0.0114	46	0.0106	66	0.0100	86	0.0101
7	0.0147	27	0.0133	47	0.0169	67	0.0118	87	0.0129
8	0.0069	28	0.0075	48	0.0119	68	0.0144	88	0.0137
9	0.0158	29	0.0138	49	0.0179	69	0.0142	89	0.0177
10	0.0086	30	0.0110	50	0.0112	70	0.0175	90	0.0176
11	0.0114	31	0.0147	51	0.0150	71	0.0120	91	0.0119
12	0.0111	32	0.0142	52	0.0131	72	0.0112	92	0.0116
13	0.0169	33	0.0148	53	0.0096	73	0.0169	93	0.0139
14	0.0098	34	0.0190	54	0.0118	74	0.0060	94	0.0102
15	0.0137	35	0.0124	55	0.0126	75	0.0134	95	0.0144
16	0.0100	36	0.0106	56	0.0092	76	0.0123	96	0.0112
17	0.0089	37	0.0132	57	0.0125	77	0.0132	97	0.0181
18	0.0179	38	0.0124	58	0.0099	78	0.0198	98	0.0135
19	0.0128	39	0.0120	59	0.0136	79	0.0120	99	0.0127
20	0.0088	40	0.0142	60	0.0106	80	0.0110	100	0.0153

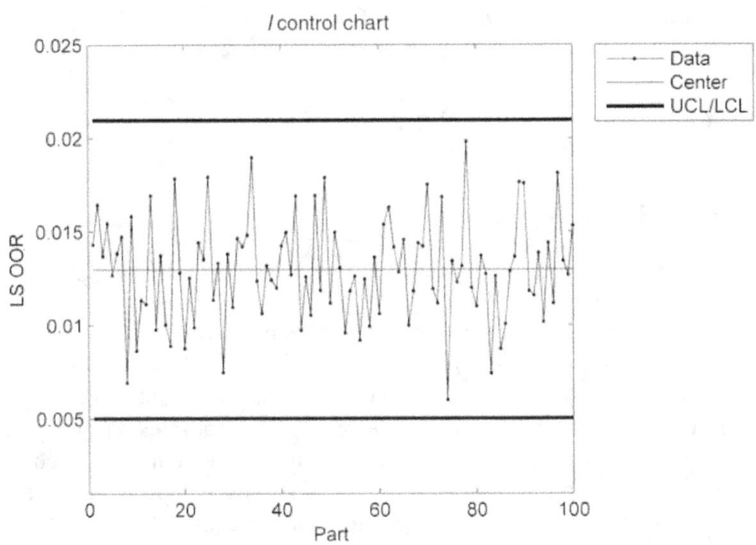

Figure 8.12 Individuals control charts of the transformed OOR values based on the LS algorithm; vertical axis scale (mm).

the *location control chart*. This approach, which was presented in Boeing (1998, pp. 89–92), consists of applying a Shewhart's control chart separately to each data point observed at a given location of the part. The rationale behind this approach is that, if the observed shape is in-control, the data observed at that specific location should stay within that interval with a given probability. On the other hand, when the process goes out-of-control, it is likely that the control interval will be violated at one or more locations.

In order to design the location control chart, the first step consists of identifying the center of each interval, i.e., the systematic pattern of the in-control shape. This reference is usually estimated as the average of all the in-control data observed at each location. Starting from the mean shape, the location control chart (LOC CC) consists of computing the upper and lower control limits at each location, by using the usual approach that place the limits at $\pm K$ standard deviations from the sample mean. According to this method, an alarm is issued when at least one point, in the whole set of data observed, exceeds the control limits.

Due to its inner simplicity, this chart can be easily applied in industrial practice (and in fact its origin is in Boeing with reference to applications in which numerous measurements of the same variable, e.g., a dimension such as thickness, are made at several locations on each manufactured part). However, since the control limits used at each location depend on the responses at that specific position only, the main disadvantage with this method is that the multivariate structure of data is completely ignored. The only form of relationship between control intervals at each location is a constraint on the false alarm, as discussed in Subsection 8.3.2.1.

8.3.2.1 Design of the Location Control Chart

The control limits for location of index p are as follows:

$$\begin{aligned} UCL(p) &= \bar{y}(p) + Z_{\alpha/2} s(p), \\ CL(p) &= \bar{y}(p), \\ LCL(p) &= \bar{y}(p) - Z_{\alpha/2} s(p), \end{aligned} \tag{8.16}$$

where $\bar{y}(p) = \frac{1}{N}\sum_{n=1}^{N} y_n(p)$ and $s(p) = \sqrt{\frac{1}{N-1}\sum_{n=1}^{N}[y_n(p) - \bar{y}(p)]^2}$ are, respectively, the sample mean and the sample standard deviation of data observed at location p, while $Z_{\alpha/2}$ represents the $(1-\alpha/2)$ percentile of the standard normal distribution. The constant $K = Z_{\alpha/2}$ of the control band is computed as a function of the required false alarm rate. However, given that P dependent control rules are simultaneously applied, the percentile of the standardized normal distribution used to compute K should be corrected. To this aim, the Bonferroni's rule for dependent events is used to attain an actual false alarm rate not greater than a predefined value. Let α' denote the upper bound of the false alarm rate the value $\alpha = \alpha'/P$ is used for designing the P control limits in Equation (8.16).

It is worth noting that different procedures can also be used, such as the Simes' modified Bonferroni procedure (Simes 1986). Colosimo and Pacella (2010) showed

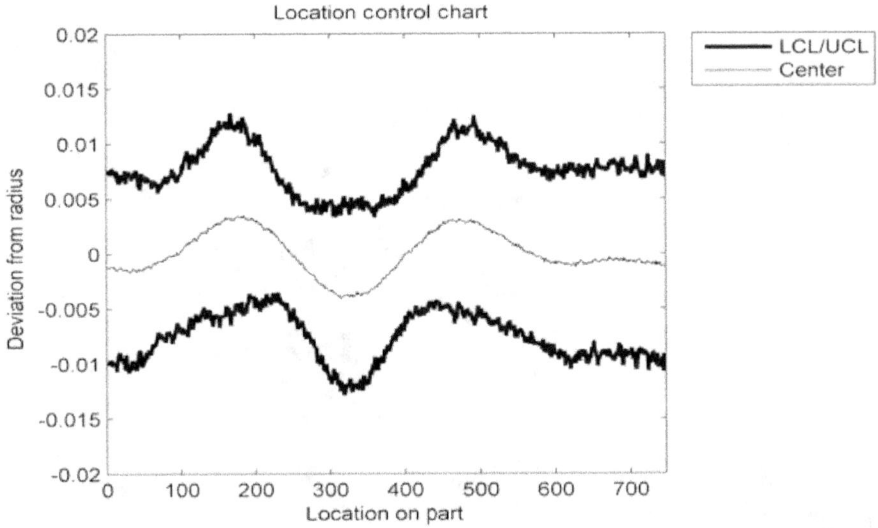

Figure 8.13 LOC CC (748 locations) with reference to the 100 simulated profiles; actual false alarm rate not greater than 0.01; vertical axis scale (mm).

that when compared with the standard Bonferroni's method, the Simes' procedure does not produce significant effects on the false alarm rate of the location control chart. Furthermore, since the Simes' procedure does not allow the graphical representation of the control region as the Bonferroni's procedure does, this latter method is the one considered in practice.

Figure 8.13 shows the average profile and the control band of the simulated roundness profiles. The Bonferroni's rule for dependent events is used to attain an actual false alarm rate not greater than a predefined value. In particular, assuming the standard value $\alpha' = 0.01$ as upper bound for the Type I error probability, the value $\alpha = 1.3369 \times 10^{-5}$ (i.e., $Z_{\alpha/2} = 4.354$) is used for designing the 748 control limits in Equation (8.16).

Figure 8.14 shows one out of the 100 profiles depicted against the control limits of the location control chart. The profile is plotted against this control region with the advantage of allowing a simple identification of the locations where problems arise. In this specific case, the profile is considered in-control. Indeed, from the visual inspection it appears that the profile behavior is close to the average common profile and there is no apparent discrepancy in the shape of the profile when compared to the center line of the location control chart.

8.4 PERFORMANCE COMPARISON

In order to allow practitioners to select a specific approach for monitoring functional data in a given situation, the present section provides a numerical comparison based on simulation of (1) the LOC CC; (2) the individuals' control chart of the geometrical

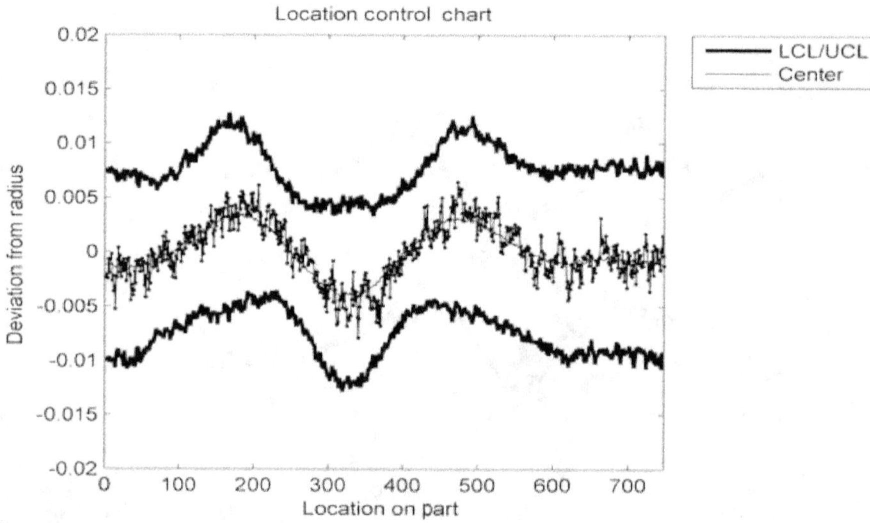

Figure 8.14 One out of the 100 profiles of the reference case study depicted against the control limits of the location control chart; vertical axis scale (mm).

errors (OOR CC); (3) the regression-based approach (REG CC); and (4) the PCA-based approach (PCA CC).

The comparison study is based on two simulation scenarios, which are obtained from the real case study of roundness profiles obtained by lathe-turning. The simulation scenarios were designed for representing two different, although realistic, productive situations. Performance is measured both as the ability to obtain a predefined false alarm rate in the design phase of the control chart (Phase I) as well as to detect unusual patterns in the functional data during the operating phase (Phase II).

For Phase I analysis, only groups of uncontaminated process samples were considered in this study. In-control profiles were obtained by simulation, while the performance of the competing methods were compared in terms of the probability of obtaining at least one statistic outside the control limits when performing control charting using the set of simulated samples. Control chart parameters estimated from these Phase-I samples were not used in the subsequent Phase-II comparison study, where performance comparison was based on the ideal assumption that the in-control parameters are known. Indeed, computer simulation was used to obtain a large data set of in-control profiles in order to estimate as closely as possible control chart's parameters.

Basically, the objective in this section is to investigate situations where each approach should be preferred to the others, thus to provide some guidelines for implementing profile monitoring in actual applications.

8.4.1 Production Scenarios Under Study

In this section, the focus is on two different production scenarios. On the one hand, the first production scenario mimics the real case study where the C-length parameter vector $\mathbf{c}_n$ changes from profile to profile according to a C-variate normal distribution of mean $\boldsymbol{\mu}$ and covariance matrix $\boldsymbol{\Sigma}$, i.e., $\mathbf{c}_n \sim N(\boldsymbol{\mu}, \boldsymbol{\Sigma})$. The actual values of the mean vector $\boldsymbol{\mu}$ and of the covariance matrix $\boldsymbol{\Sigma}$ used for the simulation of this scenario are summarized in Table 8.1.

On the other hand, the second scenario is obtained by perturbing the SARX model in Equation (8.1) with reference to the variability that characterizes the C-length parameter vector $\mathbf{c}_n$. In particular, a null matrix is considered as covariance matrix ($\boldsymbol{\Sigma} = \mathbf{0}$) for the C-variate normal distribution of model parameters. Hence, a fixed vector of parameters is used for each instance ($\mathbf{c}_n = \boldsymbol{\mu}$).

The production scenario where $\mathbf{c}_n$ may change from profile to profile is referred to as *random-effect* model. The production scenario, in which parameters $\mathbf{c}_n$ do not change, is referred to as *fixed-effect* model. It is worth noting that random-effect and fixed-effect are referred to both the large-scale and small-scale variability components in Equation (8.1).

A fixed-effect model is usually assumed in traditional approaches for profile monitoring (Woodall et al. 2004). In manufacturing, assuming that the input material is very stable and homogeneous, profile-to-profile variability of the small-scale component (spatial structure) can probably be neglected. Furthermore, assuming a process more stable and/or more controlled, also the profile-to-profile variability in the large-scale component (harmonic structure) tends to vanish. In other words, while the first scenario with random effects corresponds to a common machining process, in which natural causes of variability affects the variability in both the parameters and residuals of the model, the second scenario with fixed effects can be considered representative of a process more stable and/or more controlled in which natural causes can affect only the residuals of the model.

8.4.2 Phase I: Performance Comparison

In Phase I, a set of N process samples is analyzed in order to both evaluate the stability of the process and estimate the in-control state's parameters. During Phase I, the performances of the competing control charts are compared in terms of the probability of deciding whether or not the process is stable. This is the probability of obtaining at least one statistic outside the control limits when performing control charting using the set of N process samples.

Given a desired false alarm or Type I error rate α', assumed equal to $\alpha' = 1\%$ in the following, the competing approaches can be compared in terms of the probability of actually achieving this nominal value.

To compare the performance of the alternative approaches, we considered 10000 replicates of Phase I control charting for different values of the number of profiles used in this phase, specifically $N = 50$, 100, 150, and 200. We recorded the false alarm rate in each replicate (computed as the number of out-of-control signals divided by N) and then we stored the average false alarm rate obtained in the whole set of 10000 replicates.

Table 8.3 summarizes the actual Type I error rates for each of the competing methods in the two simulation scenarios considered in our work. For the two production

Table 8.3 Phase I Simulation Results; Average type I Error Rate and Corresponding Standard Deviations within Brackets

	Random-Effect Model				Fixed-Effect Model			
N	LOC CC	OOR CC	REG CC	PCA CC	LOC CC	OOR CC	REG CC	PCA CC
50	0.070% (0.004%)	0.815% (0.013%)	1.048% (0.014%)	1.210% (0.015%)	0.048% (0.003%)	1.124% (0.015%)	1.049% (0.014%)	0.935% (0.014%)
100	0.283% (0.005%)	0.833% (0.009%)	1.037% (0.010%)	1.495% (0.012%)	0.208% (0.005%)	1.167% (0.011%)	1.013% (0.010%)	1.022% (0.010%)
150	0.391% (0.005%)	0.836% (0.007%)	1.044% (0.008%)	1.562% (0.010%)	0.306% (0.005%)	1.167% (0.009%)	0.985% (0.008%)	1.048% (0.008%)
200	0.454% (0.005%)	0.828% (0.006%)	1.031% (0.007%)	1.607% (0.009%)	0.353% (0.004%)	1.186% (0.008%)	0.988% (0.007%)	1.065% (0.007%)

scenarios under study it can be observed that the false alarm rates produced by the LOC CC are always lower than the nominal value (1%), although the gap between actual and nominal rate decreases as the number N of profiles used in Phase I increases. This is an expected result as the Bonferroni's inequality allows one to set only an upper bound on the actual false alarm rate. Note that a reduced false alarm rate could seem an advantage at first sight. However, it means that control limits are too far from the center line, thus, resulting in the ineffective detection of out-of-control profiles when they arise.

The false alarm rates produced by the OOR CC are smaller than the nominal value in the case of random-effect model, while they are greater than the nominal value in the case of fixed-effect model. These results can be mainly ascribed to a slight departure from normality observed for the OOR values, where the departure depends on the specific scenario (random-effects or fixed-effects). It is also worth noting that, in the case of the individuals control chart on the geometrical errors, the false alarm rate of the control chart is not influenced by the number N of profiles used in Phase I.

In contrast to the previous two approaches, the actual false alarm rates obtained by the regression-based approaches are much closer to the nominal value. In fact, as shown in Table 8.3, in spite of both the specific scenario and the dimension N, there is no practical difference between the actual false alarm rate produced by the REG CC and the nominal value.

The PCA CC is a little more sensitive than the REG CC to the production scenario, since it gives a false alarm closer to the nominal value in the case of fixed-effect model but not in the case of random-effect. With reference to this last scenario, the PCA CC consists of two control charts. A multivariate control chart based on the first four retained PCs and a univariate control chart for monitoring the residuals (the Q chart). The excessive false alarm rate produced is mainly due to the Q chart. As a matter of fact, Colosimo and Pacella (2010) showed that the upper control limit for the Q statistic proposed by Nomikos and MacGregor (1995) is substantially adequate when an independently distributed error term is considered in the model, as well as in the case of correlated errors with fixed effects in the small-scale component of the model. However, such a limit is not suitable in the case of correlated errors with random effects in the small-scale component of the model.

Finally, with reference to the Phase I simulation results reported in Table 8.3, it is fair to conclude that regression-based control charts can easily be designed, since the actual false alarm rate can be achieved in practice by using an analytic computation of the control limits. Similar conclusions can also be drawn with reference to the PCA CC , but only in the specific production scenario of fixed effects. On the contrary, the design of an LOC CC can be a difficult task, since the actual false alarm rate can be different from the expected rate. In the case of the individuals control chart on the geometrical errors, we observed that a departure from normality may also make difficult the design of this control chart. However, as we are monitoring non-negative values (the OOR values), if a proper power transformation (Box–Cox transformation) can be identified, we expect that the design of this individuals control chart should be simplified.

8.4.3 Phase II: Performance Comparison

The objective in Phase II is to quickly detect any change in the process from its in-control state. The monitoring approaches are compared in terms of the average run length (ARL), where the run length is defined as the number of samples taken until an out-of-control signal is issued.

In order to evaluate performance in Phase II, occurrences of assignable causes are simulated by a total of three out-of-control conditions. These out-of-controls are simulated by spindle-motion errors (Cho and Tu 2001), which are modeled by introducing a spurious harmonic in the baseline model of roundness profile data. Each condition is then characterized by a parameter directly proportional to the severity of the out-of-control introduced in the baseline model. In particular, denoting by $y_n(p)$ the measurement of index $p = 1, \ldots, P$ on the profile of index $n = 1, 2, \ldots$, the out-of-controls are simulated according to the following three models:

half-frequency spindle-motion error, which is modeled as:

$$y_n(p) + \sqrt{\frac{2}{P}}\delta_1 \sin\left(\frac{1}{2}\theta_p\right), \tag{8.17}$$

where δ_1 is the size of the shift ($\delta_1 = 0.1, 0.15, 0.2, 0.25$);

bi-lobe out-of-control, which is simulated by incrementing the amplitude of the second harmonic in the baseline model as follows:

$$y_n(p) + \sqrt{\frac{2}{P}}\delta_2 \left[c_{1n}\cos\left(2\theta_p\right) + c_{2n}\sin\left(2\theta_p\right)\right], \tag{8.18}$$

where δ_2 is the increasing factor ($\delta_2 = 0.1, 0.2, 0.3, 0.4$);

tri-lobe out-of-control, which is simulated by incrementing the amplitude of the third harmonic with respect to the baseline model as follows:

$$y_n(p) + \sqrt{\frac{2}{P}}\delta_3 \left[c_{3n} \cos\left(3\theta_p\right) + c_{4n} \sin\left(3\theta_p\right)\right], \tag{8.19}$$

where δ_3 is the increasing factor ($\delta_3 = 0.1,\ 0.2,\ 0.3,\ 0.4$).

Data obtained under these Phase II models are also scaled (by subtracting the least squares estimation of the radius) and centered (on the least squares estimation of the center) before applying the profile monitoring method. In fact, we are assuming that centering and scaling are standard steps applied to data when the focus is on the geometrical errors (Cho and Tu 2001).

Performance comparison is based on the ideal assumption that the in-control parameters for each competing method are known. Indeed, computer simulation is used in our work to obtain a large dataset of in-control profiles in order to estimate as closely as possible the parameters of each method. For the two production scenarios, all simulations were conducted by first tuning each competing approach in order to achieve the same in-control ARL value of about 100. Hence, the performance in Phase II are related to the ability of detecting out-of-control profiles, given that all the approaches are designed to achieve the same false-alarm probability approximately equal to 1%.

Tables 8.4 and 8.5 summarize the simulation results for the two production scenarios under study. In particular, Table 8.4 refers to the case of random-effect model, while Table 8.5 refers to the case of fixed-effect model. Each table reports the ARLs estimated by computing a set of 1000 run lengths, given new profiles simulated

Table 8.4 Phase II Simulation Results for the Production Scenario with Random-Effect Model; ARLs and Corresponding Standard Deviations within Brackets (1000 trials)

	Delta	LOC CC	OOR CC	REG CC	PCA CC
Half-frequency	0.1	73.26 (2.18)	98.15 (3.12)	93.18 (2.83)	80.04 (2.56)
	0.15	50.53 (1.57)	97.89 (3.06)	78.68 (2.63)	61.07 (1.90)
	0.2	35.61 (1.16)	85.00 (2.71)	68.49 (2.16)	44.43 (1.37)
	0.25	22.49 (0.72)	63.51 (2.01)	48.50 (1.56)	29.05 (0.89)
Bi-lobe	0.1	64.08 (1.92)	72.39 (2.32)	76.41 (2.42)	64.29 (1.89)
	0.2	36.07 (1.16)	44.66 (1.42)	47.48 (1.60)	38.03 (1.22)
	0.3	24.01 (0.77)	25.69 (0.81)	29.76 (0.92)	21.21 (0.67)
	0.4	14.98 (0.45)	18.40 (0.57)	16.93 (0.53)	12.61 (0.38)
Tri-lobe	0.1	70.44 (2.16)	88.29 (2.80)	72.51 (2.27)	70.08 (2.19)
	0.2	43.66 (1.44)	67.87 (2.18)	47.72 (1.50)	37.55 (1.17)
	0.3	30.33 (0.94)	46.82 (1.49)	27.57 (0.85)	21.70 (0.67)
	0.4	19.43 (0.60)	33.81 (1.04)	17.39 (0.54)	12.60 (0.40)

Table 8.5 Phase II Simulation Results for the Production Scenario with Fixed-Effect Model; Actual ARLs and Corresponding Standard Deviations within Brackets (1000 trials)

	Delta	LOC CC	OOR CC	REG CC	PCA CC
Half-frequency	0.1	33.31 (1.00)	42.47 (1.36)	1.27 (0.02)	6.00 (0.17)
	0.15	12.26 (0.38)	19.71 (0.59)	1.00 (0.00)	1.51 (0.03)
	0.2	4.81 (0.13)	11.21 (0.33)	1.00 (0.00)	1.03 (0.01)
	0.25	2.37 (0.06)	5.83 (0.17)	1.00 (0.00)	1.00 (0.00)
Bi-lobe	0.1	72.11 (2.13)	58.40 (1.86)	6.75 (0.19)	45.89 (1.40)
	0.2	37.70 (1.21)	21.72 (0.69)	1.20 (0.02)	7.02 (0.20)
	0.3	21.47 (0.63)	8.87 (0.27)	1.00 (0.00)	1.71 (0.04)
	0.4	10.63 (0.32)	4.22 (0.12)	1.00 (0.00)	1.05 (0.01)
Tri-lobe	0.1	86.85 (2.72)	82.43 (2.71)	16.70 (0.50)	64.30 (1.94)
	0.2	60.26 (1.92)	38.99 (1.24)	2.44 (0.06)	20.18 (0.57)
	0.3	33.59 (1.04)	20.37 (0.67)	1.11 (0.01)	5.20 (0.15)
	0.4	21.56 (0.66)	10.42 (0.32)	1.01 (0.00)	1.87 (0.04)

according to a specific out-of-control model. Standard deviations estimated for such mean values are also included within brackets.

Since in industrial applications the analyst is not expected to know a priori what kind of out-of-control condition will affect the production process and how severe it will be, we consider a measure of the overall performance for each of the five competing approaches in each production scenario. To do this, we consider the mean ARL values for each competing approach in signaling a generic out-of-control condition of any severity for that production scenario. We are assuming that all out-of-controls previously considered are equally probable and that the analyst knows the model of monitored functional data (this is plausible when a retrospective phase of control charting has been accomplished).

Figures 8.15 and 8.16 graphically depicts the 95% Bonferroni's confidence intervals of the overall ARLs presented by the four competing approaches in each production scenarios considered in our study. A discussion on the performance comparisons is detailed in the following two subsections for the production scenarios with random-effect model and fixed-effect model, respectively.

8.4.3.1 Production Scenario with Random-Effect Model

From the results reported in Table 8.4, graphically summarized in Figure 8.15, it can be observed that the OOR CC presents the lowest power of detection when compared to the competing methods.

Similarly, the REG CC has a small power of detection in signaling out-of-controls. This may be mainly ascribed to the variability in the regression parameters that characterizes this production scenario with random effects for the baseline model. As a matter of fact, it should be noted that for the out-of-control conditions considered in our study (half-frequency, bi-lobe and tri-lobe), the majority of alarms released

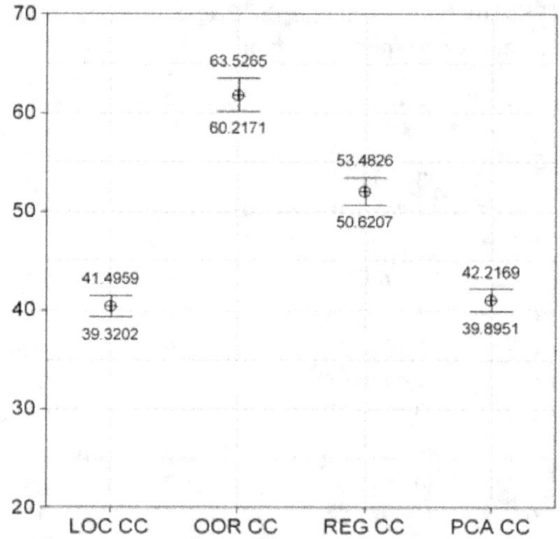

Figure 8.15 95% Bonferroni's confidence intervals of the overall ARL for the competing approaches for the production scenario with random-effect model.

by the REG CC are produced by the multivariate control chart on the vector of fitted parameters. The extra variability in the regression parameters, which naturally characterizes the vector of fitted parameters in this scenario, causes a lower detection power of the regression-based control charts, in particular of the multivariate control chart.

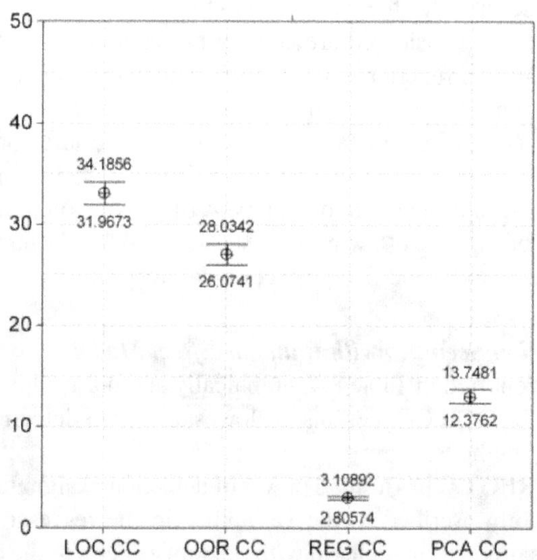

Figure 8.16 95% Bonferroni's confidence intervals of the overall ARL for the competing approaches for the production scenario with fixed-effect model.

On the other hand, it can be noted that the PCA CC presents, in many cases, outperforming performance when compared with the regression-based approach. As previously mentioned, the PCA CC consists of a multivariate control chart based on the first four retained PCs and of a univariate control chart for monitoring the residuals variability (the Q control chart).

Furthermore, from Table 8.4 it can be also noted that the LOC CC presents comparable performance to that observed for the PCA CC . In a few of cases, especially in the case of half-frequency out-of-control, surprisingly the LOC CC outperforms the PCA CC. This result shows that the simple LOC CC can be considered a valuable alternative to parametric methods for profile monitoring, at least in a production scenario with random effects.

8.4.3.2 Production Scenario with Fixed-Effect Model

In the case of a fixed-effect model as reference production scenario (results in Table 8.5 and Figure 8.16), each of the competing approach presents better performance in signaling any kind of out-of-control condition when compared with the production scenario with random-effect model. This can be easily explained by observing that no extra variability is naturally affecting the in-control profiles.

The REG CC outperforms all of the other competing methods, even if in a few cases, especially when a high severity of the out-of-control condition is considered, the regression-based and the PCA CC may have comparable performance.

Note that for the out-of-control conditions considered in our study (half-frequency, bi-lobe, and tri-lobe), the majority of alarms released by the REG CC are produced by the multivariate control chart on the vector of fitted parameters, while the PCA CC consists of a univariate control chart only (a Q control chart). As a matter of fact, when PCA is performed in the case of a fixed-effect model, no significant PCs are identified (Colosimo and Pacella 2007) as the PCA is performed after data centering and this first step consists of subtracting the mean pattern (described by the fixed-effect model) from each profile data. Thus, PCA is performed just on the error terms and hence no significant PC is correctly reported. Also, note that in this case the Q statistic is given by the sum of the squared difference between data observed at each location and the average profile at that location.

It can be noted that the LOC CC presents the lowest overall power of detection when compared to the competing methods. Also the OOR control chart does not present a better performance than those that characterize the model-based control chats (regression-based and PCA-based).

8.5 MOVING FROM 2D PROFILES TO 3D SURFACES

In the case of cylindrical features, Henke et al. (1999) presented an analytical model for aiding the interpretation of the relationship between manufacturing processes and the typical deviation from the ideal cylinder. The approach combines Chebyshev polynomials and sinusoidal functions for representing axial errors and circular errors, respectively. Subsequently, a similar approach was proposed by Zhang et al. (2005),

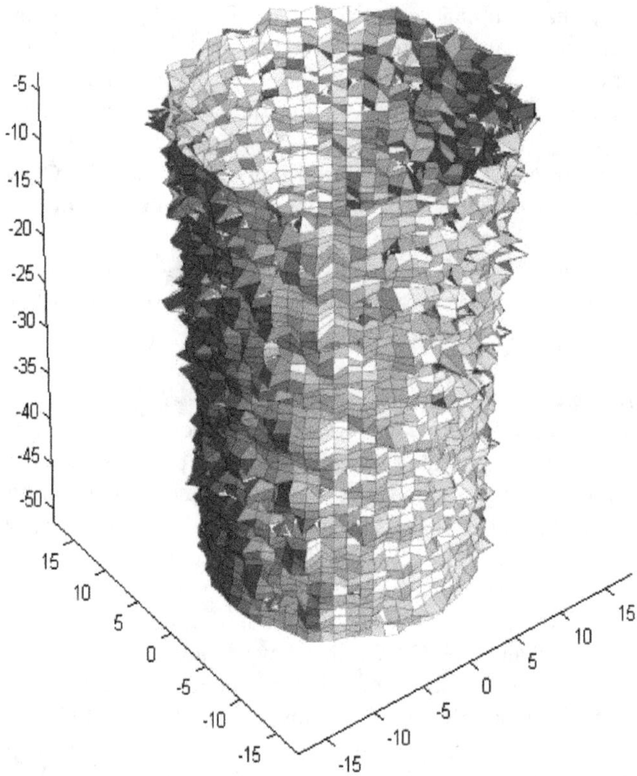

Figure 8.17 The actual shape of one out of the 100 machined cylinders.

who used the Legendre polynomials instead of the Chebyshev ones to describe the deviation along the cylinder axis, while keeping the sinusoidal functions for describing the cross-section form errors. Regression models presented in the literature on cylindrical specifications can be considered as starting reference to identify the parametric models of the large-scale pattern characterizing all the cylindrical surfaces machined. Figure 8.17 shows the shape of one real cylinder obtained by lathe-turning (Colosimo et al. 2010) while Figure 8.18 shows some typical form errors associated to machined surfaces (Henke et al. 1999 and Zhang et al. 2005).

Similarly to what was done for 2D profiles related with circularity, a parametric model for the machined cylinder can be obtained by merging a large-scale pattern

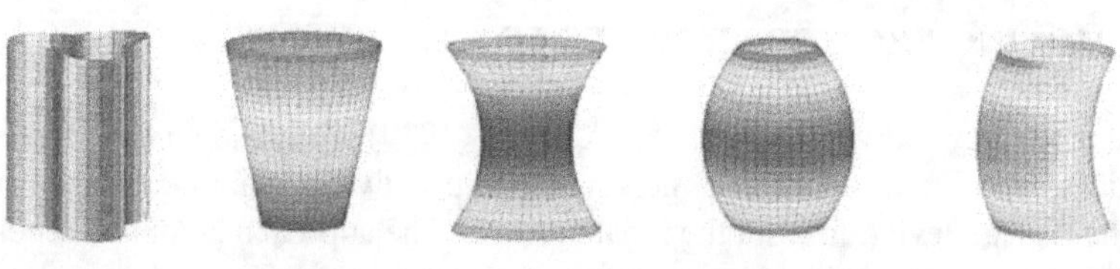

Figure 8.18 Typical form errors in manufactured cylindrical surfaces (from left to right: three-lobed, taper, hourglass, barrel, banana).

Table 8.6 The Regressor Functions as a Function of the Location Indices (ζ_q, θ_p) Shown in Figure 8.3

	Order of the Chebyshev Polynomial.	Order of the Periodic Component	Regressor
1	0	2	$T_0(\zeta_q)\cos(2\theta_p)$
2	0	2	$T_0(\zeta_q)\sin(2\theta_p)$
3	0	3	$T_0(\zeta_q)\cos(3\theta_p)$
4	0	3	$T_0(\zeta_q)\sin(3\theta_p)$
5	1	0	$T_1(\zeta_q)$
6	1	2	$T_1(\zeta_q)\cos(2\theta_p)$
7	1	2	$T_1(\zeta_q)\sin(2\theta_p)$
8	1	3	$T_1(\zeta_q)\cos(3\theta_p)$
9	1	3	$T_1(\zeta_q)\sin(3\theta_p)$
10	2	0	$T_2(\zeta_q)$
11	2	1	$T_2(\zeta_q)\cos(\theta_p)$
12	2	1	$T_2(\zeta_q)\sin(\theta_p)$
13	2	2	$T_2(\zeta_q)\cos(2\theta_p)$
14	2	2	$T_2(\zeta_q)\sin(2\theta_p)$
15	2	3	$T_2(\zeta_q)\cos(3\theta_p)$
16	2	3	$T_2(\zeta_q)\sin(3\theta_p)$

with a small-scale one. The large-scale component can be expressed as a combination of functions such as $T_i(\zeta_q)\cos(h\theta_p)$ or $T_i(\zeta_q)\sin(h\theta_p)$ i.e., as a combination of sinusoidal functions with frequency equal to h upr, multiplied by a second-type Chebyshev polynomial of order $i (i = 0, 1, 2)$, i.e., $T_0(\zeta_q) = 1$, $T_1(\zeta_q) = 2\zeta_q$, $T_2(\zeta_q) = 4\zeta_q^2 - 1$, where $\zeta_q = 2\frac{q-1}{Q-1} - 1$.

From Table 8.6, which summarizes the expression of the regressor functions, it can be observed that some of the possible combinations of the Chebyshev polynomials and the periodic functions are not included as possible regressor functions. As examples, $T_0(\zeta_q)$, representing the (least-square) cylinder radius, or $T_0(\zeta_q)\cos(\theta_p)$ and $T_0(\zeta_q)\sin(\theta_p)$, associated to the translation of the (least-square) cylinder axis, are not included in this table. Indeed, the large-scale component describes the deviation of the observed data from the ideal cylinder. By definition of form error, only the deviation from a perfect cylinder has to be modeled and this is why polynomials aimed at representing the cylinder radius or position can not be included in the cylindiricity form error model.

As for the small-scale component, i.e., the error terms, they can be considered to be uncorrelated and normally distributed. However, if the assumption of uncorrelated residuals is rejected, models able to deal with spatial correlation structures should be entailed. Colosimo et al. (2010) used approaches taken from spatial statistics to deal with the issue of possibly correlated errors. In particular, they assumed that the grid of equally spaced points observed on the machined surface is kept fixed, as it were implemented on a coordinate measuring machine that automatically performs the required measurement path. Given this fixed grid of measurement locations, each

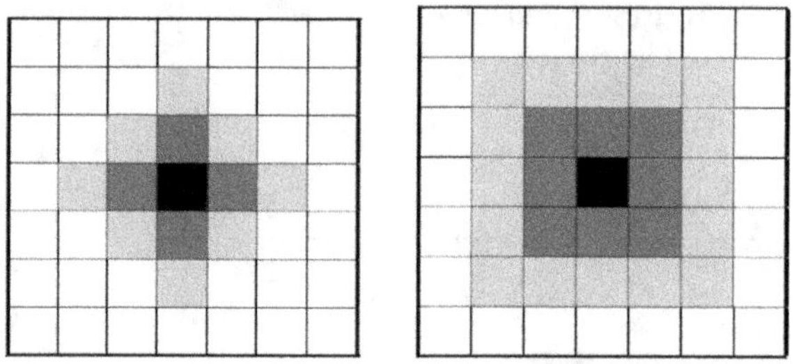

Figure 8.19 Plot of the neighbors of the rook-based (on the left) and the queen-based (on the right) contiguity: the first-order and second-order neighbors of the center point (shown in black) are represented in darker and lighter gray, respectively.

point can be related to a different set of adjacent points. Therefore, a hypothesis on the adjacency, contiguity, or spatial weight matrices $\mathbf{W}^{(s)}$ ($s = 1, \ldots, S$) has to be firstly considered.

Two traditional ways for defining these matrices are presented in the literature on spatial statistics (Cressie 1993). The first is the rook-based contiguity, where neighbors share a common border (Figure 8.19 on the left). The second is the queen-based contiguity, which defines neighbors as locations that share either a border or a vertex in their boundaries (Figure 8.19 on the right). In practice, a spatial weights matrix is rather arbitrarily selected, especially when there is no formal theoretical foundation for the extent of spatial interaction. Despite of the specific type of contiguity structure (rook or queen-based), a SAR model of proper order can be used as reference model for the noise term of the surfaces.

8.5.1 Quality Control Charts for Surface Monitoring

In order to detect out-of-control cylindrical surfaces, two alternative approaches can be considered. The first approach is the simplest one and consists in computing the OOC value associated to each machined surface and in monitoring it with an individuals control chart.

Similarly to control charts suggested in the literature on profile monitoring, the second approach consists of a multivariate control chart for monitoring the vector of estimated parameters and a univariate control chart aimed at monitoring the estimated residuals variance.

In the paper of Colosimo et al. (2010) the method for monitoring profile data developed by Colosimo et al. (2008) was extended to 3D forms (i.e., surfaces). The proposed approach combined a regression model with spatial correlated noise to univariate and multivariate control charting and was applied to real data related to cylindricity surfaces obtained by lathe-turning. The simulation study showed that the results found in (Colosimo et al. 2008) for roundness profiles are confirmed when surfaces are machined instead of profiles. In particular, the main results were that approaches based on monitoring the coefficients of the parametric model that describes the systematic pattern of machined surface outperforms the industrial practice, which is based on monitoring a synthetic indicator of form error (such as the OOCs). Furthermore, the simulation study showed that the extra effort required for modeling the spatial correlation can be worthy just when the small-scale (e.g., correlation structure and/or noise variance) characteristics of the machined surface are of interest.

8.6 CONCLUDING REMARKS

Data collected by measuring equipment can be modeled as functional data, where the quality outcome (dependent variable) is actually a function of one or more location variables (independent variables). In this framework, the present chapter has introduced different approaches for profile monitoring of geometric features. Both parametric models and nonparametric methods have been considered.

With the former kind of approaches, the in-control shape of the profiles is summarized by a parametric model, while profile monitoring is based on monitoring the parameters of this model (regression-based and PCA CC). On the other hand, with the latter kind of approaches, one can monitor the discrepancies between observed profiles and a baseline profile established using historical data. In this chapter, the LOC CC has been considered to this aim. A Shewhart's control chart on the geometrical errors has been also included in the comparison study, as this kind of approach is still the most representative of industrial practice.

By comparing the Phase I performance, which refers to the ability to obtain a predefined false alarm rate when uncontaminated process samples are considered, we can conclude that the extra effort required by an REG CC is worthwhile. Indeed, a regression-based control chart can easily be designed, since the actual false alarm rate can be achieved in practice using an analytic computation of the control limits, despite the specific production scenario considered and the number of historical data available. On the contrary, the actual false alarm rate presented by the other approaches for profile monitoring can be very different from that expected.

By comparing the overall performance in Phase II, which refers to the ability to signal a generic out-of-control condition of any severity for any type of production scenario (with random-effect and fixed-effect model), we can conclude that the extra effort required by the regression-based and PCA CC is also worthwhile. In fact, both the regression-based and PCA CC are more effective in signaling a generic change in the functional data. However, while the PCA CC shows superior robustness to change of the productive scenario, the REG CC may produce poor performance in signaling out-of-controls in the case of a production scenario with a random-effects model.

Furthermore, although the LOC CC is less effective in Phase II than the parametric approaches, the performance observed using this method is comparable to those produced by the competing approaches in some cases. Given the simplicity of the location control chart, its use in practice can be justified in these production situations, even if attention should be paid to properly designing (Phase I) this tool. On the other hand, the OOR CC is not suitable for profile monitoring and hence this approach is not recommended for actual applications.

Finally, in this chapter the method for monitoring 2D profile data has been also extended to 3D forms (i.e., surfaces). In this direction, there is some work (Colosimo et al. 2010) and we think that profile monitoring research will focus more and more on this topic in the near future. Although a specific geometry was used in (Colosimo et al. 2010), any 3D shape can be monitored with a similar approach, given that the regression terms and the spatial weight matrices are appropriately selected.

ACKNOWLEDGMENTS

This research work was partially funded by the Italian Ministry of University and Research (MIUR). The authors also thank the Machining Tool and Production Systems Research Laboratory (MUSP) in Piacenza (Italy) for providing the actual real case studies.

REFERENCES

Boeing Commercial Airplane Group, Quality Assurance Department, 1998. Advanced quality system tools, AQS D1-9000-1. <http:// www.boeingsuppliers.com/supplier/d1-9000-1.pdf> [Accessed March 2011].

Chang, T. C. and Gan, F. F. (2006) Monitoring linearity of measurement gauges. *Journal of Statistical Computation and Simulation*, **76**(10), 889–911.

Cho, N. W. and Tu, J. F. (2001) Roundness modeling of machined parts for tolerance analysis. *Precision Engineering*, **25**(1), 35–47.

Cho, N. W. and Tu, J. F. (2002) Quantitative circularity tolerance analysis and design for 2D precision assemblies. *International Journal of Machine Tools & Manufacture*, **42**(13), 1391–1401.

Colosimo, B. M. and Pacella, M. (2007) On the use of principal component analysis to identify systematic patterns in roundness profiles. *Quality and Reliability Engineering International*, **23**(6), 707–725.

Colosimo, B. M., Pacella, M., and Semeraro, Q. (2008) Statistical process control for geometric specifications: On the monitoring of roundness profiles. *Journal of Quality Technology*, **40**(1), 1–18.

Colosimo, B. M., Mammarella, F., and Petrò, S. (2010) Quality control of manufactured surfaces. In: *Frontiers of Statistical Quality Control 9*, edited by H. J. Lenz and P. T. Wilrich. Springer, New York.

Colosimo, B. M. and Pacella, M. (2010) A comparison study of control charts for statistical monitoring of functional data. *International Journal of Production Research*, **48**(6), 1575–1601.

Cressie, N. A. C. (1993) *Statistics for Spatial Data*, Revised edition. John Wiley & Sons, Inc., New York.

Ding, Y., Zeng, L., and Zhou, S. (2006) Phase I analysis for monitoring nonlinear profiles in 755 manufacturing processes. *Journal of Quality Technology*, **38**(3), 199–216.

Forbes, B. (1994) *Algorithms for Geometrical Tolerance Assessment. Report DITC 228/94.* National Physical Laboratory, UK.

Gupta, S., Montgomery, D. C., and Woodall, W. H. (2006) Performance evaluation of two methods for online monitoring of linear calibration profiles. *International Journal of Production Research*, **44**(10), 1927–1942.

Henke, R. P., Summerhays, K. D., Baldwin, J. M., Cassou, R. M., and Brown, C. W. (1999) Methods for evaluation of systematic geometric deviations in machined parts and their relationships to process variables. *Precision Engineering*, **23**(4), 273–292.

Jackson, J. E. (2003) *A User's Guide to Principal Components*. John Wiley & Sons, Inc., New York.

Jensen, W. A., Birch, J. B., and Woodall, W. H. (2008) Monitoring correlation within linear profiles using mixed models. *Journal of Quality Technology*, **40**(2), 167–183.

Jin, J. and Shi, J. (1999) Feature-preserving data compression of stamping tonnage information using wavelets. *Technometrics*, **41**, 327–339.

Jin, J. and Shi, J. (2001) Automatic feature extraction of waveform signals for in-process diagnostic performance improvement. *Journal of Intelligent Manufacturing*, **12**, 257–268.

Jolliffe, I. T. (2002) *Principal Component Analysis*, 2nd Edition. Springer, Springer Series in Statistics, New York.

Kang, L. and Albin, S. L. (2000) On-line monitoring when the process yields a linear profile. *Journal of Quality Technology*, **32**, 418–426.

Kim, K., Mahmoud, M. A., and Woodall, W. H. (2003) On the monitoring of linear profiles. *Journal of Quality Technology*, **35**, 317–328.

LeSage, J. P. (1999) *The Theory and Practice of Spatial Econometrics*. Available at http://www.spatialeconometrics.com/.

Mahmoud, M. A. and Woodall, W. H. (2004) Phase I analysis of linear profiles with calibration applications. *Technometrics*, **46**, 377–391.

Mahmoud, M. A., Parker, P. A., Woodall, W. H., and Hawkins, D. M. (2007) A change point method for linear profile data. *Quality & Reliability Engineering International*, **23**(2), 247–268.

Montgomery, D. C. (2004) *Introduction to Statistical Quality Control*, 5th Edition. John Wiley & Sons, Inc., New York.

Moroni, G. and Pacella, M. (2008) An approach based on process signature modeling for roundness evaluation of manufactured items. *Journal of Computing and Information Science in Engineering*, **8**(2), 021003-1–021003-10.

Nomikos, P. and MacGregor, J. F. (1995) Multivariate SPC charts for monitoring batch processes. *Technometrics*, **37**, 41–59.

Ramsay, J. O. and Silverman, B. W. (2005) *Functional Data Analysis*, 2nd Edition. Springer, Springer Series in Statistics, New York.

Simes, R. J. (1986) An improved Bonferroni procedure for multiple tests of significance. *Biometrika*, **73**(3), 751–754.

Stover, F. S. and Brill, R. V. (1998) Statistical quality control applied to ion chromatography calibrations. *Journal of Chromatography, A*, **804**, 37–43.

Walker, E. and Wright, S. P. (2002) Comparing curves using additive models. *Journal of Quality Technology*, **34**, 118–129.

Whittle, P. (1954) On stationary processes in the plane. *Biometrika*, **41**, 434–449.

Williams, J. D., Woodall, W. H., Birch, J. B., and Sullivan, J. H. (2006) Distribution of Hotelling's T^2 statistic based on the successive differences estimator. *Journal of Quality Technology*, **38**, 217–229.

Williams, J. D., Woodall, W. H., and Birch, J. B. (2007) Statistical monitoring of nonlinear product and process quality profiles. *Quality and Reliability Engineering International*, **23**(8), 925–941.

Woodall, W. H., Spitzner, D. J., Montgomery, D. C., and Gupta, S. (2004) Using control charts to monitor process and product quality profiles. *Journal of Quality Technology*, **36**(3), 309–320.

Woodall, W. H. (2007) Current research on profile monitoring. *Produção*, **17**(3), 420–425.

Young, T. M., Winistorfer, P. M., and Wang, S. (1999) Multivariate control charts of MDF and OSB vertical density profile attributes. *Forest Products Journal*, **49**, 79–86.

Zhang, H. and Albin, S. (2009) Detecting outliers in complex profiles using a 2 control chart method. *IIE Transactions*, **41**(4), 335–345.

Zhang, X. D., Zhang, C., Wang, B., and Feng, S. C. (2005) Unified functional tolerancing approach for precision cylindrical components. *International Journal of Production Research*, **43**(1), 25–47.

Zou, C. L., Zhang, Y. J., and Wang, Z. J. (2006) A control chart based on a change-point model for monitoring linear profiles. *IIE Transactions*, **38**(12), 1093–1103.

Zou, C. L., Tsung, F. G., and Wang, Z. J. (2008) Monitoring profiles based on nonparametric regression methods. *Technometrics*, **50**(4), 512–526.

CHAPTER 9

Correlation and Autocorrelation in Profiles

Willis A. Jensen

Medical Products Division, W. L. Gore & Associates, Inc., Flagstaff, AZ, USA

Jeffrey B. Birch

Virginia Polytechnic Institute, Virginia State University, Blacksburg, VA, USA

INTRODUCTION

The previous chapters in this book have presented methods for which the profiles are independent of each other and/or when the measurements within the profile are independent of each other. In reality, oftentimes the measurements are correlated with each other. Following the time series literature, when referring to the measurements within a single variable (profile), we refer to this lack of independence as autocorrelation. In contrast, when this lack of independence occurs across multiple variables (profiles), we refer to it simply as correlation. In this chapter, we address methods for monitoring profiles when the profiles themselves are not independent of each other (between-profile correlation) and when the measurements within a profile are not independent of each other (within-profile autocorrelation or more simply, autocorrelation). Throughout this chapter we will distinguish between methods for between-profile correlation (BPC) and within-profile autocorrelation (WPA). BPC is present when the profile measurements are correlated with measurements from other profiles and can result from a variety of reasons. Likewise, WPA or autocorrelation can be due to a wide variety of causes. For example, if the profiles or measurements within a profile are collected at time intervals that are very small, then the results of one measurement may be related to the results of the preceding measurement.

Statistical Analysis of Profile Monitoring, First Edition. Edited by Rassoul Noorossana, Abbas Saghaei and Amirhossein Amiri.

This dependence over time is often referred to as serial correlation. Alternatively, the measurements within a profile may be autocorrelated because they are measurements on the same physical object. This correlation due to physical distance is often referred to as spatial correlation. Here we will not distinguish between the causes of the correlation because serial and spatial correlation can occur both between and within profiles.

Note that when we refer to BPC or WPA, we are referring to the correlation in the errors or data values themselves. We are not referring to the correlation in the parameter estimators that are used to represent the profile with fewer values than the original data itself. It is entirely possible to have independent profiles with independent error terms within a profile and still have parameter estimators that are correlated with each other (see Kim et al. 2003) for a discussion of the correlation of parameter estimators for simple linear profiles). It is for that reason, that a multivariate T^2 statistic is often employed in profile monitoring, because it can properly take into account the correlation in the parameter estimators.

To illustrate one way that WPA can be present in a profile, consider the following model for a simple linear profile that was discussed in Chapter 2:

$$y_j = \beta_0 + \beta_1 x_j + \varepsilon_j, \tag{9.1}$$

where y_j and x_j are vectors containing the responses and profile values, respectively. If there is no WPA (or it is negligible), then we assume that $\varepsilon_j \sim N(0, \sigma^2 I)$, i.e., the error terms are independent and identically distributed (i.i.d.) observations from a normal (Gaussian) distribution. On the other hand, if we want to allow for and model the WPA, then we can relax the restriction on the error terms to model the autocorrelation. In other words, we allow $\varepsilon_j \sim N(0, R)$, where R is some positive definite covariance matrix. Off-diagonal elements of the covariance matrix that are not zero allow us to model the autocorrelation between observations within a profile. While there is a tremendous amount of flexibility in the structure of the covariance matrix to allow for different types of autocorrelation, some structures are more commonly used than others. For example, an AR(1) structure can be set for the covariance matrix of the errors. Some examples of the various structures can be found in Littell et al. (1996). These structures are often chosen to be simpler forms so as to reduce the number of parameters needed to model the data.

Consider Figure 9.1, which shows a single randomly generated profile that can be fit by a simple linear regression (SLR) model discussed in Chapter 2. Using the model shown in Equation (9.1), we set the intercept to be 3, the slope equal to 2, and randomly generated deviations to follow a normal distribution with a mean of zero and a variance of one. The exception to Equation (9.1) is that now the errors are correlated with the previous point within the profile. That is, this profile follows an AR(1) structure with a correlation coefficient, $\rho = 0.9$. This high level of correlation leads to the cyclical pattern in the points up and down around the fitted regression line. It should be noted that Figure 9.1 has $n = 50$ points which make up the profile. It is easier to determine if autocorrelation within a profile is present with larger amounts of data per profile. If n is smaller, and separate SLR models are fit to each profile, this cyclical pattern will not show up in a visual display of the profile. Notice that in this

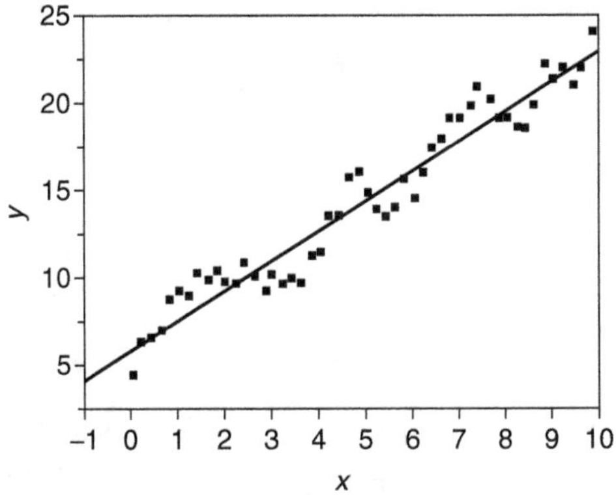

Figure 9.1 Randomly generated profile showing the presence of autocorrelation.

preceding example, the autocorrelation does not refer to the correlation between the response variable and the explanatory variable, which is captured by the slope term in the SLR model.

In this chapter, we first cover some methods for handling WPA and BPC before presenting an example based on lumber profiles. We then discuss some related issues for profiles with BPC. Finally, we spend some time in the last portion discussing some open research questions and issues that have not been addressed in the literature. We feel that there is much to be done in this area.

9.1 METHODS FOR WPA FOR LINEAR MODELS

Profiles that exhibit autocorrelation can be thought of as the profile monitoring extension of control charts for autocorrelated data that make heavy use of time series methods for serial correlation. Some examples of standard control charts for autocorrelated data can be found in the review papers of Montgomery and Mastrangelo (1991) and Psarakis and Papaleonida (2007). Thus, it is natural to extend these time series methods to handle the WPA that exists within the profiles.

For linear profiles that can be fit with an SLR model, several approaches have been proposed for both Phase I and Phase II applications. These approaches apply to the parameter estimates that remain after the data reduction that occurs from fitting the profile. These approaches are covered in Chapter 2 and include the T^2 chart, the EWMA/R approach of Kang and Albin (2000) and the EWMA-3 approach of Kim et al. (2003).

Soleimani et al. (2009) extended the work covered in Chapter 2 by considering an SLR model with correlated errors within a profile. In other words, they used an autoregressive model of order 1 (AR(1)) for the error terms and thus rewrote Equation (9.1) as follows:

$$y_j = \beta_0 + \beta_1 x_j + \varepsilon_j, \tag{9.2}$$

with $\varepsilon_j \sim N(0, R)$ where R can be written as follows:

$$R = \sigma^2 \begin{bmatrix} 1 & \rho & \rho^2 & \cdots & \rho^n \\ \rho & 1 & & & \vdots \\ \rho^2 & & \ddots & & \rho^2 \\ \vdots & & & 1 & \rho \\ \rho^n & \cdots & \rho^2 & \rho & 1 \end{bmatrix}. \tag{9.3}$$

This matrix notation to represent the WPA makes it simpler to consider a variety of types of autocorrelation. Soleimani et al. (2009) found that the best control chart approach for this scenario follows the approach recommended by Kim et al. (2003), which consists of first centering the profiles by subtracting the mean value of the x_i. This has the effect of making the estimators of the slope and intercept independent and allows the construction of separate charts for the intercept and slope as was done by Kim et al. (2003) for simple linear profiles with uncorrelated errors. This results in three charts: one to monitor the intercept, another to monitor the slope, and the third to monitor the error variance. This was referred to as the EWMA-3 approach by Kim et al. (2003). This use of multiple EWMA charts also simplifies the interpretation of an out-of-control signal because the user knows which portion of the profile (slope, intercept, or error variance) is causing the signal.

The simulation studies of Soleimani et al. (2009) showed that in general, the autocorrelation in the errors leads to the decrease in ARL performance of the chart. These simulation study results match the conclusion of Jensen et al. (2006), who found that the frequency of false alarms increases (ARL decreases) when there is autocorrelation present for standard Shewhart charts. As a result of this increasing frequency of false alarms, it is necessary to adjust the control limits for autocorrelated data. In addition, because of the autocorrelation that is present, larger amounts of Phase I data will be needed to ensure that the estimators used for Phase II application will be adequate. In addition, the simulation studies showed that the EWMA-3 approach was uniformly superior to other approaches. The approach of Soleimani et al. (2009) could be extended for any type of autocorrelation structure in the errors by fitting the appropriate time series model, although only an AR(1) structure was considered by Soleimani et al. (2009).

Staudhammer et al. (2007) considered linear profiles of lumber manufacturing where the profile consists of measurements along a board. For a single board, four different lasers (top right, top left, bottom left, bottom right) are set up to measure the distance from the laser (thickness) to the board as the board is being cut, with many measurements taken by each laser along the length of the board. Because the measurements within a board were taken very close together in time and space, they exhibit strong autocorrelation. Variation in these measurements indicates potential sawing defects. Researchers were aware of a variety of known defect types. As a result, the proposed methods of Staudhammer et al. (2007) were designed to address specific defect types that can occur in the profile; thus, the proposed approaches are not broadly applicable. We will use this lumber profile data in an example later in this chapter.

Jensen et al. (2008) considered a linear mixed model (LMM) with WPA. It is important to note that when using an LMM, the WPA can be induced either through random effects or through the autocorrelation of the error terms. Jensen et al. (2008) included in their model both random effects as well as simple correlation structures for the error terms like those considered by Soleimani et al. (2009). If we rewrite Equation (9.2) for the jth profile with autocorrelated errors in matrix form, we have the following:

$$y_j = X_j \beta_j + \varepsilon_j, \tag{9.4}$$

where X_j is an n by 2 matrix with a column of 1's in the first column and a second column corresponding to the values in x_j. $\boldsymbol{\beta}_j$ is a 2 by 1 vector containing the slope and intercept parameters. Note that this equation in (9.4) can be extended to include various types of linear models, such as models with multiple variables (x) as well as polynomial models. The LMM of Jensen et al. (2008) can be given by rewriting the model in (9.4) for the jth profile as follows:

$$y_j = X_j \beta + Z_j b_j + \varepsilon_j, \tag{9.5}$$

where Z_j is a matrix containing the columns of X_i that have corresponding random effects, which are contained in the vector, $b_j \sim$ MN(0, D), and D is a variance–covariance matrix of the random effects. Jensen et al. (2008) restricted Z_j to be a subset or equivalent to X_j, which does not limit the applicability of the LMM. The simulation studies of Jensen et al. (2008) showed that under certain conditions when there is autocorrelation in the data, there is little difference between the LMM approach and the SLR approach when using a T^2 statistic to determine outlying profiles. In other words, reduction of the data by either approach gives more or less the same ability to detect the outlying profiles. Amiri et al. (2010) give a case study example of automotive profiles that can be modeled with a second-order LMM and an AR(1) correlation structure in the errors.

9.2 METHODS FOR BPC FOR LINEAR MODELS

In contrast to the methods discussed in the previous section, we discuss here profiles where the measurements are correlated across profiles. As in Section 9.1, we restrict our attention here to profiles that can be fit with a linear model. Noorossana et al. (2008) considered an SLR model with the correlation introduced in the observations between the profiles. When this correlation is introduced, it is transferred to the estimators of slope and intercept, so they proposed in a Phase II control chart application to monitor the residual values of the fitted slope and intercepts. This results in two EWMA charts, one for the residual slope value and the other for the residual intercept value.

Noorossana et al. (2008) supplemented these two EWMA charts with another EWMA chart to monitor the error variance. This EWMA-3 approach is similar to that used by Kim et al. (2003) and Soleimani et al. (2009). The simulation studies of Noorossana et al. (2008) showed the EWMA-3 approach to be superior to other approaches.

Kazemzadeh et al. (2010) extended the results of Noorossana et al. (2008) to consider BPC that follows an AR(1) structure for polynomial profiles. Thus, rather than fitting an SLR model like that shown in Equation (9.2), a second-order polynomial is fit to the profiles. Simulation studies were done to compare a T^2 chart versus an EWMA/R chart and no single approach performs best for different out-of-control scenarios. The simulation studies of Kazemzadeh et al. (2010) did not include the EWMA-3 method in its comparison, so it is not clear whether or not this would be better than those considered.

9.3 METHODS FOR WPA AND BPC FOR OTHER (NONLINEAR) MODELS

Jensen and Birch (2009) considered the application of nonlinear mixed models (NLMM) to profile monitoring. In general, it is not common to include modeling of the autocorrelated errors in an NLMM because of computational difficulties, although autocorrelation can still be accounted for by the inclusion of the random effects. As shown by Jensen and Birch (2009), the NLMM is a superior procedure in the presence of autocorrelated errors to a procedure that fits a separate non-linear regression model to each profile. Thus, even though a mixed model (inclusion of random effects) does not specifically model the autocorrelation in the errors, it still makes more effective the Phase I control chart procedure. In our review, we could find no literature that considered nonlinear models and autocorrelation in the errors.

Qiu et al. (2010) presented recent work on correlated profiles, which is similar in concept to the work done by Noorossana et al. (2008) and Kazemzadeh et al. (2010). For Phase II applications, they proposed the use of nonparametric methods in conjunction with mixed models in order to adequately model more complicated profiles. The mixed model introduces correlation in measurements across the profiles, although they did not explicitly introduce autocorrelated error terms. The methods of Qiu et al. (2010) could be extended to include both the random effects and an autocorrelation structure in the errors. They adapted the LMM method of Jensen et al. (2008) for Phase II analysis. When there is no model misspecification and a parametric model works well, there is little difference between a nonparametric and a parametric approach. Qiu et al. (2010) found that in Phases II analysis, a mixed model combined with a nonparametric approach works better than a purely parametric mixed model approach, in the case where a parametric approach is subject to model misspecification.

Example 9.1 Lumber profiles. To illustrate the concept of profile monitoring for autocorrelated data, we present an example here based on the lumber profile data discussed in Staudhammer et al. (2007). A total of 110 boards are available in

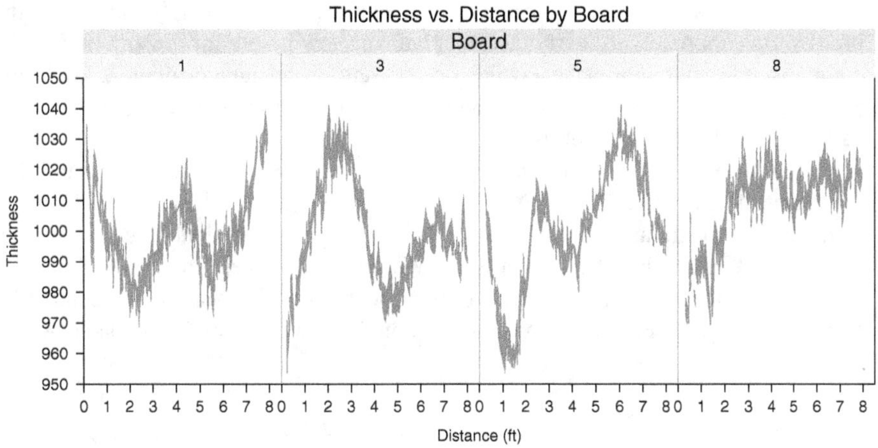

Figure 9.2 Raw data profiles for the lumber data.

the data set. However, these boards are cut with different saw types, which results in differences in the profiles, thus only the boards ($n = 41$) from one saw type (band saw) are used here. In addition, four profiles are available for each board corresponding to different locations. We use only a single profile here corresponding to the laser from the top of the board on one side. Each board has nearly 3000 measurements for the profile, thus giving us approximately 120,000 total data points in the data set. To reduce the computational burden, we also cut the measurements for a profile in half by using every other point in the profile, which has a negligible effect on an approach that accounts for the autocorrelation. The raw data profiles for several boards are shown in Figure 9.2. The distance in feet refers to the point along the board that was measured; thus, each board is roughly 7–8 feet in length. The thickness is measured in thousandths of an inch.

Similar to the pattern displayed by the autocorrelated data in Figure 9.1, Figure 9.2 clearly shows the autocorrelation present within each profile. While each profile could be fit by a more sophisticated model, such as smoothing splines or other nonparametric methods, the general pattern of the profiles varies widely, making it more difficult to find a single model that works for all the profiles. In addition, Staudhammer et al. (2007) noted that it is desirable to have the profiles in Figure 9.2 resemble a flat, straight line with a slope of zero and an intercept of 1000. A flat line indicates the absence of sawing defects. The patterns are a function of the strong autocorrelation and the many different patterns can be summarized nicely by considering the autocorrelation structure of the errors. We first illustrate the Phase I analysis using the first 30 boards as the "historical" data set.

9.4 PHASE I ANALYSIS

Our review of the literature earlier in this chapter did not cover any Phase I methods for autocorrelated profiles to be fit with SLR models.

Thus, we first consider an approach that ignores the autocorrelation within the profile. This approach is clearly incorrect but we show it here for illustration and for comparison to later approaches, which do account for the autocorrelation. We fit the SLR lines to each profile and obtain an estimate of the slope, intercept, and residual variability (MSE) for each profile.

Prior to fitting the SLR line, we first center the data by subtracting the mean value for the distance for each profile. As noted by Kim et al. (2003), this has the advantage of making the estimators for the slope and intercept independent; thus, a separate Shewhart chart can be fit for each term. We then supplement these two Shewhart charts with another univariate chart to monitor the MSE, as was recommended by Kim et al. (2003). Figure 9.3 shows the control charts for this approach, which

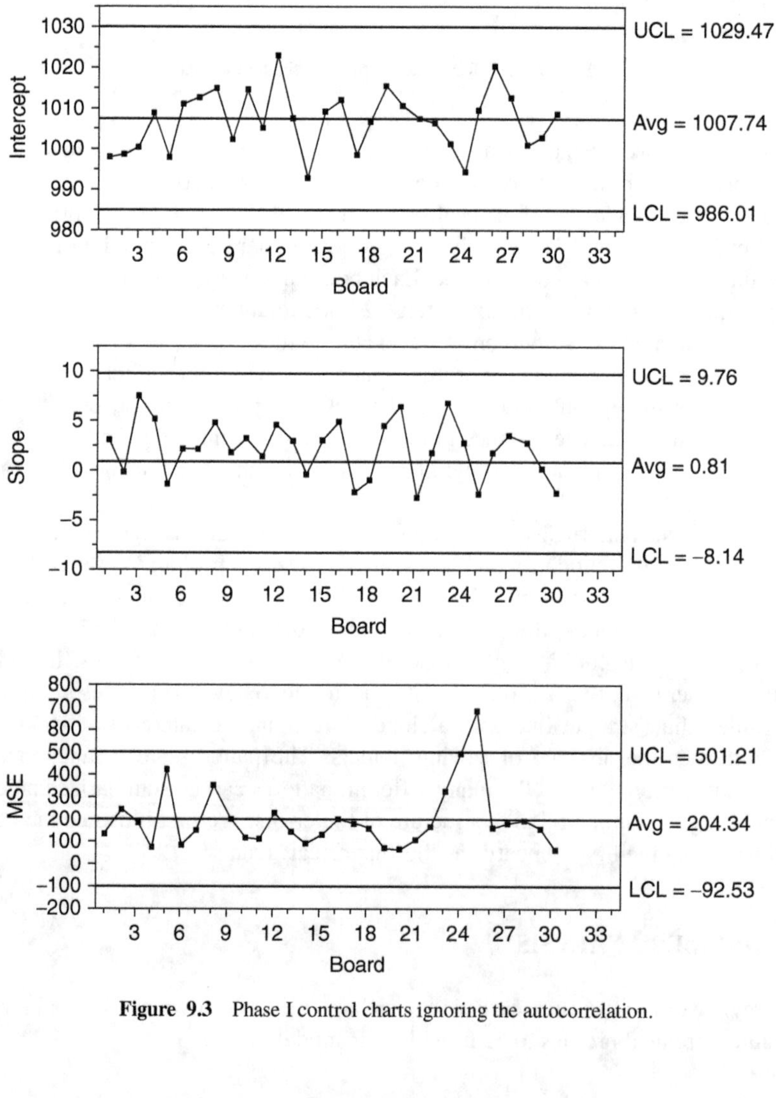

Figure 9.3 Phase I control charts ignoring the autocorrelation.

ignores the autocorrelation. In general, the 30 boards are in-control as evidenced by no signals on either of the charts for the intercept or slope. Only one board, number 25, is deemed an outlier, because its MSE value is unacceptably high. It would be removed prior to the Phase II analysis.

As a second approach, we consider the lumber data and now account for the autocorrelated errors. To do so, we fit an LMM with a random effect for the slope and intercept as illustrated by Jensen et al. (2008). A time series analysis on the individual profiles, not shown here, shows that an AR(1) model does quite well in explaining a large portion of the variability within a profile. Thus, the AR(1) structure for the error terms is fit in conjunction with the LMM. Even though the data were centered as was before, the predicted random effects for the slope and intercept are not independent of each other, thus a T^2 chart (top chart in Figure 9.4) will be used to jointly monitor

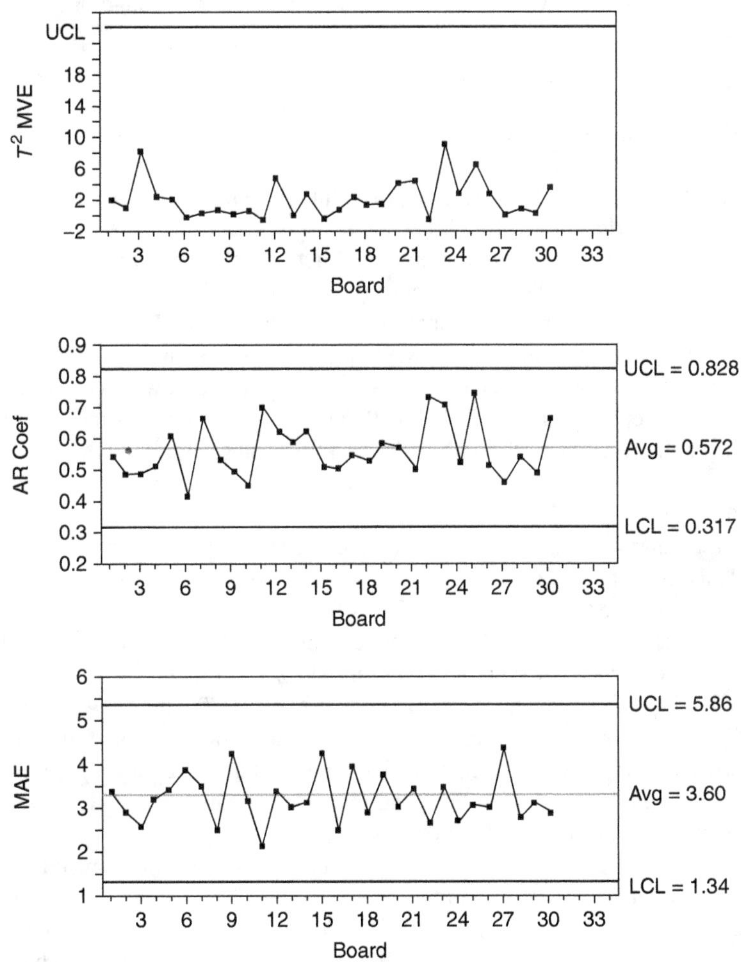

Figure 9.4 Phase I control charts accounting for the autocorrelation.

the intercept and slope random effects. Based on the recommendation of Jensen et al. (2007), we use the chart based on the minimum volume ellipsoid (MVE). The upper control limit (UCL) was given in Jensen et al. (2007) as 24.31. We supplement this chart with two univariate charts to monitor the autocorrelation within the profile. The first (middle chart on Figure 9.4) chart contains the estimated AR coefficients of the first lagged time period for each profile. A difference in the coefficient between the profiles would be an indication that the strength of the autocorrelation is different for that particular profile. The second (bottom chart on Figure 9.4) chart is of the mean absolute error (MAE), which is the sum of the absolute values of the deviations of the individual measurements from its predicted value, based on the fitted time series model. The MAE is analogous to the MSE that was used when no autocorrelation is present. Large MAE values indicate profiles that are not fit very well by the AR(1) error structure. It is clear that all 30 profiles are in-control, in contrast to the original analysis ignoring the autocorrelation that found one profile to be out-of-control.

9.5 PHASE II ANALYSIS

To illustrate the Phase II analysis with the lumber data, we treat the remaining 11 profiles (of the original 41 profiles) as the Phase II data set to show how the various charts will correctly diagnose the signals. First, we show the updated charts from Figure 9.3 that ignore the autocorrelation. For Phase II, we fix the control limits that result from excluding board number 25, which had a high MSE value. We then compute the intercepts, slopes, and MSE values for the 11 new profiles. Figure 9.5 shows the resulting control charts. We see that there are now three signals on the chart for intercepts (board numbers 33, 35, and 38), no signals for the slope and two signals for the MSE (board numbers 32 and 36). These signals indicate an out-of-control process, which should be investigated prior to producing more boards on this particular saw.

For the charts that account for the autocorrelation, we perform a similar Phase II analysis. As was noted by Amiri et al. (2010), we use the LMM estimates from Phase I to compute the new random effects for slope and intercept for the additional 11 profiles. These new random effects are computed using the estimated fixed effects and estimated variance–covariance matrices shown in Equation (9.5). The random effects for slope and intercept are combined into a single T^2 chart as was done earlier. In addition, we compute the same AR coefficient and MAE values to supplement the T^2 chart. The updated control charts with the Phase II data are shown in Figure 9.6. A signal occurs on the T^2 chart at board number 33, indicating a need for investigation into the out-of-control state of the process.

We note in this example that there is not a large difference in the results for methods that account for the WPA and those that do not. This example demonstrates, for profiles adequately represented by linear models, that accounting for strong autocorrelation within the profiles, has a negligible effect on the results obtained by ignoring this autocorrelation. This result agrees with the conclusion of Jensen et al. (2008).

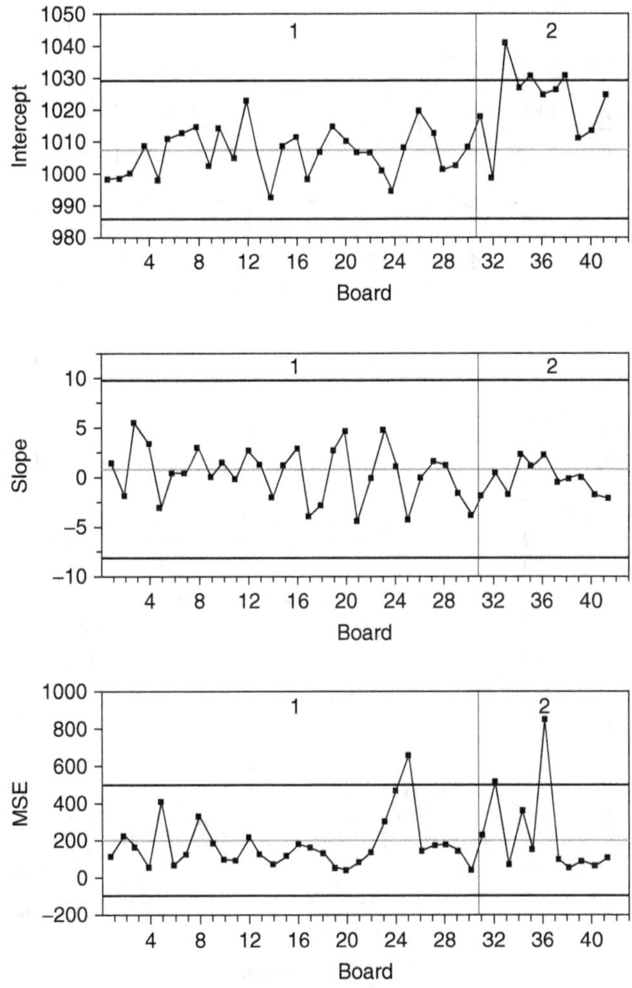

Figure 9.5 Phase II control charts ignoring the autocorrelation.

We believe that this is because the current approaches may not completely capture all the salient features of the data.

9.6 RELATED ISSUES: RATIONAL SUBGROUPING AND RANDOM EFFECTS

We wish to discuss here two related issues to BPC: (1) rational subgrouping and (2) random effects. As discussed in Montgomery (2003, pp. 143–146), rational subgrouping signifies a type of grouping where the differences that are due to assignable

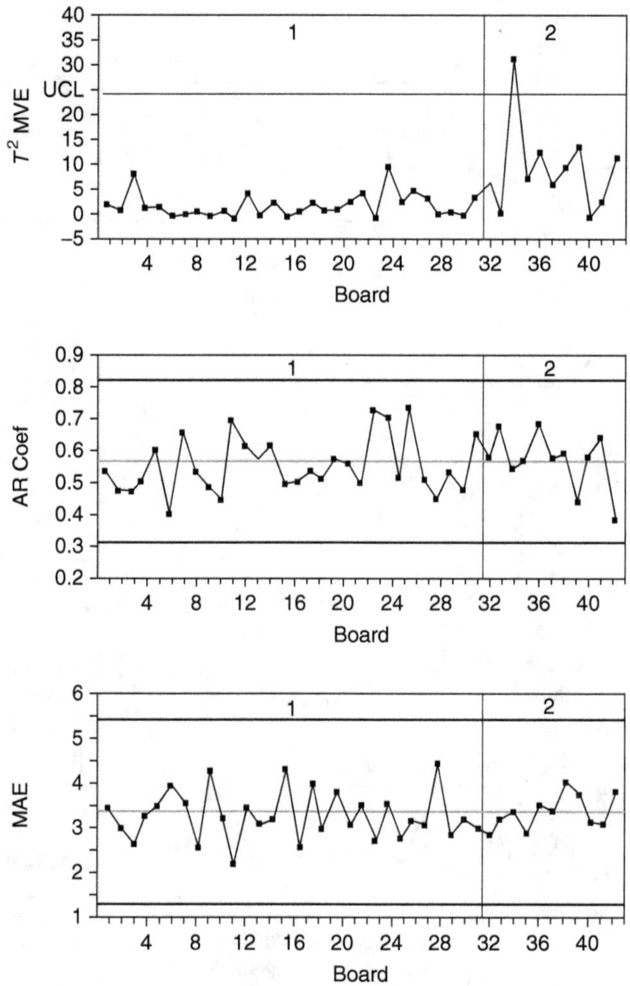

Figure 9.6 Phase II control charts accounting for the autocorrelation.

or special causes within the group are minimized. In addition, the differences between the groups due to special causes are maximized. In other words, if several profiles form a natural collection of profiles where the profiles are collected under similar conditions, they can be grouped. A common way to form rational subgroups is to collect observations that are close together in time. As such, the observations within a group will be similar to each other. The rational subgroups should be formed such that the variability between groups is maximized and the variability within a group is minimized. Nelson (1988, p. 73) defined items within a subgroup as those for which "only random effects are responsible for the observed variation."

To illustrate the concept of rational subgrouping in profile monitoring, consider for example an oven used in a manufacturing process where thermocouples are placed in several places within the oven to record the temperature. Suppose that six thermocouples are placed in six different locations in the oven, and the temperature of the oven is recorded every second for a specified period of time after the oven is turned on and set to a particular temperature value. The resulting temperature profiles show an increase in the temperature during the warm-up period before stabilizing at the set temperature. However, there are differences due to the location of the thermocouples within an oven, representing within-oven variability that will be present every time the oven is started up. Over time, it is necessary to determine if there are differences in the way the oven is performing, not necessarily to determine differences between the six profiles within an oven. If we desire to monitor the temperature profiles and obtain a set of six profiles for each start-up run (which often corresponds to a day), then the six temperature profiles can form a rational subgroup, if the differences between the profiles can be considered as a random effect. Alternatively, the profiles can exhibit systematic differences, such as one thermocouple being closer to the door of the oven and consequently show lower temperatures than the other profiles. The six profiles are correlated with each other (because they all come from the same oven run), although they are not treated by the methods for correlated profiles that we discuss here. Rather these six temperature profiles that form a group (with differences either from random effects or systematic differences) are better handled by the multivariate profile methods discussed in Chapter 7.

Another example to illustrate this concept of rational subgrouping can be found in Staudhammer et al. (2007), who considered profiles of pieces of lumber cut with various types of saws. As each laser produces its own profile of measurements, the data collection results in a set of four profiles for each board. Because the profiles come from the same board and defects in the board can show up in multiple profiles, it is natural to consider as a rational subgroup, the four profiles that are correlated with each other. Example 9.1 only considered one of the four profiles for each board and did not include all the data across the multiple profiles.

In addition, BPC can occur even when there is no natural grouping, simply because profiles exhibit some correlation with each other over time. The approaches of Noorossana et al. (2008) and Kazemzadeh et al. (2010) address this scenario. In addition to these approaches, there are other ways to introduce BPC. For example, the LMM approach of Jensen et al. (2008) made the assumption that the random effects were uncorrelated, i.e., the variance–covariance matrix of the random effects, D, is a diagonal matrix. One approach not considered in the literature is to allow D to follow some correlation structure, such as the AR(1) structure considered for the errors. This would allow the profiles to be correlated with each other, in addition to the errors within a profile to be autocorrelated with each other.

The second issue related to BPC is that of random effects. It should be noted that fitting an LMM or NLMM (i.e., introducing random effects) does not require the inclusion of autocorrelated errors. A mixed model approach is simply a modeling approach that reduces the profile data to a smaller set of parameter estimates that contain the necessary information about the profile to determine whether or not it is an outlying profile. Introducing random effects into the model allows for the assumption that the profiles are a random sample from a population of profiles that follows some distribution. Thus, rather than fitting a separate model to each profile, a single model is fit to all the profiles, with the deviation of each profile from the average profile being considered the random effect(s). While profiles may be considered a random sample from a population of profiles, and thus independent profiles, the inclusion of random effects allows for the pooling of information across profiles. This pooling can be valuable when the profile data does not come in equally spaced intervals along the profile. Pooling is also advantageous when some profiles have missing observations or otherwise have an unequal number of observations at different points along the profile. Inclusion of random effects in models is simply a modeling and data reduction technique. It is not a control chart technique like the T^2 or EWMA charts mentioned here and in other chapters. Thus, nearly any Phase I or Phase II control chart method may be applied to profiles modeled with random effects.

9.7 DISCUSSION AND OPEN QUESTIONS

As noted in Section 9.6, it is important to recognize that in profile monitoring there are methods of data reduction and methods of process monitoring for Phase I and Phase II. Data reduction methods include SLR, polynomial models, and LMM for profiles that fall within the category of linear models as well as nonlinear models or NLMM for other types of profiles. Any of these data reduction approaches can also allow for specific modeling of the autocorrelation in the errors. In Table 9.1, we summarize the studies that have been completed in the literature according to six different data reduction schemes. Blank cells correspond to areas where no research has been completed to the best of our knowledge. Phase I methods include the multivariate T^2 statistics based on the successive differences estimator mentioned by Sullivan and Woodall (1996) and the robust methods discussed in Jensen et al. (2007). Phase II methods include the T^2 chart, the EWMA/R approach of Kang and Albin (2000), and the EWMA-3 approach of Kim et al. (2003).

In our review of the existing literature as covered in this chapter, which is summarized in Table 9.1, it is clear that there are many unanswered questions related to correlation and autocorrelation in profile monitoring. Because of the many blank spaces in Table 9.1, we believe that addressing BPC and WPA in profile monitoring is a promising area of research.

In addition, very little work has been done for autocorrelation structures other than an AR(1) structure. The literature shown in Table 9.1 has extensively focused on AR(1) structures. The data reduction and process monitoring methods shown here can be extended to cover other types of autocorrelation structures, such as more complex ARIMA time series type models as well as more complex error structures that allow for different types of spatial correlation as discussed in Littell et al. (1996).

We also note that there are other methods of introducing BPC that have not been considered in the literature on profile monitoring. For example, generalized estimating equations (GEE) have been proposed in other contexts as a way to introduce correlation between longitudinal data, but have yet to be introduced into profile monitoring. An overview to GEE can be found in Hardin and Hilbe (2003).

Table 9.1 Summary of Completed Research for Profiles with WPA and BPC

	WPA		BPC	
Data Reduction Scheme	Phase I Methods	Phase II Methods	Phase I Methods	Phase II Methods
1. Simple linear regression (SLR)		Soleimani et al. (2009)		Noorossana et al. (2008)
2. Polynomial regression with fixed effects				Kazemzadeh et al. (2010)
3. Linear mixed model (LMM)	Jensen et al. (2008), Amiri et al. (2010)			
4. Nonlinear model				
5. Nonlinear mixed model (NLMM)	Jensen and Birch (2009)			
6. Non-parametric methods				Qiu et al. (2010)

From a broader perspective, profile monitoring involves trying to capture the salient features of the profile in a small number of parameters, without losing the essential information about the profile. This data reduction allows the practitioner to monitor the corresponding process in an effective way, resulting in good decisions being made about the state of the process. When compared to profiles without autocorrelation, autocorrelated profiles have additional salient features, such as autocorrelation or partial autocorrelation values. It is not clear how to best account for those additional salient features and thus we encourage additional research in this area to provide better tools for these types of situations.

ACKNOWLEDGMENT

The authors wish to thank Dr. Christina Staudhammer for providing the lumber profile data used in the example.

REFERENCES

Amiri, A., Jensen, W. A., and Kazemzadeh, R. B. (2010) A case study on monitoring polynomial profiles in the automotive industry. *Quality and Reliability Engineering International*, **26**(5), 509–520.

Hardin, J. and Hilbe, J. (2003) *Generalized Estimating Equations*. Chapman and Hall/CRC, London.

Jensen, W. A. and Birch, J. B. (2009) Profile monitoring via nonlinear mixed models. *Journal of Quality Technology*, **41**(1), 18–34.

Jensen, W. A., Birch, J. B., and Woodall, W. H. (2007) High breakdown estimation for phase I multivariate control charts. *Quality and Reliability Engineering International*, **23**(5), 615–629.

Jensen, W. A., Birch, J. B., and Woodall, W. H. (2008) Monitoring correlation within linear profiles using mixed models. *Journal of Quality Technology*, **40**(2), 167–183.

Jensen, W. A., Jones-Farmer, L. A., Champ, C. W., and Woodall, B. W. (2006) Effects of parameter estimation on control chart properties: A literature review. *Journal of Quality Technology*, **38**(4), 349–364.

Kang, L. and Albin, S. L. (2000) On-line monitoring when the process yields a linear profile. *Journal of Quality Technology*, **32**(4), 418–426.

Kazemzadeh, R. B., Noorossana, R., and Amiri, A. (2010) Phase II monitoring of autocorrelated polynomial profiles in AR(1) processes. *International Journal of Science and Technology, Scientia Iranica*, **17**(1), 12–24.

Kim, K., Mahmoud, M. A., and Woodall, W. H. (2003) On the monitoring of linear profiles. *Journal of Quality Technology*, **35**(3), 317–328.

Littell, R. C., Milliken, G. A., Stroup, W. W., and Wolfinger, R. D. (1996) *SAS System for Mixed Models*. SAS Institute, Inc., Cary, NC.

Montgomery, D. C. (2003) *Introduction to Statistical Quality Control*, 3rd Edition. John Wiley & Sons, New York.

Montgomery, D. C. and Mastrangelo, C. M. (1991) Some statistical process control methods for autocorrelated data. *Journal of Quality Technology*, **23**(3), 179–204.

Nelson, L. S. (1988) Control charts: Rational subgroups and effective applications. *Journal of Quality Technology*, **20**(1), 73–75.

Noorossana, R., Amiri, A., and Soleimani, P. (2008) On the monitoring of autocorrelated linear profiles. *Communications in Statistics-Theory and Methods*, **37**(3), 425–442.

Psarakis, S. and Papaleonida, G. E. A. (2007) SPC procedures for monitoring autocorrelated processes. *Quality Technology and Quantitative Management*, **4**(4), 501–540.

Qiu, P., Zou, C., and Wang, Z. (2010) Nonparametric profile monitoring by mixed effects modeling (with discussions). *Technometrics*, **52**(3), 265–277.

Soleimani, P., Noorossana, R., and Amiri, A. (2009) Simple linear profiles monitoring in the presence of within profile autocorrelation. *Computers and Industrial Engineering*, **57**(3), 1015–1021.

Staudhammer, C., Maness, T. C., and Kozak, R. A. (2007) New SPC methods for identifying lumber manufacturing defects with real-time laser range sensor data. *Journal of Quality Technology* **39**(3), 224–240.

Sullivan, J. H. and Woodall, W. H. (1996) A comparison of multivariate control charts for individual observations. *Journal of Quality Technology* **28**(4), 398–408.

CHAPTER 10

Nonparametric Profile Monitoring

Changliang Zou

Department of Statistics, School of Mathematical Sciences,
Nankai University, Tianjin, China

INTRODUCTION

The approaches introduced in the previous chapters focus on linear profiles that have been particularly popular in the practice application. However, there are many situations in practice in which the profile cannot be represented easily by a parametric model (linear or nonlinear). For example, a piecewise polynomial model is required to characterize the curve of vertical density profile (VDP; Williams et al. 2007a). In practical applications, engineers may want to avoid spending much time on fitting a complicated parametric model. The methods for profile monitoring with nonparametric regression will be quite useful when the relationship is too complicated to be described parametrically.

In this chapter, we focus on monitoring a general profile that can be well represented by a regression function. To be specific, assume that for the jth random sample collected over time, we have the observations $(\mathbf{X}_j, \mathbf{y}_j)$, where $\mathbf{y}_j = (y_{1j}, \ldots, y_{n_j j})^T$ is an n_j-variate response vector and $\mathbf{X}_j$ is an $n_j \times p$ regressor matrix. It is assumed that when the process is in statistical control, the underlying model is

$$y_{ij} = g\left(\mathbf{x}_j^{(i)}, \boldsymbol{\beta}\right) + \varepsilon_{ij}, \quad i = 1, \ldots, n_j \quad j = 1, 2, \cdots, \tag{10.1}$$

where $\mathbf{x}_j^{(i)}$ denotes the ith row of $\mathbf{X}_j$, $\boldsymbol{\beta}$ is a q-dimensional parameter vector, and the $\boldsymbol{\varepsilon}_{ij}$s are i.i.d. normal random variables with means of zero and variances of σ_j^2. The function g can be either linear or nonlinear and it possesses some degree of smoothness, such as continuity or the existence of derivatives.

Statistical Analysis of Profile Monitoring, First Edition. Edited by Rassoul Noorossana, Abbas Saghaei and Amirhossein Amiri.

Assuming a nonlinear model can well characterize the profile curve, the parametric control schemes, such as Kim et al.'s (2003) multi-charts and Zou et al.'s (2007) MEWMA scheme may be applied by obtaining the least squares estimators of the parameters of each profile. This is usually accomplished by employing the Gauss–Newton procedure and iterating until convergence. Note that, unlike linear regression, the exact small-sample distribution of parameter estimators in nonlinear regression is unobtainable. That said, these parameter-based charts may still be utilized because the estimated parameters are usually asymptotically normally distributed. Nevertheless, such a "rough" control scheme will not only deteriorate the properties of in-control (IC) average run length (ARL), but also have a significant effect on the out-of-control (OC) performance. Moreover, when the process is out of control, nonconvergence or slow convergence often occurs, thus resulting in extensive computational efforts or even a failure in monitoring and detection. This makes the practical implementation of parametric methods infeasible and inconvenient. Besides, the parametric methods for monitoring model (10.1) with both linear and nonlinear regressions have an inherent problem; that is, they are based on the assumption that the IC and OC models have the same form but only the parameters may be different. Although the IC model can be determined before the beginning of the on line monitoring, the OC model often cannot be specified easily, especially when the IC curves have complicated forms, such as various nonlinear models. The parametric monitoring methods are generally powerful when matched with the specific OC model for which they were designed, but they can have very poor ARL performance with other types of OC models.

In this chapter, we will discuss several nonparametric profile monitoring schemes to resolve the issues mentioned above to some extent. In Section 10.2, we introduce a multivariate EWMA procedure with the generalized likelihood ratio (GLR) test based on nonparametric regression to monitor changes in both the regression relationship and the variation of the profile on line. This scheme can be regarded as a nonparametric counterpart of Zou et al.'s (2007) MEWMA scheme. In Section 10.3, a self-staring type procedure using sequential change-point formulation with GLR tests will be presented. Its control limits are determined by a bootstrap procedure. This chart can be implemented without any knowledge about the error distributions, as long as a few IC profiles are available beforehand. Finally, in Section 10.3 we focus on nonparametric profile monitoring when within-profile data are spatially or serially correlated and different profiles have different design points. The method is based on nonparametric mixed-effects (NME) modeling and EWMA process monitoring.

10.1 MONITORING PROFILES BASED ON NONPARAMETRIC REGRESSION

10.1.1 Model and Assumption

To facilitate the presentation, we choose to use the one-dimensional covariant case of model (10.1). That is, the underlying model is

$$y_{ij} = g\left(x_{ij}\right) + \varepsilon_{ij}, \quad i = 1, \ldots, n_j \quad j = 1, 2, \ldots, \tag{10.2}$$

where x_{ij} denotes the value of the regressor for the ith observation in the jth profile. Note that, in model (10.12), we suppress the parameter $\boldsymbol{\beta}$ because we focus on the nonparametric regression method. The n_js are taken to be equal (denoted as n) and the explanatory variable, $x_{1j}, x_{2j} \ldots, x_{nj}$, is assumed to be fixed for different j (denoted as $\mathbf{x} = \{x_1, x_2, \ldots, x_n\}$). In the nonparametric regression context, this case is the so-called common fixed design. This is usually the case in practical calibration applications in industrial manufacturing and is also consistent with the settings in previous chapters. Without loss of generality, we assume that $x_1 \leq x_2 \leq \ldots \leq x_n$ and x_i vary in the interval [0, 1]. Otherwise, we can obtain this form through some appropriate linear transformations and permutation. This assumption facilitates the technical arguments and eases the exposition. In practice, it is not necessary to make such a transformation in advance. In the later subsections, we introduce how to use one single chart to monitor the regression function and variation simultaneously.

10.1.2 The Generalized Likelihood Ratio Test for Profile Data

To monitor a general profile model (10.12), the regression function, g, and the standard deviations, σ, should be controlled simultaneously. An effective nonparametric monitoring approach for g can be designed based on the GLR statistics proposed by Fan et al. (2001). Suppose that $\{y_i, x_i\}_{i=1}^n$ is a random profile sampled from model (10.2). Consider the simple null hypothesis-testing problem:

$$H_0 : g = g_0 \quad \sigma = \sigma_0 \qquad \longleftrightarrow \qquad H_1 : g \neq g_0 \quad \sigma = \sigma_0. \tag{10.3}$$

According to the normality assumption of the error, the logarithm of the likelihood function is given by

$$-n \ln\left(\sqrt{2\pi}\sigma\right) - \frac{1}{2\sigma^2}\sum_{i=1}^{n}(y_i - g(x_i))^2 .$$

Then, the maximum likelihood under H_0 can be expressed by

$$l_0 = -n \ln\left(\sqrt{2\pi}\sigma_0\right) - \frac{1}{2\sigma_0^2}\sum_{i=1}^{n}(y_i - g_0(x_i))^2 .$$

Fan et al. (2001) suggest to replace the unknown function g under H_1 with a reasonable nonparametric estimator that leads to the logarithm of likelihood function H_1:

$$l_1 = -n \ln\left(\sqrt{2\pi}\sigma_0\right) - \frac{1}{2\sigma_0^2}\sum_{i=1}^{n}(y_i - \hat{g}(x_i))^2 ,$$

and the GLR test statistic

$$lr = -2(l_0 - l_1) = \frac{1}{\sigma_0^2}\left[\sum_{i=1}^{n}(y_i - g_0(x_i))^2 - (y_i - \hat{g}(x_i))^2\right]. \tag{10.4}$$

Large values of lr lead to a rejection of the null hypothesis.

In Fan et al. (2001), the local linear smoother (see Fan and Gijbels 1996) is used to estimate, g, and we also follow this approach. The local linear estimator for any given point, x, takes the form $\hat{g}(x) = \sum_{i=1}^{n} W_{ni}(x)y_i$, where

$$W_{ni}(x) = U_{ni}(x)\Big/\sum_{j=1}^{n} U_{nj}(x),$$

$$U_{nj}(x) = K_h(x_j - x)\left[m_{n2}(x) - (x_j - x)m_{n1}(x)\right],$$

$$m_{nl}(x) = \frac{1}{n}\sum_{j=1}^{n}(x_j - x)^l K_h(x_j - x), \quad l = 1, 2,$$

and $K_h(\cdot) = K(\cdot/h)/h$ with K being a symmetric probability density function and h a bandwidth. Here, to ease the exposition, we define an $n \times n$ smoothing matrix, $\mathbf{W}$, as

$$\mathbf{W} = (\mathbf{w}_n(x_1), \mathbf{w}_n(x_2), \ldots, \mathbf{w}_n(x_n))^T,$$

where $\mathbf{w}_n(x_i) = (W_{n1}(x_i), W_{n2}(x_i), \ldots, W_{nn}(x_i))^T$. Then, the lr in model (10.4) can be reexpressed in vector–matrix notation as

$$lr = \frac{1}{\sigma_0^2}\left[(\mathbf{y} - \mathbf{g}_0)^{\otimes} - (\mathbf{y} - \mathbf{W}\mathbf{y})^{\otimes}\right],$$

where $\mathbf{g}_0 = (g_0(x_1), g_0(x_2), \ldots, g_0(x_n))^T$, $\mathbf{y} = (y_1, y_2, \ldots, y_n)^T$, and $\mathbf{A}^{\otimes}$ means $\mathbf{A}^T\mathbf{A}$.

The following Wilks phenomenon is revealed by Fan et al. (2001): under some regular conditions, the asymptotic null distributions of model (10.4) are independent of nuisance functions and approximately follow a scaled chi-square distribution. Hence, we now can develop a control chart based on the lr statistics. However, the dependence of the small-sample distribution of lr on g_0 makes it difficult and infeasible to construct the control chart for monitoring the general profile. To overcome this kind of difficulty, we consider transforming each profile dataset, $\{y_i, x_i\}_{i=1}^n$, to $\{y_i - g_0(x_i), x_i\}_{i=1}^n$ first, because the function, g_0, is known in advance (or estimated from some historical data). Now, the testing problem (10.3) is equivalent to the problem $H_0 : g = 0 \quad \sigma = \sigma_0 \longleftrightarrow H_1 : g \neq 0 \quad \sigma = \sigma_0$. Denote $z_i = (y_i - g_0(x_i))/\sigma_0$ and $\mathbf{z} = (z_1, z_2, \ldots, z_n)^T$. Then, the GLR statistics will be

$$lr_z = \mathbf{z}^{\otimes} - (\mathbf{z} - \mathbf{W}\mathbf{z})^{\otimes} = \mathbf{z}^T\mathbf{V}\mathbf{z}^T,$$

where $\mathbf{V} = \mathbf{W}^T + \mathbf{W} - \mathbf{W}^{\otimes}$. Because the GLR test is indeed general, simple and powerful for nonparametric testing problems based on the regression function estimation, next we integrate it with the MEWMA procedure for monitoring the general profile.

10.1.3 The Control Scheme for Monitoring a General Profile

With the jth random profile collected over time, we have the observations set $\{\mathbf{x}, \mathbf{y}_j\}$. Because both the regression relationship and variance need to be monitored, a nonparametric test for the σ_j is also necessary. Note that, under certain conditions, we have the following well-known fact:

$$\hat{\sigma}_j^2 = \frac{1}{n}\left(\mathbf{z}_j - \mathbf{W}\mathbf{z}_j\right)^{\otimes} = \sigma_j^2 + O_p(n^{-1/2}) + O_p((nh)^{-1}),$$

regardless of if the regression function has changed. Thus, we can utilize the statistic $\left(\mathbf{z}_j - \mathbf{W}\mathbf{z}_j\right)^{\otimes}$ to conduct an appropriate test for a possible change in variance, in which a very large (small) value indicates an increase (decrease) in the variance of the process.

In adopting the GLR test and the foregoing test for variance for charting, we consider developing an EWMA-type chart, which is known to be more sensitive to small and moderate shifts than a Shewhart-type chart is. At first, we transform the $\hat{\sigma}_j^2$ to a normal random variable. That is,

$$\tilde{\sigma}_j = \Phi^{-1}\left\{\psi\left(n\hat{\sigma}_j^2; \mathbf{I} - \mathbf{V}\right)\right\},$$

where $\Phi^{-1}(\cdot)$ is the inverse of the standard normal cumulative distribution function (CDF) and $\psi(\cdot; \mathbf{A})$ is the CDF of random variable $n\hat{\sigma}_j^2$ when the process is IC. The advantage of this transformation is that the distribution of $\tilde{\sigma}_j$ would be symmetric, so that the EWMA control chart is not only relatively easy to construct but is also sensitive to decreases in the variance. An accurate and fast method for approximating the $\psi(\cdot; \mathbf{A})$ is provided in Appendix A.

Next, we denote $\mathbf{u}_j$ as $(\mathbf{z}_j^T, \tilde{\sigma}_j)^T$, which is an $(n+1)$-variate random vector, and $\mathbf{\Sigma} = \begin{pmatrix} \mathbf{V} & \mathbf{0} \\ \mathbf{0} & 1 \end{pmatrix}$, which is a $(n+1)$-dimensional symmetric matrix. The proposed EWMA charting statistic can then be defined as

$$\mathbf{e}_j = \lambda \mathbf{u}_j + (1-\lambda)\mathbf{e}_{j-1} \quad j = 1, 2, \ldots, \tag{10.5}$$

where $\mathbf{e}_0$ is a $(n+1)$-dimensional starting vector and λ is a weight, ($0 < \lambda \leq 1$), that regulates the magnitude of the smoothing. The chart signals if

$$Q_j = \mathbf{e}_j^T \mathbf{\Sigma} \mathbf{e}_j > L\frac{\lambda}{2-\lambda}, \tag{10.6}$$

where $L > 0$ is chosen to achieve a specified IC ARL. Hereafter, we call this nonparametric control scheme a NEWMA chart for brevity. The form of the NEWMA chart is analogous to the MEWMA charts used in Zou et al. (2007) for parametric monitoring of a linear profile.

One issue is that $\mathbf{z}_j$ and $\tilde{\sigma}_j$ are usually dependent, but we have not considered their correlation in the matrix Σ. In fact, putting $\mathbf{z}_j$ and $\tilde{\sigma}_j$ together makes the resulting $\mathbf{u}_j$ an obscure joint distribution with normal marginal distributions. However, we may expect the proposed schemes to still be effective, because the ignorance of the correlation of $\mathbf{z}_j$ and $\tilde{\sigma}_j$ does not affect the testing of the regression relationship and the variance to any important extent, and the simulation results also concur.

Remark 1 Here, two possible extensions of the NEWMA chart are discussed to indicate its versatility. In practical applications, engineers may use the Phase-I IC samples directly rather than fitting a linear or nonlinear regression model before starting the Phase-II monitoring. In such a case, one may use the local linear estimator of g_0 based on the IC profile samples, say $\hat{g}_0$, as a replacement for g_0 in the NEWMA chart. There is no need to have equal design points for each IC profile sample, but the number and positions of the design points should be properly chosen in order to describe the regression function well. Moreover, the number of IC samples should be large enough to result in a sufficient signal-to-noise ratio so that the properties of the IC ARL can be obtained. Note that one could collect a larger number of observations in each profile for Phase-I analysis than for Phase-II monitoring. That is to say, we may expend more measurement efforts on the Phase-I analysis so that the underlying regression model can be estimated accurately and, correspondingly, desired IC run-length behavior can be achieved, while a large n may not be necessary in Phase II because the OC condition could usually be effectively captured by using relatively small or moderate n. Based on extensive simulations and theoretical properties of the local linear smoother, in the Phase-I analysis, we recommend using at least 40 IC profile samples, and the number of observations in each profile should not be fewer than 50. Of course, this is a general guide, and the engineers need to take the engineering knowledge about the specific profile into consideration in practical applications, such as the smoothness and variation of the profile.

Another natural extension of NEWMA charts is to deal with profiles with multidimensional regressors. For this case, one may use a multivariate local linear regression estimator in place of the univariate one, and the NEWMA chart can still be used. The technical details of the multivariate local linear regression estimator can be found in Section 7.8 of Fan and Gijbels (1996) and the references therein.

10.1.4 Guideline on Design and Implementation

This subsection provides guidelines on how to design and implement the NEWMA scheme. Several practical issues will be discussed including the choices of the kernel function, $K(\cdot)$, the bandwidth, h, the smoothing weight, λ, and the determination of the control limits, L.

On choosing the smoothing weight, λ: In general, a smaller λ leads to a quicker detection of smaller shifts (Lucas and Saccucci 1990). The starting vector, $\mathbf{e}_0$, is chosen to be the zero vector. In fact, the fast initial response (Lucas and Saccucci 1990) can also be extended to the NEWMA chart.

On choosing the sample size and regressor positions: For the NEWMA chart to perform well, a relatively large sample size, n, of the profile is needed, especially with complicated profiles. This is because, due to its flexibility, the nonparametric smoother absorbs considerably more degrees of freedom than do parametric approaches. However, this has become a less significant limitation with advances in electronic, sensor, and information technologies. New instruments can capture more information, and a large amount of data becomes available at one time. Also, we recommend that engineers carefully choose the regressor, x_is (i.e., the design point positions), in terms of the IC profile model. Since the shifts often tend to appear yet are difficult to detect in the regions of the profile that have more curvature, we suggest that a good design should have a higher concentration of points at xs where $g_0(x)$ has a sharp peak.

On choosing the kernel function, $K(\cdot)$: As noted by Fan et al. (2001), the construction of the GLR test does not depend on the special structures of the smoothing procedure, although proofs of the theorems may need minor modification. For the kernel function, there are many kernels to meet the requirements, such as Uniform, Epanechnikov, Quadratic, and Gaussian kernels. We observe that the performance of NEWMA is mostly unaffected by the choice of the kernel according to our simulations. For simplicity, the Epanechnikov kernel,

$$K_E(u) = \frac{3}{4}(1 - u^2)I(|u| \leq 1),$$

is recommend in practice.

On choosing the bandwidth, h: In the context of nonparametric regression estimation, the optimal bandwidth, h, is usually determined by minimizing the asymptotic mean square error of the estimator. Frequently used bandwidth selection techniques are data-driven methods, such as the least square cross-validation (CV) and generalized cross-validation (GCV) (see Fan and Gijbels 1996). However, the data-driven bandwidth methods that are well suited for producing visually smooth estimates of the underlying curves may not in general be appropriate for our online monitoring problem.

If an observed profile of data indeed comes from the IC model, then the optimal bandwidth for local fitting of the $\mathbf{Z}$ should be close to 1 (according to our assumption that $x_i \in [0, 1]$), and data-driven bandwidth selectors will lead to a large bandwidth. However, the distributional properties of GLR statistics rely implicitly on the assumption that $h \to 0$. Moreover, we usually do not have specific information about the OC model so that we cannot choose an optimal h for the OC condition before we start Phase-II monitoring. Roughly speaking, one would expect the size of the optimal bandwidth to be proportional to the smoothness of the underlying function. In other words, a very smooth difference between IC and OC regression models requires larger bandwidths than do less smooth differences, all other factors being equal.

However, we expect the value of h to be less important since the performance of the chart is indeed insignificantly affected by h. The amount of smoothing applied will affect the power of the test, but we have observed in many simulation results (not reported here but available from the authors) that the observed significance changes little over a wide range of the values of h. Hence, we recommend using the following empirical bandwidth formula,

$$h_E = c \times \left(\frac{1}{n} \sum_{i=1}^{n} (x_i - \bar{x})^2 \right)^{1/2} n^{-1/5}, \tag{10.7}$$

where $\bar{x} = \sum_{i=1}^{n} x_i$ and c is a constant. Empirically, c can be any value in the interval [1.0,2.0]. Note that, here, $n^{-1/5}$ is the order of the classical optimal bandwidth for curve estimation and $\frac{1}{n}\sum_{i=1}^{n}(x_i - \bar{x})^2$ is a measure of the sparseness of the design points, which is also involved in an asymptotic optimal h formula in the context of nonparametric estimation (see Fan and Gijbels 1996 for details).

Remark 2 It can be verified that using model (2.6) to determine the bandwidth guarantees that the design of the NEWMA chart is affine-invariant to covariate $\mathbf{x}$. That is to say that, the smoothing matrix, $\mathbf{W}$, remains unchanged if we do the same linear transformation for $\mathbf{x}$.

On determining the control limits, L: Finally, there is a vital issue remaining to be addressed, which is how to determine the control limit, L, after all of the preceding functions and parameters have been chosen. We note that the NEWMA chart is a single two-sided chart. Thus, only one control limit, L, needs to be determined. As mentioned in Section 10.2.2, the L is free of the IC model, $g_0(x)$. Hence, after n, $\mathbf{x}$, λ, $K(\cdot)$, h and the desired IC ARL are fixed, the values of L can be determined using Monte Carlo simulations by generating a random series, $\mathbf{z}_j$. Although the simulations involve calculations in the quadratic form, considering that the matrix $\boldsymbol{\Sigma}$ is fixed, the computation task is trivial by virtue of the massive computing and data storage capabilities of modern computers. Determining the control limit of such a nonparametric scheme is just a little more time-consuming than that of a parametric approach, but much more efficiency will be gained when their assumption on OC models is violated.

10.1.5 Diagnostic Aids in Profile Monitoring

In the practice of quality control, in addition to detecting a process change quickly, it is also critical to determine when the process change occurs and to identify the kind of change that occurs in a profile after an OC signal is triggered. A diagnostic aid to locating the change point in the process and to isolating the type of changes in the profile will help an engineer to identify and eliminate the root cause of the problem quickly and easily. In this subsection, we discuss the diagnosis of a general profile and provide the following operational steps for systematic diagnosis.

Step 1: Identify the location of the change point. This is the first and critical step in our diagnostic procedure, because we need it to separate the OC profiles from the entire observed profile samples so that we can make an accurate inference about this type of change. Here, the generalized maximum likelihood estimator of the change-point statistic is used. We assume that an OC signal is triggered at the profile, k, by the NEWMA chart. Our suggested estimator of the change point, τ, of the step shift is given by

$$\hat{\tau} = \arg \max_{0 \leq t < k} \{lr_{t,k}\}, \tag{10.8}$$

where $lr_{t,k}$ is the GLR statistic. The expressions of $lr_{t,k}$ and the involved deductions are given in Appendix B. Zou et al.'s (2008) present an asymptotic result on the consistency of the change-point estimator (2.7), which ensures that the estimator is asymptotically effective.

Step 2: Determine if the variance is stable. After obtaining the estimate of the change point, we firstly suggest to test if the variance of the profile has changed based on $(k - \hat{\tau})$ OC samples. This is because once the variation of the process has shifted, we usually cannot make an accurate inference on the regression relationship. By using an extension of the variance estimator introduced in Section 10.2.2 to the multiple samples case, we may apply the following test statistic and acceptance region

$$\psi_1^{-1}\left(\frac{\alpha}{2}; \mathbf{I} - \mathbf{V}, k - \hat{\tau}\right) < \sum_{j=\hat{\tau}+1}^{k} \mathbf{z}_j^T(\mathbf{I} - \mathbf{V})\mathbf{z}_j < \psi_1^{-1}\left(1 - \frac{\alpha}{2}; \mathbf{I} - \mathbf{V}, k - \hat{\tau}\right), \tag{10.9}$$

where $\psi_1(\cdot; \mathbf{A}, l)$ is the CDF of the sum of l independent random variables $\mathbf{z}^T\mathbf{A}\mathbf{z}$ and correspondingly $\psi_1^{-1}(\alpha; \mathbf{A}, l)$ is the α percentile of the $\psi_1(\cdot; \mathbf{A}, l)$ distribution. $\psi_1(\cdot; \mathbf{A}, l)$ can be easily approximated by $\psi(\cdot; \mathbf{A})$ and use of the additive property of chi-squar random variables. If the null hypothesis is rejected, an increase in the variation corresponds to greater inaccuracies in the manufacturing process or increased measurement error, and a decrease in the variation about the curve would indicate an improvement in the process.

Step 3: Determine if the regression function has changed. If the stability of the variance is ensured in Step 2, we then need to diagnose whether the regression relationship in the profile has changed. Again, the GLR test can be adopted and simple derivations lead to the following test statistic,

$$\frac{1}{(k - \hat{\tau})}\left(\sum_{j=\hat{\tau}+1}^{k} \mathbf{z}_j\right)^T \mathbf{V}\left(\sum_{j=\hat{\tau}+1}^{k} \mathbf{z}_j\right) > \psi^{-1}(1 - \alpha; \mathbf{V}). \tag{10.10}$$

where $\psi^{-1}(\alpha; \mathbf{A})$ is the α percentile of the distribution of quadratic form $\mathbf{z}^T\mathbf{A}\mathbf{z}$. If we accept the null hypothesis that there is no change in the regression function, we may conclude that a false alarm is triggered by NEWMA. Otherwise, we may proceed to the next step.

Step 4: Further diagnosis by graphical analysis. At this point, the engineer may care about the following problems: which part of the regression curve changes; what is the OC regression function; and how great is the difference between the IC and OC models? It is obvious that using the traditional hypothesis testing method to

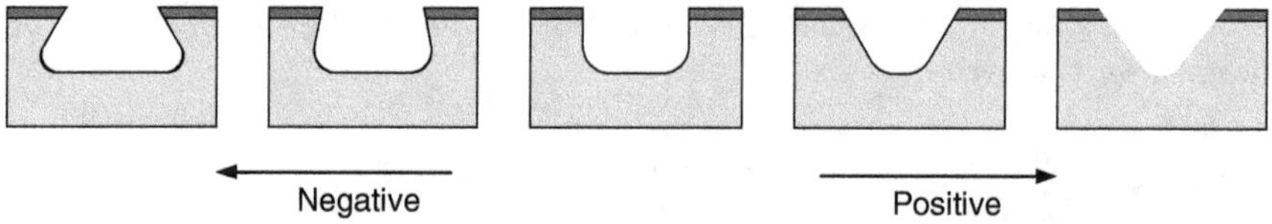

Figure 10.1 Illustrations of various etching profiles from a DRIE process.

get the answer seems inappropriate and infeasible. However, some simple methods, such as graphical analysis, may shed light on these problems. We suggest plotting the nonparametric smoothing curve of the average of $(k - \hat{\tau})$ sample profiles, say $\frac{1}{k-\hat{\tau}} \sum_{j=\hat{\tau}+1}^{k} \mathbf{W}\mathbf{y}_j$, and the IC profile together.

On the basis of that, a priori engineering knowledge and experience can then be fused to offer a visual and practical interpretation of the foregoing problems. Such graphical analysis has been commonly adopted in various industrial applications with the emergence of massive computing and data storage capabilities. We will illustrate this with a semiconductor manufacturing example in Section 10.2.6.

10.1.6 An Application

Here, we apply the proposed monitoring scheme to a deep reactive ion etching (DRIE) process in semiconductor manufacturing, which is very critical to the quality of the output wafer and requires careful control and monitoring on a run-to-run basis. In the DRIE process, the desired profile is one with smooth and vertical sidewalls, which is called the anisotropic profile. Ideally, the sidewalls of a trench are perpendicular to the bottom of the trench with certain degrees of smoothness around the corners. Various shapes of profiles, such as positive and negative profiles, which are due to underetching and overetching, are considered to be unacceptable (cf., Figure 10.1). More detailed discussion about the DRIE example may be found in Zou et al. (2007) and the references therein.

In practice, it is not necessary to monitor the entire DRIE profile because the two sides of each profile are usually symmetrical. Here, similar to the approach by Zou et al. (2007), we focus on one side of the profile (e.g., the left side), and then rotate the side of the profile by 45 degrees along a reference point according to a prescribed chosen coordinate as in Figure 10.2. The transformed nonlinear profile in Figure 10.2 will be used to demonstrate the proposed monitoring scheme.

It can be seen that each side of the DRIE profile may be constituted by three key parts: (1) the smooth and straight sidewall (corresponding to (i) in Figure 10.2), (2) the smooth and curly corner (corresponding to (ii) in Figure 10.2), and (3) the flat bottom (corresponding to (iii) in Figure 10.2). Here, we only select the segments near the corner for our investigation, since engineers consider that those segments contain enough critical information to distinguish the OC conditions in the process.

Zou et al. (2007) found that the part (ii) profile may be adequately described by the following quadratic polynomial model

$$y_{ij} = 0.62 \cdot x_i^2 + \varepsilon_{ij}, \quad i = 1, \ldots, n, \tag{10.11}$$

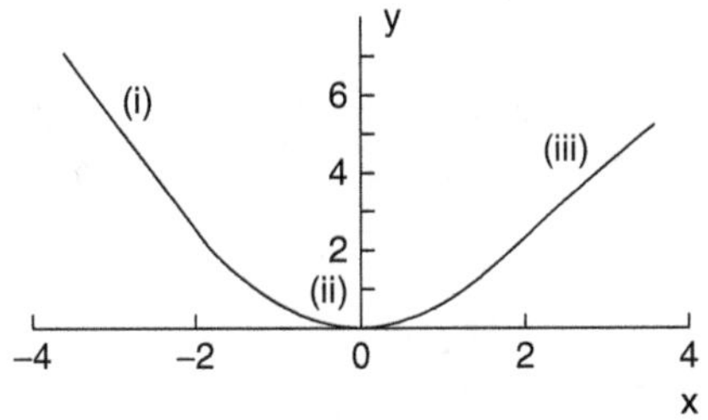

Figure 10.2 Illustration of modeling the DRIE profile.

and the dataset was tabulated in the Appendix of Zou et al. (2007), which contains five IC profile samples and nine OC profile samples. Here, we apply our proposed NEWMA scheme to monitoring this dataset. For each profile, $x_i, i = 1, \ldots, 11$ are fixed as equally spaced values, −2.5, (0.5), 2.5. Detailed implementation steps of the proposed scheme are as follows.

Step 1. Choose the kernel function, $K(\cdot)$, bandwidth h, smoothing weight, λ, and the desired IC ARL following the guidelines in Section 10.2.4. Here, we adopt the Epanechnikov kernel, and h is determined by Equation (10.7) with $c = 1.5$. According to the values of the design points, x_is, we can obtain $L = 18.09$ given that ARL = 370 and $\lambda = 0.2$. Consequently, $L\frac{\lambda}{2-\lambda}$ will be 2.01. Then, we can construct the NEWMA control chart as in Figure 10.3.

Step 2. Start monitoring the process and sequentially compute the $\mathbf{z}_j$ and $\mathbf{e}_j$ vectors. Consequently, compute the plot statistic, Q_j, in (10.6) and compare it with the control limit, $L\frac{\lambda}{2-\lambda}$. In Figure 10.3, we can see that the NEWMA chart signals at the 14th sample, just as effectively as the parametric MEWMA chart of Zou et al. (2007) does.

Step 3. Then, by computing the values of $lr_{j,14}$ for $j = 0, 1, \ldots, 13$, we can find that its maximum occurs at $j = 6$ when $lr_{6,14}$ is 18.60. By computing the test statistics given in models (10.9) and (10.10), respectively, we can obtain the corresponding

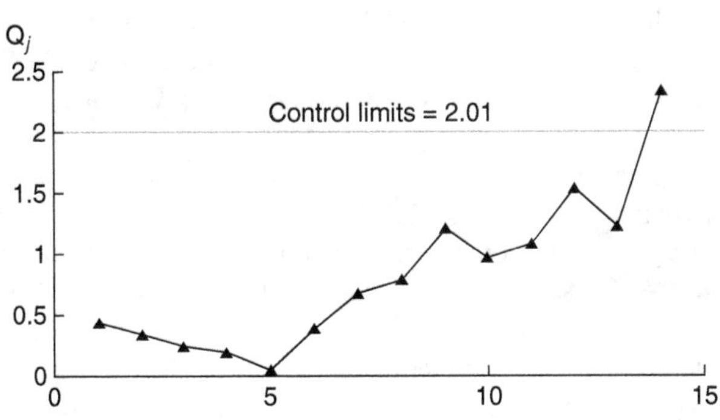

Figure 10.3 The NEWMA chart for monitoring the part (ii) of the DRIE example.

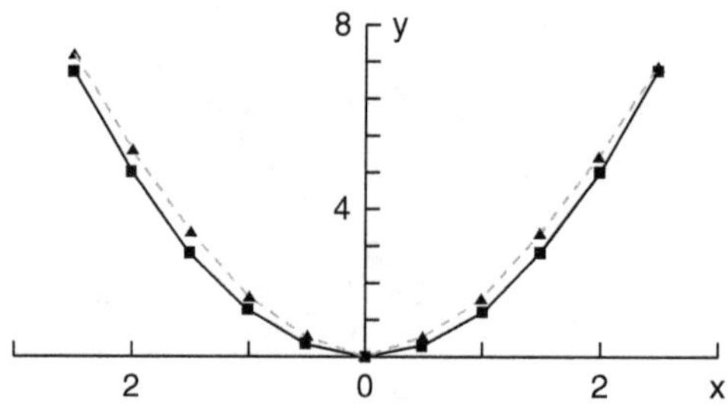

Figure 10.4 Graphical diagnosis of the part (ii) of the DRIE example. The solid and the dotted lines represent the IC and the estimated OC models, respectively.

p-values of 0.209 and 0.002, which indicate that the variance is stable but there may be a shift in the regression function.

Step 4. Finally, we plot the local linear smoothing curve of the average of the last eight sample profiles and the IC profile model together as shown in Figure 10.4. From this figure, we can see that the OC profile model presents an apparent negative trend, which may be due to overetching of the DRIE process.

Next, in order to demonstrate the effectiveness of our proposed approach, we consider monitoring the whole transformed DRIE profile (i.e., all three parts) shown in Figure 10.2. Obviously, these three parts cannot be readily viewed as a polynomial or a general linear model, so existing parametric methods cannot be applied directly. Thus, we choose to apply the proposed NEWMA chart to monitor such a profile.

As discussed in Section 10.2.3, considering the efficiency of monitoring and diagnosing inferences in using the NEWMA chart, 35 design points, x_is, are chosen, which are -3.6, (0.3), -1.8, -1.5, (0.15), 1.5 and 1.8, (0.3), 3.6. Note that the corner requires more design points (smaller spaced values) than do the other two parts of the profile. This setting is consistent with our mentioned principle on the choice of design points in Section 10.2.4. Here, we collect dimensional readings (seventy design points) by the scanning electron microscope from $m = 18$ anisotropic profiles that are known to be in control. The local linear smoothing curve of the average of the these 18 IC profiles is plotted in Figure 10.5. This curve can be deemed as an estimate ($\hat{g}_0$) of the true IC profile model (g_0) and thus we do not need to model this profile as some piecewise polynomial (or other nonlinear model) (see Remark 1). In addition, the estimate of variance is 0.409. From here, we can start the monitoring and diagnosing procedure.

Similar to the preceding monitoring case, we use ARL $= 370$, $\lambda = 0.2$, and $c = 1.5$ and consequently the control limit is 2.30. For ease of illustration, we generate six IC profiles by adding random errors to the $\boldsymbol{g}_0$ estimated by those 18 anisotropic profiles. We also obtain three OC sample profiles that are classified as inferior profiles based on engineering knowledge. In this example, we artificially assume that we first monitor the six simulated IC samples and then obtain the three OC profiles. Accordingly, we construct NEWMA and plot the test statistics, Q_j, one by one as shown in Figure 10.6.

We can see that the NEWMA chart signals after the third OC sample is monitored. Then, by looking at the values of $lr_{j,9}$ for $j = 0, 1, \ldots, 8$, we find that its maximum

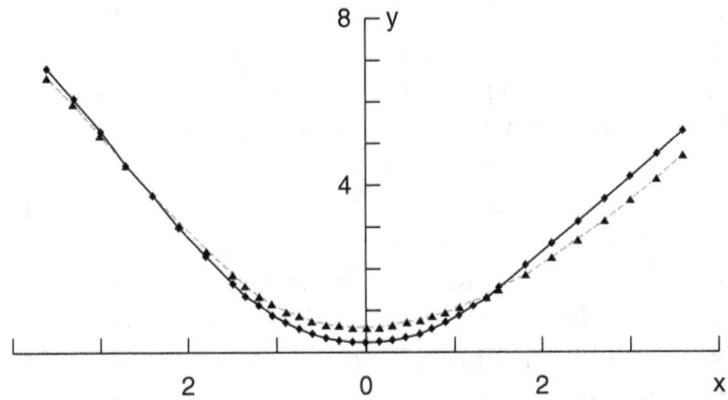

Figure 10.5 Graphical diagnosis of the whole DRIE profile. The solid and the dotted lines represent the IC and the estimated OC models, respectively.

occurs at $j = 6$ with $lr_{6,9} = 41.09$. This maximum accurately indicates the change-point location, τ, of the shift. In addition, by computing the test statistics given in models (10.9) and (10.10), respectively, we can obtain the corresponding p-values of 0.633 and 10^{-6}, which significantly indicate that the regression function has changed. Finally, the local linear smoothing curves of the average of these three sample profiles are shown in Figure 10.5 coupled with the IC curve. From this figure, we can see that the OC profile model indicates an apparent positive trend that may be due to underetching of the DRIE process, so we would suggest that the ICP machine settings and the DRIE process conditions should be reexamined.

10.2 NONPARAMETRIC PROFILE MONITORING USING CHANGE-POINT FORMULATION AND ADAPTIVE SMOOTHING

10.2.1 Motivation

The NEWMA control chart introduced in the previous section has a number of model assumptions. In certain applications, these assumptions may not all hold. In such

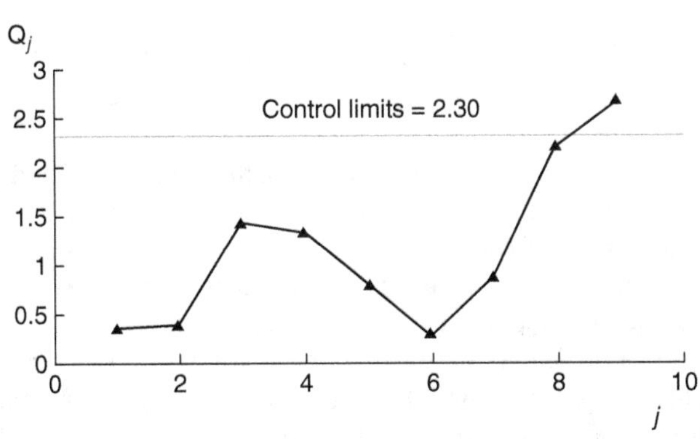

Figure 10.6 Thd NEWMA chart for the DRIE example.

cases, its performance would be unsatisfactory, which is briefly discussed below. First, NEWMA makes explicit use of the IC true regression function g and the error variance σ^2 in model (10.2). In practice, both g and σ^2 are often unknown. Therefore, they need to be estimated from IC data. If such data are of small to moderate size, then there would be considerable uncertainty in the parameter estimates, which in turn would distort the IC run-length distribution of the control chart. Even if the control limit of the chart is adjusted properly to attain a desired IC run-length behavior, its OC run length would still be severely compromised. Second, the NEWMA chart assumes that the error distribution, say F, is Normal. In practice, however, F is often unknown. In such cases, it remains challenging to design the NEWMA chart properly. Third, numerical examples in Zou et al. (2008) demonstrate that NEWMA depends heavily on the choice of a bandwidth used in smoothing profile data. However, the proper choice of this parameter is not discussed thoroughly. Fourth, the NEWMA chart has another challenge that its best performance can be achieved only after it is "tuned" to the shift magnitude which is often unknown in practice.

In this section, we introduce a new control chart proposed by Zou et al. (2009) that addresses all issues raised above about the NEWMA chart. The new control chart adopts the on line change-point detection approach. It handles sequential profile readings by simultaneously updating parameter estimates and checking for OC conditions. An adaptive procedure for selecting the bandwidth parameter is incorporated to the construction of the control chart so that it can adapt to the unknown smoothness of the difference between the IC and OC regression functions, which remarkably improves its robustness to various OC profile conditions. Furthermore, a bootstrap procedure is used to determine the control limits of the proposed chart without any knowledge of F, as long as a few, say m_0, IC profiles are available beforehand. These m_0 IC profiles are mainly for estimating the error distribution. Thus, m_0 does not need to be large.

10.2.2 Change-Point Model and Generalized Likelihood Ratio Test

To ease presentation, we assume that model (10.2) has an unknown but stable error variance over time. The change-point model can be expressed as

$$y_{ij} = \begin{cases} g(x_i) + \varepsilon_{ij}, & i = 1, \ldots, n, \quad \text{if} \quad 1 \le j \le \tau \\ g_1(x_i) + \varepsilon_{ij}, & i = 1, \ldots, n, \quad \text{if} \quad \tau < j \le t, \end{cases} \tag{10.12}$$

where τ is the unknown change point, $g \neq g_1$ are the unknown IC and OC regression functions, and ε_{ij} are i.i.d. errors with an unknown distribution F of mean 0 and unknown variance σ^2.

To check whether a possible change point occurs at $\tau = k$, a two-sample GLR test for testing the null hypothesis that g and g_1 are the same can be derived in a similar way to that in Fan et al. (2001). For ease of exposition, we can think of F as a Normal distribution, although this is not necessary in either asymptotic theory or practical use of the proposed chart. The major idea in deriving the GLR test is to replace the unknown functions g and g_1 by their nonparametric estimators constructed from profile data when we define the GLR statistic. To be specific, the generalized log-likelihood functions under the IC and OC conditions are, respectively,

$$l_0 = -nt \ln\left(\sqrt{2\pi}\sigma\right) - \sum_{j=1}^{t}\left[\frac{1}{2\sigma^2}\sum_{i=1}^{n}\left(y_{ij} - \hat{g}^{(0)}(x_{ij})\right)^2\right]$$

$$l_1 = -nt \ln\left(\sqrt{2\pi}\sigma\right) - \sum_{j=1}^{k}\left[\frac{1}{2\sigma^2}\sum_{i=1}^{n}\left(y_{ij} - \hat{g}^{(1)}(x_{ij})\right)^2\right]$$
$$- \sum_{j=k+1}^{t}\left[\frac{1}{2\sigma^2}\sum_{i=1}^{n}\left(y_{ij} - \hat{g}_1(x_{ij})\right)^2\right],$$

where $\hat{g}^{(0)}(\cdot)$ denotes the local linear kernel estimator (LLKE) of g based on the pooled t profiles, and $\hat{g}^{(1)}(\cdot)$ and $\hat{g}_1(\cdot)$ denote the LLKEs of g and g_1 based on the first k and the remaining $t-k$ profiles, respectively. After some mathematical manipulations, the GLR statistic up to time point t is defined by

$$T_{h,k,t} = -2(l_0 - l_1) = \frac{k(t-k)}{t\sigma^2}\left(\bar{\mathbf{y}}_{0,k} - \bar{\mathbf{y}}_{k,t}\right)^T \mathbf{V}_h \left(\bar{\mathbf{y}}_{0,k} - \bar{\mathbf{y}}_{k,t}\right), \tag{10.13}$$

where

$$\bar{\mathbf{y}}_{i,m} = \frac{1}{m-i}\sum_{j=i+1}^{m}\mathbf{y}_j, \quad \mathbf{V}_h = \mathbf{W}_h^T + \mathbf{W}_h - \mathbf{W}_h^{\otimes},$$

$$\mathbf{W}_h = (\mathbf{w}_n(x_1), \ldots, \mathbf{w}_n(x_n))^T, \quad l = 1, 2,$$

where the definition of $\mathbf{w}_n(\cdot)$ is as the same as in Section 10.2.2. Obviously, the test statistic (10.13) is a two-sample counterpart of the one-sample GLR test statistic in Fan et al. (2001). Since σ^2 is assumed unknown here, we replace it by the following consistent nonparametric estimator originally suggested by Hall and Marron (1990):

$$\hat{\sigma}_t^2 = \frac{1}{t(n-df)}\sum_{j=1}^{t}(\mathbf{y}_j - \mathbf{W}_h\mathbf{y}_j)^{\otimes}, \tag{10.14}$$

where $df = \mathrm{tr}(\mathbf{V}_h)$.

Like many other smoothing-based tests, performance of the test (10.13) depends upon the smoothing bandwidth h. Selection of h such that the testing power is optimal remains an open problem in this area. It is widely recognized that the optimal h for nonparametric curve estimation is generally not optimal for testing (see, e.g., Hart 1997). A uniformly most powerful test usually does not exist because nonparametric regression functions have infinite dimensions. For the lack-of-fit testing problem, Horowitz and Spokoiny (2001) suggest choosing a single h based on the maximum of a studentized conditional moment test statistic over a sequence of smoothing parameters, and prove that the resulting test would have certain optimality properties. Based on these results, Guerre and Lavergne (2005) suggest choosing h from a sequence of pre specified values and demonstrate that their method has a number of favorable properties. Because this method is easy to use and has good performance in various cases, we use it here for choosing h. Let $\mathcal{H}_n$ be a set of admissible smoothing parameter values defined to be the following geometric grid:

$$\mathcal{H}_n = \{h_j = h_{\max} a^{-j} : h_j \geq h_{\min},\ j = 0, \ldots, J_n\}, \tag{10.15}$$

where $0 < h_{\min} < h_{\max}$ are the lower and upper bounds and $a > 1$ is a parameter. Clearly, the number of values in $\mathcal{H}_n$ is $J_n \leq \log_a(h_{\max}/h_{\min})$. Following Guerre and Lavergne (2005), we select h to be

$$\tilde{h} = \arg\max_{h \in \mathcal{H}_n} \left\{(T_{h,k,t} - \mu_h) - (T_{h_0,k,t} - \mu_{h_0}) - \gamma_n v_{h,h_0}\right\},$$

where $\gamma_n > 0$ is a chosen penalty parameter, μ_h is the mean of $T_{h,k,t}$, and $v^2_{\tilde{h},h_0}$ is the variance of $T_{h,k,t} - T_{h_0,k,t}$. After accommodating the chosen bandwidth, the testing statistic becomes

$$\tilde{T}_{k,t} = (T_{\tilde{h},k,t} - \mu_{\tilde{h}})/v_{h_0}, \tag{10.16}$$

where $v^2_{h_0}$ is the variance of $T_{h_0,k,t}$, $\mu_{\tilde{h}} = \sum_{i=1}^n V_{\tilde{h}}^{(ii)}$, and $V_{\tilde{h}}^{(ij)}$ denotes the (i, j) th element of the matrix $\mathbf{V}_{\tilde{h}}$. For given h and h_0, it can be shown that the following estimators of v^2_h and $v^2_{\tilde{h},h_0}$ are consistent:

$$\hat{v}^2_h = 2\sum_{i=1}^n \sum_{j=1}^n \left[V_h^{(ij)}\right]^2, \quad \hat{v}^2_{h,h_0} = 2\sum_{i=1}^n \sum_{j=1}^n \left[V_h^{(ij)} - V_{h_0}^{(ij)}\right]^2.$$

Some statistical properties of the adaptive GLR test statistic $\tilde{T}_{k,t}$, including asymptotic null distribution and consistency under contiguous alternatives, are given in the Appendix of Zou et al. (2009).

In practice, the true change-point τ is often unknown. To test whether there is a shift in the regression function g for profiles up to time point t, we consider the following adaptive GLR test statistic:

$$\tilde{T}_{\hat{t}} = \max_{1 \leq k < t} \tilde{T}_{k,t}, \tag{10.17}$$

which is the maximum over all possible split points (i.e., binary segmentation). The maximizer $\hat{\tau}$ is used as an estimator of τ.

10.2.3 Phase II SPC

The change-point detection procedure (10.13)–(10.17) is developed for a collection of profiles with fixed sample size. In this part, it is adapted for online Phase-II SPC. Phase-II linear profile monitoring by the change-point approach has been discussed by Zou et al. (2006). In this subsection, we discuss nonparametric profile monitoring using the change-point detection procedure introduced in the previous subsection.

Assume that there are m_0 IC profiles available. In many applications, one could collect a larger number of observations for each profile in Phase-I analysis than in Phase-II monitoring, because of the additional care commonly taken in Phase-I analysis, while the OC condition could usually be effectively captured by using relatively small n. To make a distinction, hereafter we use n_0 to denote the number of observations in each of the m_0 IC profile and n ($n \leq n_0$) the number of observations in each future profile. Moreover, let $\mathbf{x}_{\text{IC}} = \{x_1, \ldots, x_n, x_{n+1}, \ldots, x_{n_0}\}$ be the set of design points in the m_0 IC profiles and the corresponding response variables are $\{y_{1j}, \ldots, y_{nj}, y_{(n+1)j}, \ldots, y_{n_0 j}\}$ for $j = 1, \ldots, m_0$. Our proposed Phase-II SPC procedure is described as follows:

- After the $(t - m_0)$th Phase-II profile has been obtained, where $t > m_0$, calculate

$$lr_{m_0,t} = \max_{m_0 \leq k < t} \tilde{T}_{k,t}. \tag{10.18}$$

- If $lr_{m_0,t} > h_{m_0,t,\alpha}$, where $h_{m_0,t,\alpha}$ is some suitable control limit (see the next subsection for discussion about its selection), then an OC signal is trigged. After the signal, the systematic diagnostic procedure described in Section 10.2.5 can be used for locating the mean profile change.
- If $lr_{m_0,t} \leq h_{m_0,t,\alpha}$, then the monitoring process continues by obtaining the $(t+1)$th Phase-II profile and by repeating the previous two steps.

The above sequential scheme differs from NEWMA in an obvious way that the IC regression function and the error variance in each observed profile can be both unknown in the former scheme while they are assumed known in the latter one. Furthermore, in NEWMA, we need to choose the procedure parameter λ, besides the smoothing bandwidth, which is not necessary in the change-point approach. Hereafter, we refer it as adaptive change-point (ACP) procedure.

It should be noted that computing model (10.18) involves estimation of the error variance σ^2 for each t. When t is given, formula (10.14) gives a consistent estimator of σ^2. From tth profile to $(t+1)$th profile, this estimator can be easily updated by the following formula:

$$\hat{\sigma}_{t+1}^2 = [(t + m_0)(n - df)\hat{\sigma}_t^2 + (\mathbf{y}_t - \mathbf{W}_{h_b}\mathbf{y}_t)^{\otimes}]/[(t + 1 + m_0)(n - df)], \tag{10.19}$$

where h_b is a pre-specified bandwidth. Section 10.3.5 provides some practical guidelines about selection of h_b. This parameter can also be selected before Phase-II monitoring from the m_0 IC profiles by certain data-driven procedures, such as the least square CV and the GCV.

In practice, it might be more convenient to plot the normalized statistic $lr_{m_0,t}/h_{m_0,t,\alpha}$ over t in a control chart. In such cases, the normalized control limit is a constant 1. Besides $h_{m_0,t,\alpha}$, this chart has a number of other parameters. Selection of all these parameters are discussed in the next subsections.

10.2.4 A Bootstrap Procedure for Determining Control Limits

From the construction of $lr_{m_0,t}$, it is easy to check that this statistic does not depend on the IC regression function g. Therefore, $h_{m_0,t,\alpha}$ should not depend on g either. In this part, a bootstrap procedure for determining $h_{m_0,t,\alpha}$ based on the m_0 IC profiles is described below.

First, the m_0 IC profiles are averaged and then the averaged data are smoothed by the local linear kernel smoothing. The resulting $m_0 \times n_0$ residuals are defined by

$$\hat{e}_{ij} := y_{ij} - \mathbf{W}_{n_0}(x_i)\bar{\mathbf{y}}_{0,m_0}, \quad j = 1, \ldots, m_0, \quad i = 1, \ldots, n_0,$$

where $\mathbf{W}_{n_0}(x_i)$ is the smoothing operator defined immediately below (10.13), using $\mathbf{x}_{\mathrm{IC}}$ (instead of $\mathbf{x}$) and h_b. Second, generate a bootstrap profile $\{(x_i, y_i^*), i = 1, \ldots, n\}$ by defining $y_i^* = e_i^*$, for $i = 1, \ldots, n$, where e_i^* is drawn from $\{\hat{e}_{ij}, i = 1, \ldots, n_0, j = 1, \ldots, m_0\}$ with replacement. Third, for each given t value, by this resampling procedure, simulate the whole monitoring process, including generating the first m_0 IC profiles (at design points $\mathbf{x}$) and all future profiles and then computing the corresponding $lr^*_{m_0,t}$ value. Fourth, repeat step three B times. Then, for a given false alarm probability α, which corresponds to the IC ARL $1/\alpha$, the control limits $h_{m_0,t,\alpha}$ can be approximated by values satisfying

$$\Pr\left(lr^*_{m_0,t} > h_{m_0,t,\alpha} \middle| lr^*_{m_0,i} \leq h_{m_0,i,\alpha}, 1 \leq i < t\right) = \alpha, \quad \text{for } t > 1,$$

$$\Pr\left(lr^*_{m_0,1} > h_{m,1,\alpha}\right) = \alpha.$$

Based on our numerical experience, $h_{m_0,t,\alpha}$ would gradually converge to a constant when t increases. Therefore, we suggest computing the first about $1/(2\alpha)$ control limits and then using the last one of this sequence to approximate the remaining control limits. In addition, for computing each $h_{m_0,t,\alpha}$, about 10,000 bootstrap replications should be good enough to obtain reliable approximations. For instance, if IC ARL is 200, then we need to compute the first 100 control limits, and requires about 16,500 bootstrap sequences such that there are about 10,000 sequences left for computing the 100th control limit $h_{m_0,100,\alpha}$. Numerical accuracy of this procedure can be found in Zou et al. (2009).

10.2.5 Practical Guidelines

On choosing m_0, n_0, and n: From description of the ACP procedure, we can see that its control limits are computed from the m_0 IC profiles (each with n_0 observations) by the bootstrap procedure. Therefore, both m_0 and n_0 should not be too small. If the IC error distribution F is known, then the control limits can be computed directly from F. In practice, F is often unknown and most practitioners would accumulate a few IC profiles before starting monitoring, which is a major motivation for us to propose the current SPC procedure. Based on our numerical experience, to describe the error distribution reasonably well in various cases, we recommend using $n_0 > 50$ and $m_0 \geq 8$. Note that the recommended m_0 here is much smaller than the recommended m_0 in Section 10.2.4, where $m_0 \geq 40$ is suggested. That is mainly because the m_0 IC profiles are used for estimating both the IC regression function g and the error variance σ^2 in the latter case, while g does not need to be estimated in the former case. Regarding the choice of n and design points positions, one may refer to Section 10.2.4 for a detailed discussion.

On choosing $\gamma_n, a, h_{\max}$, and $h_{\min}$ in model (10.16): Theoretically speaking, these quantities should satisfy certain conditions to obtain the corresponding asymptotic results. See the Appendix in Zou et al. (2009) for detailed discussion. Based on simulations, we observe that performance of this chart is hardly affected by these parameters. By both theoretical arguments and numerical studies, we recommend the choices that $a = 1.4$, $\gamma_n = 2.5\sqrt{\ln(J_n + 1)}$, J_n could be 4, 5, or 6, $h_{\max} = n^{-1/7}$, and $h_j = h_{\max} a^{-j}$, $j = 1, \ldots, J_n$.

On computation: To implement the proposed method efficiently, we suggest recording a recursive array of the following running total sum of $\mathbf{y}_t$:

$$\mathbf{s}_t = \mathbf{s}_{t-1} + \mathbf{y}_t.$$

Then, computation in model (10.13) can be simplified by using $\bar{\mathbf{y}}_{0,k} - \bar{\mathbf{y}}_{k,t} = \mathbf{s}_k/k - (\mathbf{s}_t - \mathbf{s}_k)/(t-k)$. Considering the facts that $\mathbf{V}_h$ can be calculated before monitoring and the estimators of the error variance are also calculated in a recursive way (cf., model (10.19)), the computational task involved in our proposed procedure is actually quite simple.

10.2.6 Application to Monitoring a Deep Reactive Ion Etching Process

In this subsection, we revisit the DRIE example illustrated in Section 10.2.6 by applying ACP to the DRIE dataset (cf., Figure 10.1).

In the dataset, the first 18 profiles are known to be IC. In order to demonstrate the effectiveness of ACP, we only take the first eight IC profiles for implementing the bootstrap procedure and discard the remaining ten. In each of these eight IC profiles, profile dimensional readings are collected at 70 design points (i.e., $n_0 = 70$), which satisfies the requirements of the bootstrap procedure. Hence, the desired IC performance could be well approximated even if we do not make assumptions on the error distribution. As in Section 10.2.6, IC ARL is fixed at 370. All other parameters of the ACP chart are chosen as the recommendations in Section 10.3.5. Then, the ACP chart is used for monitoring Phase-II profiles, each of which has $n = 35$ observations. As detailed in Section 10.2.6, there are nine Phase-II profiles in the dataset, and the last three are classified as inferior profiles on the basis of engineering knowledge. Table 10.1 tabulates the statistics $lr_{m_0,t}$, $\hat{\sigma}_t$, the bootstrap approximated control limits

Table 10.1 Results of Various Charts for Monitoring the DRIE Dataset. Note That Profile Monitoring Starts at $t = 9$, Since the First Eight Profiles Are Treated as IC Ones

t	$lr_{m_0,t}$	$\hat{\sigma}_t$	$h_{m_0,t,\alpha}$	$lr_{m_0,t}/h_{m_0,t,\alpha}$	$\hat{\tau}$
9	0.740	0.431	6.526	0.113	8
10	0.924	0.420	6.816	0.136	8
11	2.493	0.423	7.112	0.351	8
12	1.957	0.420	7.194	0.272	8
13	1.126	0.418	7.437	0.151	8
14	1.573	0.416	7.741	0.203	8
15	2.423	0.411	8.136	0.298	14
16	6.771	0.411	8.487	0.798	14
17	9.053	0.408	8.702	1.040	14

$h_{m_0,t,\alpha}$, the ratio $lr_{m_0,t}/h_{m_0,t,\alpha}$, and the estimated change-point $\hat{\tau}$. The corresponding control chart based on the ratio charting statistic $lr_{m_0,t}/h_{m_0,t,\alpha}$ is shown in Figure 10.7, where the control limit is 1. As a comparison, corresponding results of the fixed change point (FCP), self-starting NEWMA (SSN), and "naive" multivariate EWMA approach (MEW) charts are also presented in this table. We choose $\lambda = 0.2$ in the SSN and MEW charts. Their control limits are given in the last row of the table.

From Table 10.1 and Figure 10.7, it can be seen that the ACP chart signals a shift after the 17th profile is monitored, which matches the NEWMA chart in Section 10.2.6. As a by-product, the current chart gives a change-point estimate of $\hat{\tau} = 14$. In NEWMA, a separate diagnostic method needs to be used, which gives the exactly same result about the shift position.

10.3 NONPARAMETRIC PROFILE MONITORING BY MIXED-EFFECTS MODELING

10.3.1 Motivation

The foregoing nonparametric profile monitoring requires a fundamental assumption that random errors within a profile are i.i.d., which is often invalid in

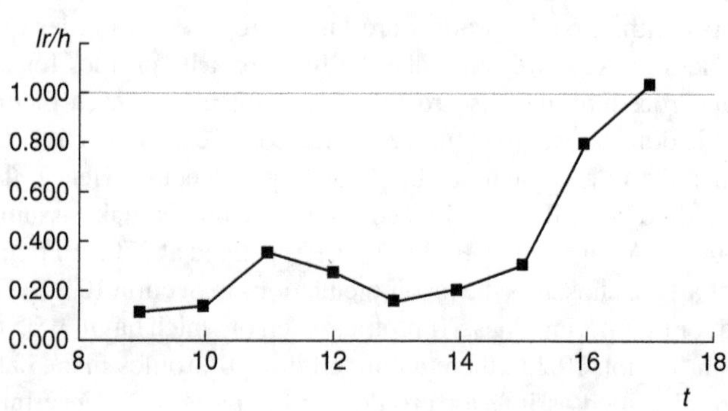

Figure 10.7 The ACP control chart for Phase-II monitoring of the DRIE process. Note that profile monitoring starts at $t = 9$, since the first eight profiles are treated as IC ones.

applications. In practice, within-profile data are often spatially or serially correlated. For instance, within-profile data of the VDPs considered by Walker and Wright (2002) and Williams et al. (2007b) are spatially correlated, since the density measurements are taken in intervals that are close to each other along the vertical depth of a particleboard. As another example, within-profile data in the DRIE example considered in Sections 10.2 and 10.3 exhibit obvious serial correlation over time. As demonstrated in the following subsections, when within-profile correlation exists, proper setup of the profile model becomes challenging and estimation of certain IC process parameters becomes difficult. Such correlation would adversely affect the IC and OC properties of the control charts that ignore its existence.

A recent research by Jensen et al. (2008) discusses linear profile monitoring in Phase-I analysis, using linear mixed-effects (LME) modeling (cf., e.g., Laird and Ware 1982). This approach can accommodate certain within-profile correlation. But, besides the fundamental difference between the Phase-I linear profile monitoring considered in that paper and the Phase-II nonparametric profile monitoring considered here, their approach assumes that both the fixed and random effects terms in their model are linear and that the covariance matrix of the random errors follows certain parametric form, such as the autoregressive and compound symmetry forms. While parametric methods are useful in certain applications, questions will always arise about adequacy of these parametric model assumptions and about potential impact of model mis-specifications on profile monitoring performance. See Hart (1997) for related discussion. In addition, design points are assumed fixed from profile to profile in the literatures of profile monitoring. In practice, however, different profiles often have different design points (i.e., the so-called unbalanced design cases). In some cases, they might be even random (i.e., the random design cases). Phase-II profile monitoring in such cases is particularly challenging, which is also discussed in this section.

To properly describe within-profile correlation, Qiu et al. (2010) propose to use a NME model (Wu and Zhang 2002) in Phase-I analysis, which allows a flexible variance–covariance structure. Based on estimated variance structure from the Phase-I analysis, they propose a novel Phase-II control chart for monitoring nonparametric profiles, which can accommodate within-profile correlation and arbitrary design. The chart is based on local linear kernel smoothing of profile data and on the EWMA weighting scheme as well. It incorporates properly both the exponential weights used in the EWMA scheme at different time points and the heteroscedasticity of observations within each profile into the local linear kernel smoother. Numerical results show that it performs well in applications. In the next several subsections, we detailly describe the methodology in Qiu et al. (2010) and demonstrate it through two real-data examples.

10.3.2 Nonparametric Mixed-Effects Modeling

Mixed-effects modeling is often used in longitudinal data analysis (cf., e.g., Laird and Ware 1982). It has become a basic tool for accommodating possible correlation among observed data. NME modeling for analyzing longitudinal data has been discussed in the literatures and here we follow this framework for modeling within-profile correlation in Phase-I analysis after a set of IC data is available. In what follows, we use the term "profile" throughout, but it should be noted that, in the mixed-effects modeling literature, it is often referred to as "cluster" or "subject." To simplify presentation, we choose to discuss cases with a single covariate, but this discussion can be easily generalized to various cases with multivariate covariates. In the IC dataset, assume that there are m profiles and the ith profile has n_i observations for $i = 1, 2, \ldots, m$. Then, the NME model can be written as

$$y_{ij} = g(x_{ij}) + f_i(x_{ij}) + \varepsilon_{ij}, \quad \text{for } j = 1, 2, \ldots, n_i, \quad i = 1, 2, \ldots, m, \tag{10.20}$$

where g is the population profile function (i.e., the fixed-effects term), f_i is the random-effects term describing the variation of the ith individual profile from g, $\{x_{ij}, y_{ij}\}_{j=1}^{n_i}$, i is the ith sample collected for the ith profile, and ε_{ij}s are i.i.d. random errors with mean 0 and variance σ^2. In model (10.20), it is routinely assumed that the random-effects term f_i and the errors ε_{ij} are independent of each other, and f_i is a realization of a mean 0 process with a common covariance function

$$\gamma(x_1, x_2) = E[f_i(x_1) f_i(x_2)].$$

Without loss of generality, we further assume that $x_{ij} \in [0, 1]$ for all i and j.

Model (10.20) is fairly flexible and includes many common correlation structures as its special cases. For instance, if $f_i(x_{ij}) = \alpha_i$ and α_i is a mean 0 random variable, then within-profile correlation would have the compound symmetry form. If Corr $(f_i(x_1), f_i(x_2)) = \rho(|x_1 - x_2|; \alpha)$, for some correlation function ρ and a coefficient α, then the correlation structure includes the nonhomogeneous Ornstein–Uhlenbeck process and the Gaussian correlation model. When the design points are equally spaced and unchanged between profiles, this model can also be used for describing the autoregressive correlation structure. Because of its flexibility, this NME model requires a relatively large size of profiles for model estimation and calibration, compared to its parametric counterparts. Thanks to fast progress in sensor and information technologies, automatic data acquisition is commonly used in industry and large amounts of IC data are often available. The NME model (10.20) allows us to make use of such data without imposing a parametric model form.

10.3.3 Estimation in NME Modeling

For constructing a Phase-II control chart when within-profile correlation is present and described by the NME model (10.20), it is required to know the IC g, γ, and σ^2 (see discussion in Section 10.3.6 below). In this part, we discuss estimation of these quantities when an IC dataset is available.

In the literature, there are some discussions about statistical analysis of correlated data under various settings and assumptions. Among others, Wu and Zhang (2002) propose a method for estimating model (10.20) by combining LME modeling and local linear kernel smoothing (cf., Fan and Gijbels 1996). They demonstrate that their estimator of g, which is referred to as local linear mixed-effects (LLME), is often more efficient than alternative estimators in the sense of mean square errors.

Furthermore, by using their approach, it is fairly easy to obtain consistent estimators of γ and σ^2, which is important for constructing a Phase-II control chart in the current study. For these reasons, we adopt Wu and Zhang's method here, which is briefly described below.

For a given point $s \in [0, 1]$, LLMEs of $g(s)$ and $f_i(s)$ are obtained by minimizing the following penalized local linear kernel likelihood function:

$$\sum_{i=1}^{m} \left\{ \frac{1}{\sigma^2} \sum_{j=1}^{n_i} [y_{ij} - z_{ij}^T(\boldsymbol{\beta} + \boldsymbol{\alpha}_i)]^2 K_h\left(x_{ij} - s\right) + \boldsymbol{\alpha}_i^T \mathbf{D}^{-1} \boldsymbol{\alpha}_i + \ln |\mathbf{D}| + n_i \ln(\sigma^2) \right\}, \tag{10.21}$$

where $K_h(\cdot) = K(\cdot/h)/h$, K is a symmetric density kernel function, h is a bandwidth, $z_{ij}^T = (1, x_{ij} - s)$, $\boldsymbol{\beta}$ is a deterministic two-dimensional coefficient vector, and $\boldsymbol{\alpha}_i \sim (0, \mathbf{D})$ is a two-dimensional vector of the random effects. Minimization of model (10.21) is accomplished by the following iterative procedure:

Step 1. Set the initial values for $\mathbf{D}$ and σ^2, denoted as $\mathbf{D}_{(0)}$ and $\sigma^2_{(0)}$.

Step 2. At the kth iteration, for $k \geq 0$, compute estimates of $\boldsymbol{\beta}$ and $\boldsymbol{\alpha}_i$ by solving the so-called mixed-model equation (cf., Wu and Zhang 2002), and the resulting estimates are denoted as

$$\hat{\boldsymbol{\beta}}^{(k)} = \left\{ \sum_{i=1}^{m} \mathbf{Z}_i^T \boldsymbol{\Sigma}_i \mathbf{Z}_i \right\}^{-1} \left\{ \sum_{i=1}^{m} \mathbf{Z}_i^T \boldsymbol{\Sigma}_i \mathbf{y}_i \right\} \tag{10.22}$$

$$\hat{\boldsymbol{\alpha}}_i^{(k)} = \{\mathbf{Z}_i^T \mathbf{K}_i \mathbf{Z}_i + \sigma^2_{(k)} [\mathbf{D}_{(k)}]^{-1}\}^{-1} \mathbf{Z}_i^T \mathbf{K}_i (\mathbf{y}_i - \mathbf{Z}_i \hat{\boldsymbol{\beta}}^{(k)}), \tag{10.23}$$

where $\mathbf{Z}_i = (z_{i1}, \ldots, z_{in_i})^T$, $\mathbf{y}_i = (y_{i1}, \ldots, y_{in_i})^T$, $\boldsymbol{\Sigma}_i = (\mathbf{Z}_i \mathbf{D}_{(k)} \mathbf{Z}_i^T + \sigma^2_{(k)} \mathbf{K}_i^{-1})^{-1}$, and $\mathbf{K}_i = \text{diag}\{K_h(x_{i1} - s), \ldots, K_h(x_{in_i} - s)\}$.

Step 3. Based on $\hat{\boldsymbol{\beta}}^{(k)}$ and $\hat{\boldsymbol{\alpha}}_i^{(k)}$, update $\mathbf{D}$ and σ^2 by

$$\mathbf{D}_{(k+1)} = \frac{1}{m} \sum_{i=1}^{m} \hat{\boldsymbol{\alpha}}_i^{(k)} [\hat{\boldsymbol{\alpha}}_i^{(k)}]^T \tag{10.24}$$

$$\sigma^2_{(k+1)} = \frac{1}{m} \sum_{i=1}^{m} \frac{1}{n_i} [\mathbf{y}_i - \mathbf{Z}_i(\hat{\boldsymbol{\beta}}^{(k)} + \hat{\boldsymbol{\alpha}}_i^{(k)})]^T \mathbf{K}_i [\mathbf{y}_i - \mathbf{Z}_i(\hat{\boldsymbol{\beta}}^{(k)} + \hat{\boldsymbol{\alpha}}_i^{(k)})]. \tag{10.25}$$

Step 4. Repeat Steps 2 and 3 until the following condition is satisfied:

$$\|\mathbf{D}_{(l)} - \mathbf{D}_{(l-1)})\|_1 \,/\, \|\mathbf{D}_{(l-1)}\|_1 \leq \epsilon,$$

where ϵ is a pre-specified small positive number (e.g., $\epsilon = 10^{-4}$), and $||\mathbf{A}||_1$ denotes the sum of absolute values of all elements of $\mathbf{A}$. Then, the algorithm stops at the lth iteration.

As a side note, similar to estimation of LME models, nonconvergence of the above iterative procedure can occasionally happen, although we found that the frequency of nonconvergence is negligible in all our simulation studies, except certain extreme cases such as the ones when m or n_is are too small. To reduce the frequency of nonconvergence, in the literature, it is suggested to use good initial values for $\mathbf{D}$ and σ^2. A simple but effective method is to set $\mathbf{D}_{(0)}$ to be the identity matrix and

$$\sigma^2_{(0)} = \frac{1}{m}\sum_{i=1}^{m}\frac{1}{n_i}\sum_{j=1}^{n_i}[y_{ij} - \hat{g}^{(P)}(x_{ij})]^2,$$

where $\hat{g}^{(P)}(x_{ij})$ is the standard LLKE constructed from the pooled data.

After obtain estimates of β and α_i using the above algorithm, we can define

$$\hat{g}(s) = \mathbf{e}_1^T\hat{\beta}(s), \quad \hat{f}_i(s) = \mathbf{e}_1^T\hat{\alpha}_i(s),$$

and

$$\hat{\gamma}(s_1, s_2) = \frac{1}{m}\sum_{i=1}^{m}\hat{f}_i(s_1)\hat{f}_i(s_2), \text{ for any } s_1, s_2 \in [0, 1], \tag{10.26}$$

where $\boldsymbol{e}_1 = (1, 0)^T$. Note that the variance estimator from the above iterative procedure depends on s. Since σ^2 is a population parameter that does not depend on s, we suggest estimating it by

$$\hat{\sigma}^2 = \frac{1}{m}\sum_{i=1}^{m}\frac{1}{n_i}\sum_{j=1}^{n_i}[y_{ij} - \hat{g}(x_{ij}) - \hat{f}_i(x_{ij})]^2, \tag{10.27}$$

which is similar to model (10.14).

Proposition 1 in Qiu et al. (2010) establish the consistency of the estimators of g, γ, and σ^2, which is important for the Phase-II profile monitoring problem discussed in the following subsections. We refer it for details.

10.3.4 Phase-II Nonparametric Profile Monitoring with NME

In this part, we present a Phase-II nonparametric profile monitoring scheme in the general case when within-profile data might be correlated and the design points within and between profiles are arbitrary. This is a challenging task due to the following two major reasons. First, because the within-profile data might be correlated, estimation of the profile function g involves a considerable amount of computation if the NME modeling is also used in Phase-II SPC, as described in Section 10.3.3. However, a good online control chart should maintain a reasonable efficiency while be effective in detecting profile shifts. Second, in cases when the design points $\mathbf{x}_i$ are unchanged from profile to profile, a method comes to our mind is to first averages observed responses y_{ij}s across individual profiles and then detects potential profile shifts using a GLR test statistic. This idea cannot be applied to the current problem directly because the response is observed at different design points in different profiles. One immediate alternative is to estimate g from individual profile data at a given set of points in [0, 1]. But the resulting estimates would be inefficient since they are constructed from individual profile data instead of from all observed data.

To overcome the above difficulties, at any point $s \in [0, 1]$, consider the following weighted local likelihood (Qiu and Zou 2010):

$$WL(a, b; s, \lambda, t) = \sum_{i=1}^{t}\sum_{j=1}^{n_i}[y_{ij} - a - b(x_{ij} - s)]^2 K_h\left(x_{ij} - s\right)(1-\lambda)^{t-i}/\nu^2(x_{ij}),$$

where λ is a weighting parameter and $\nu^2(x) = \gamma(x, x) + \sigma^2$ is the variance function of the response. Note that $WL(a, b; s, \lambda, t)$ combines the exponential weighting scheme used in EWMA at different time points through the term $(1-\lambda)^{t-i}$ and LLKE. At the same time, it takes into account the heteroscedasticity of observations by using $\nu^2(x_{ij})$. Then, the LLKE of $g(s)$, defined as the solution to a of the minimization problem $\min_{a,b} WL(a, b; s, \lambda, t)$, has the expression

$$\hat{g}_{t,h,\lambda}(s) = \sum_{i=1}^{t}\sum_{j=1}^{n_i} U_{ij}^{(t,h,\lambda)}(s) y_{ij} \Bigg/ \sum_{i=1}^{t}\sum_{j=1}^{n_i} U_{ij}^{(t,h,\lambda)}(s), \tag{10.28}$$

where

$$U_{ij}^{(t,h,\lambda)}(s) = \frac{(1-\lambda)^{t-i} K_h(x_{ij} - s)}{\nu^2(x_{ij})}\left[m_2^{(t,h,\lambda)}(s) - (x_{ij} - s) m_1^{(t,h,\lambda)}(s)\right],$$

$$m_l^{(t,h,\lambda)}(s) = \sum_{i=1}^{t}(1-\lambda)^{t-i}\sum_{j=1}^{n_i}(x_{ij} - s)^l K_h(x_{ij} - s)/\nu^2(x_{ij}), \quad l = 0, 1, 2. \tag{10.29}$$

From models (10.28) and (10.29), we can see that $\hat{g}_{t,h,\lambda}(s)$ makes use of all available observations up to the current time point t, and different profiles are weighted as in a conventional EWMA chart (i.e., more recent profiles get more weights and the weights change exponentially over time).

Following the convention in Phase-II analysis, we assume that the IC regression function, denoted as g_0, and the variance function $\nu^2(\cdot)$ are both known. In practice, they need to be estimated from an IC dataset, as described in Section 10.3.3. Let $\xi_{ij} = [y_{ij} - g_0(x_{ij})]$, for all i and j, and $\hat{\xi}_{t,h,\lambda}(s)$ be the estimator defined in model (10.28) after y_{ij} are replaced by ξ_{ij}. After this transformation, the IC distribution of $\hat{\xi}_{t,h,\lambda}(s)$ would not depend on g_0 as in NEWMA.

When the process is IC, $|\hat{\xi}_{t,h,\lambda}(s)|$ should be small. So, a natural statistic that can be used for SPC would be

$$T_{t,h,\lambda} = c_{0,t,\lambda}\int \frac{[\hat{\xi}_{t,h,\lambda}(s)]^2}{\nu^2(s)}\Gamma_1(s)ds,$$

where

$$c_{t_0,t_1,\lambda} = a^2_{t_0,t_1,\lambda}/b_{t_0,t_1,\lambda}, \quad a_{t_0,t_1,\lambda} = \sum_{i=t_0+1}^{t_1} (1-\lambda)^{t_1-i} n_i, \quad b_{t_0,t_1,\lambda} = \sum_{i=t_0+1}^{t_1} (1-\lambda)^{2(t_1-i)} n_i,$$

and Γ_1 is some pre-specified density function. In the expression of $T_{t,h,\lambda}$, quantities $c_{0,t,\lambda}$ and $\nu(\cdot)$ are used for unifying its asymptotic variance. In practice, we suggest using the following discretized version:

$$T_{t,h,\lambda} \approx \frac{c_{0,t,\lambda}}{n_0} \sum_{k=1}^{n_0} \frac{\left[\hat{\xi}_{t,h,\lambda}(s_k)\right]^2}{\nu^2(s_k)}, \tag{10.30}$$

where $\{s_k, k = 1, \ldots, n_0\}$ are some pre-specified i.i.d. design points from Γ_1. Then, the chart triggers a signal if

$$T_{t,h,\lambda} > L,$$

where $L > 0$ is a control limit chosen to achieve a specific IC ARL, denoted as ARL_0. Hereafter, this chart is referred to as the mixed-effects nonparametric profile control (MENPC) chart. Some asymptotic properties regarding the distribution of $T_{t,h,\lambda}$ in the IC and OC conditions can be found in Qiu et al. (2010).

10.3.5 Some Computational Issues

Although computing power gets improved fast and it is computationally trivial to do nonparametric function estimation for individual profiles, for online process monitoring, which generally handles a large amount of profiles, fast implementation is important and some computational issues worth our careful examination. For the proposed chart, computing the test statistic $T_{t,h,\lambda}$ by formulas (10.28)–(10.30) directly requires a considerable amount of computing time and a substantial amount of storage space as well to save all past profile observations. In this part, we provide updating formulas for computing $T_{t,h,\lambda}$, which can greatly simplify the computation and reduce the storage requirement. Let

$$\tilde{m}_l^{(t,h)}(s) = \sum_{j=1}^{n_k} (x_{tj} - s)^l K_h(x_{tj} - s)/\nu^2(x_{tj}), \quad l = 0, 1, 2,$$

$$\tilde{q}_l^{(t,h)}(s) = \sum_{j=1}^{n_k} (x_{tj} - s)^l K_h(x_{tj} - s) y_{tj}/\nu^2(x_{tj}), \quad l = 0, 1.$$

Then, $m_l^{(t,h,\lambda)}(s)$ in model (10.29) can be recursively updated by

$$m_l^{(t,h,\lambda)}(s) = (1-\lambda) m_l^{(t-1,h,\lambda)}(s) + \tilde{m}_l^{(t,h)}(s), \quad l = 0, 1, 2,$$

where $m_l^{(0,h,\lambda)}(s) = 0$ for $l = 0, 1, 2$. Let $q_l^{(t,h,\lambda)}(s)$, for $l = 0, 1$, be two working functions defined by the recursive formula

$$q_l^{(t,h,\lambda)}(s) = (1-\lambda)q_l^{(t-1,h,\lambda)}(s) + \tilde{q}_l^{(t,h)}(s), \quad l = 0, 1,$$

where $q_l^{(0,h,\lambda)}(s) = 0$ for $l = 0, 1$. Then, we have

$$\hat{g}_{t,h,\lambda}(s) = \left[M^{(t,h,\lambda)}\right]^{-1}\left\{(1-\lambda)^2 M^{(t-1,h,\lambda)}\hat{g}_{t-1,h,\lambda} + \left[\tilde{q}_0^{(t,h)} m_2^{(t,h,\lambda)} - \tilde{q}_1^{(t,h)} m_1^{(t,h,\lambda)}\right]\right.$$
$$\left. + (1-\lambda)\left[q_0^{(t-1,h,\lambda)}\tilde{m}_2^{(t,h)} - q_1^{(t-1,h,\lambda)}\tilde{m}_1^{(t,h)}\right]\right\}, \tag{10.31}$$

where $M^{(t,h,\lambda)}(s) = m_2^{(t,h,\lambda)}(s)m_0^{(t,h,\lambda)}(s) - [m_0^{(t,h,\lambda)}(s)]^2$. On the right-hand side of the above equation, dependence on s in each function is not made explicit in notation for simplicity, which should not cause any confusion.

Using the above updating formulas, implementation of the MENPC chart can be briefly described as follows. At time point t, we first compute quantities $\tilde{m}_l^{(t,h)}(s)$, for $l = 0, 1, 2$, and $\tilde{q}_l^{(t,h)}(s)$, for $l = 0, 1$, at n_0 pre-determined s locations. Then, $m_l^{(0,h,\lambda)}(s_k)$, for $l = 0, 1, 2$, and $q_l^{(0,h,\lambda)}(s_k)$, for $l = 0, 1$, are updated by the above formulas. Finally, $\hat{g}_{t,h,\lambda}(s)$ is computed from model (10.31), and the test statistic $T_{t,h,\lambda}$ is computed by $\hat{g}_{t,h,\lambda}(s)$, after y_{ij} is replaced by ξ_{ij}. This algorithm only requires $O(n_0 n_i h)$ operations for monitoring the ith profile, which is in the same order as the computation involved in conventional local linear kernel smoothing. If n_i and n_0 are both large, we could further decrease the computation to the order of $O(n_i h)$, by using the updating algorithm (Fan and Gijbels 1996). Obviously, using the proposed updating formulas, computer storage does not grow sequentially with time t.

10.3.6 Certain Practical Guidelines

On the sizes of m and n_i: The number of IC profiles should be large enough to generate accurate estimates of IC g, γ, and σ^2. This has become a less significant issue nowadays because a large amount of IC data is often available due to advances in data acquisition techniques. Empirically speaking, to attain desirable IC distributional properties, we recommend using IC data with $n_i \geq 40$ and $m \geq 80$, although more systematic future research is required to determine the size of a necessary IC dataset.

On choosing the bandwidth: In estimation of the NME model (10.20) using the iterative procedure described in Section 10.3.3 for Phase-I analysis, people often use data-driven bandwidth selection techniques, such as CV and GCV. With respect to Phase-II online monitoring, like many other smoothing-based tests, performance of the MENPC chart depends on selection of the bandwidth parameter h used in model (10.28). Following the recommendation in Section 10.2.4, we suggest using the following empirical bandwidth formula,

$$h_E = \begin{cases} c_1 n^{-1/5} \left(\sum_{i=1}^{n} (x_i - \bar{x})^2 / n \right)^{1/2} & \text{for balanced design} \\ c_2 [\tilde{n}(2-\lambda)/\lambda]^{-1/5} \sqrt{\mathrm{Var}(x)} & \text{for random design,} \end{cases} \tag{10.32}$$

where $\bar{x} = \sum_{i=1}^{n} x_i$ is the mean of the n design points in the balanced design case, $\tilde{n}$ and $\mathrm{Var}(x)$ are the averaged number of design points and the variance of design points within a profile, respectively, in the random design case, which can be estimated from the IC data used in the Phase-I analysis, and c_1 and c_2 are two constants. Empirically, c_1 and c_2 can be any values in interval [1.0,2.0]. A smaller bandwidth for the random design case is suggested above because the actual number of observations used in the MENPC chart at each time point is about $c_{0,t,\lambda}$, which is roughly $\tilde{n}(2-\lambda)/\lambda$.

On choosing λ: Traditionally, a larger λ leads to a quicker detection of larger shifts. However, in the mixed-effect modeling, efficiently estimate the common profile function g requires to use observations across a number of different profiles, due mainly to the existence of random effects. From Theorems 1 and 2 in Qiu et al. (2010), we can see that the effective number of profiles used in the MENPC chart at each time point is asymptotically $(2-\lambda)/\lambda$. So, to estimate g properly, $(2-\lambda)h/\lambda$ should be large enough (cf., Condition (C8)-II in Appendix A of Qiu et al. 2010). Consequently, λ cannot be chosen too large. Otherwise, even for a shift of large magnitude, $T_{t,h,\lambda}$ may not be able to detect it quickly, due to a large bias in estimating g. Empirically, we suggest choosing $\lambda \in [0.02, 0.1]$ if h_E in model (10.32) is used.

On choosing $\{s_k, k = 1, 2, \ldots, n_0\}$: Based on our numerical experience, selection of $\{s_k, k = 1, 2, \ldots, n_0\}$ does not affect performance of the MENPC chart much, as long as n_0 is not too small and s_k s cover all the key parts of g_0 (e.g., peaks/valleys or oscillating regions) well. By numerical examples, we find that results would hardly change when $n_0 \geq 40$.

10.3.7 Real-Data Applications

In this subsection, we demonstrate MENPC by applying it to a dateset from a manufacturing process of aluminum electrolytic capacitors (AECs). This process transforms raw materials, such as anode aluminum foil, cathode aluminum foil, guiding pin, electrolyte sheet, plastic cover, aluminum shell and plastic tube, into AECs that are appropriate for use in low leakage circuits and are well adapted to a wide range of environmental temperatures. The whole manufacturing process consists of a sequence of operations, including clenching, rolling, soaking, assembly, cleaning, aging, and classifying. Before packing, a careful quality monitoring step is required by sampling from a batch of products.

Regarding quality of AECs, the most important characteristic is dissipation factor (DF), which can be automatically measured by an electronic device. However, it is known that DF measurements would change significantly with environmental temperature, and there is a specific requirement about the adaptability of AECs to

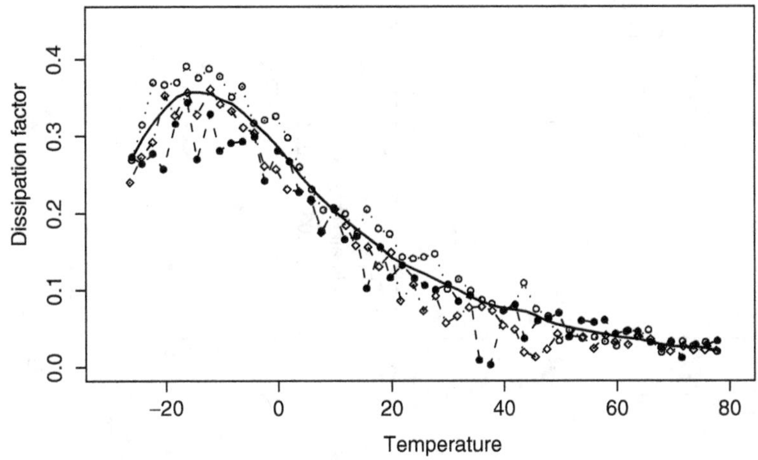

Figure 10.8 Three AEC profiles (lines connecting points with three different symbols) and the NME estimate (solid curve) of the IC profile function.

the temperature. In order to monitor the adaptability, engineers put a sampled AEC in a container. Then, the container's temperature is controlled, and the temperature is supposed to stably increase from −26°F to 78°F. In this process, measurements of DF and the actual temperature inside the container are taken at 53 equally spaced time points. The actual temperature inside the container is reported by a temperature sensor. So, for each sampled AEC, a set of 53 observations of the pair (temperature, DF), which corresponds to (x, y) in model (10.20), is obtained for monitoring the adaptability of the AEC to the temperature. Figure 10.8 shows three AEC profiles along with an NME estimate of the IC profile function (see related discussion below). It should be noted that the actual temperature inside a container would fluctuate around its nominal level at each observation time. Therefore, actual temperature readings of different containers at a given observation time are all different, although the differences are usually small. For this dataset, profile monitoring charts requiring deterministic and fixed design points in different profiles would be difficult to use.

The entire AEC dataset contains 144 profiles, and each profile has $n = 53$ observations. We use the first 96 profiles to calibrate the proposed model and the remaining ones to test the model. A calibration sample of this size might be smaller than one would like to fully determine the IC distribution, but suffices to illustrate the use of the method in a real-world setting. Since the DF measurements are taken in consecutive time intervals, the AEC data exhibit a considerable amount of positive serial autocorrelation, which is confirmed by our analysis described below.

We first fit model (10.20) to the calibration sample of the first 96 profiles of the data by the iterative procedure (10.22)–(10.25), using the suggested initial values of $\mathbf{D}$ and σ^2 given in Section 10.3.3 and the CV bandwidth selection procedure suggested by Wu and Zhang (2002). The resulting IC profile estimate $\hat{g}$ is displayed in Figure 10.8 by the solid curve. From (10.26) and (10.27), we can also compute the estimated

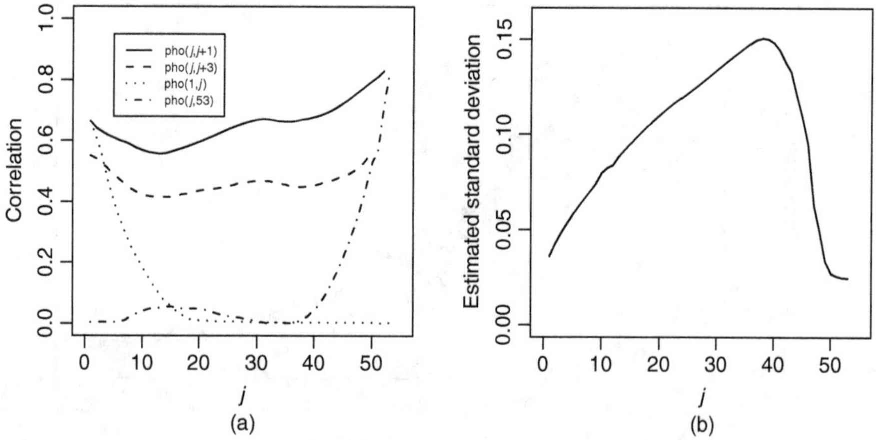

Figure 10.9 (a) Solid, dashed, dotted, and dash-dotted curves represent estimated within-profile correlations $\hat{\rho}(x_j^*, x_{j+1}^*)$, $\hat{\rho}(x_j^*, x_{j+3}^*)$, $\hat{\rho}(x_1^*, x_j^*)$, and $\hat{\rho}(x_j^*, x_{53}^*)$, for $j = 1, 2, \ldots, 53$, where $\{x_j^*, j = 1, 2, \ldots, 53\}$ are 53 equally spaced points in the design interval $[-26, 78]$. (b) Estimated standard deviation $\hat{\upsilon}(x_j^*)$ of the response variable y at x_j^* for $j = 1, 2, \ldots, 53$.

correlation of two observations of the response variable y at any two points s_1 and s_2 in the design interval

$$\hat{\rho}(s_1, s_2) = \hat{\gamma}(s_1, s_2) \Big/ [\hat{\upsilon}(s_1)\hat{\upsilon}(s_2)],$$

where $\hat{\gamma}(s_1, s_2)$ is defined in model (10.27), $\hat{\upsilon}^2(s) = \hat{\gamma}(s, s) + \hat{\sigma}^2$ is the estimated variance of y at s, and $\hat{\sigma}^2$ is defined in model (10.27). Let $x_j^* = 2(j-1) - 26$, for $j = 1, 2, \ldots, 53$, be 53 equally spaced points in the design interval $[-26, 78]$, which denote the nominal temperature levels used in taking DF measurements of the sampled AECs. The estimated correlations $\hat{\rho}(x_j^*, x_{j+1}^*)$, $\hat{\rho}(x_j^*, x_{j+3}^*)$, $\hat{\rho}(x_1^*, x_j^*)$, and $\hat{\rho}(x_j^*, x_{53}^*)$, for $j = 1, 2, \ldots, 53$, are shown in Figure 10.9(a). From the plot, we can see that correlation within AEC profiles is substantial; thus, it should not be ignored. Figure 10.9(b) shows the estimated standard deviation $\hat{\upsilon}(x_j^*)$ of the response variable y at x_j^*, for $j = 1, 2, \ldots, 53$, from which heteroscedasticity of the response variable y at different positions of x is clearly seen. Therefore, the proposed MENPC chart would be more appropriate to use in this case. In addition, we can obtain an estimate of the error standard deviation σ to be 0.016, by formula (10.27), which is much smaller than $\hat{\upsilon}(x_j^*)$, especially when $j \in [12, 50]$. This result implies that the random-effects term in model (10.20) describes a substantial amount of random variation in the data. Next, we construct the proposed MENPC chart for Phase-II profile monitoring, using the estimated IC parameters computed from the IC data. The IC ARL is fixed at 200, and λ is chosen to be 0.1. For simplicity, we choose $n_0 = n = 53$ and $\{s_k, k = 1, 2, \ldots, n_0\}$ to be equally spaced in the design interval $[-26, 78]$ of the explanatory variable. The control limit is computed to be 18.24 by simulation. The charting statistics $T_{t,h,\lambda}$, for $t = 97, \ldots, 144$, are shown in Figure 10.10 along with the control limit by the

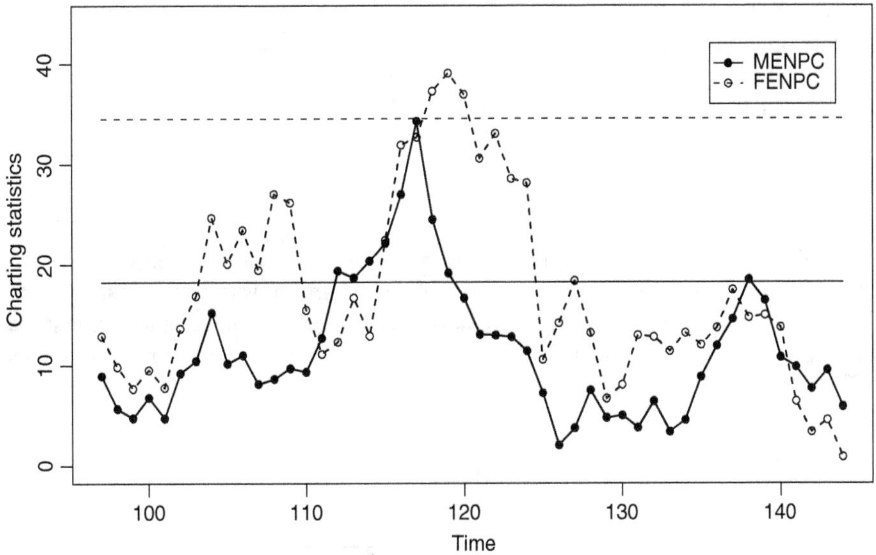

Figure 10.10 MENPC and FENPC control charts for monitoring the AEC process. The solid and dashed horizontal lines indicate their control limits, respectively.

solid curve and solid horizontal line, respectively. It is challenging to compare the proposed method with alternative methods, since there is no obvious comparable method in the literature. Here, we consider the control chart based on fixed-effects modeling for monitoring nonparametric profiles as an alternative method, denoted as fixed-effects nonparametric profile chart (FENPC). In this approach, f_i in model 10.20 is assumed to be zero, and consequently $\nu^2(x) = \sigma^2$ is used in the construction of $T_{t,h,\lambda}$ (cf., expressions 10.28–10.30). Note that the FENPC chart can be regarded as a generalization of the NEWMA chart by Zou et al. (2008). The corresponding FENPC chart, using the same λ and h as those in the MENPC chart, is also presented in the figure along with its control limit 34.52 by the dashed curve and the dashed horizontal lines. From the plot, it can be seen that the MENPC chart gives a signal of profile shift around the 112th time point, and remains above the control limit for several profiles until the 120th profile. This result confirms a marked step-change that seems to have occurred around the 108th profile. The process seems to have been adjusted around the 119th profile; thus, the MENPC charting statistic goes back below its control limit afterward. As a comparison, the FENPC chart does not give a signal until the 118th profile.

APPENDIX A: APPROXIMATE THE DISTRIBUTIONS OF QUADRATIC FORMS LIKE $\mathbf{z}^T\mathbf{A}\mathbf{z}$

Here, we recommend using the method that matches the first three moments of the distribution of $n\hat{\sigma}_j^2$ with those of a chi-square distribution (Imhof 1961), which is demonstrated to be quite accurate for approximating the distribution of the quadratic forms. Our numerical simulation also shows that this method not only eases the computation but it is also quite accurate and effective for our monitoring problem.

The sth cumulant of $\mathbf{z}^T\mathbf{A}\mathbf{z}$ is given by (Imhof 1961):

$$\kappa_s = 2^{s-1}(s-1)!\text{tr}(\mathbf{A}^s).$$

This is used to match the first three moments of the distribution of $\mathbf{z}^T\mathbf{A}\mathbf{z}$ with those of a chi-square distribution. That is, given the mean, variance, and skewness of the distribution of $\mathbf{z}^T\mathbf{A}\mathbf{z}$ (using the relationship between cumulant and moment), we find c_1, c_2, and c_3 such that a $c_1\chi^2_{c_2} + c_3$ distribution has the same mean, variance, and skewness. Simple calculations yield

$$\begin{aligned} c_1 &= \sqrt{\text{tr}(\mathbf{A}^2)/\text{tr}(\mathbf{A}^3)} \\ c_2 &= \text{tr}(\mathbf{A}^3) \\ c_3 &= \text{tr}(\mathbf{A}) - \sqrt{\text{tr}(\mathbf{A}^2)\cdot\text{tr}(\mathbf{A}^3)}. \end{aligned}$$

This $c_1\chi^2_{c_2} + c_3$ distribution is then used as an approximate null distribution to calculate a p -value in the usual manner.

APPENDIX B: THE EXPRESSION OF $lr_{t,k}$ IN MODEL (10.8)

Denote $g_i(\cdot)$ and σ_i^2 as the regression function and variance of the ith profile sample $\{\mathbf{x}, \mathbf{y}_i\}$, respectively. Then, after k samples have been collected, we care about the following null hypothesis of no change,

$$H_0: g_1 = g_2 = \cdots = g_k = g_0 \quad \text{and} \quad \sigma_1^2 = \sigma_2^2 = \cdots = \sigma_k^2 = \sigma_0^2,$$

against the following alternative of one change

$$\begin{aligned} H_0: g_0 = g_1 = g_2 = \cdots = g_{t^*} \neq g_{t^*+1} = \cdots = g_k \quad &\text{or} \\ \sigma_0^2 = \sigma_1^2 = \sigma_2^2 = \cdots = \sigma_{t^*}^2 \neq \sigma_{t^*+1}^2 = \cdots = \sigma_k^2, & \end{aligned}$$

where $1 \le t^* < k$.

The logarithm of the likelihood function is given by

$$-\frac{1}{2}\sum_{j=1}^{k}\left[n\ln(2\pi\sigma_j^2) + \frac{1}{\sigma_j^2}(\mathbf{y}_j - \mathbf{g}_j)^{\otimes}\right],$$

where $\mathbf{g}_j = (g_j(x_1), \ldots, g_j(x_n))^T$. If the data are collected under IC conditions, i.e., under the null hypothesis, the value of the logarithm of the likelihood function is

$$l_0 = -\frac{1}{2}\sum_{j=1}^{k}\left[n\ln(2\pi\sigma_0^2) + \frac{1}{\sigma_0^2}(\mathbf{y}_j - \mathbf{g}_0)^{\otimes}\right].$$

Assuming that a change occurs after t, then the corresponding logarithm of generalized likelihood is

$$l_1 = -\frac{1}{2}\sum_{j=1}^{t}\left[n\ln(2\pi\sigma_0^2) + \frac{1}{\sigma_0^2}(\mathbf{y}_j - \mathbf{g}_0)^{\otimes}\right] - \frac{(k-t)n}{2}\left[\ln\left(\frac{2\pi}{(k-t)n}\sum_{j=t+1}^{k}(\mathbf{y}_j - \hat{\mathbf{g}})^{\otimes}\right) + 1\right],$$

where $\hat{\mathbf{g}}$ is the local linear estimator based on the profile samples $\{\mathbf{x}, \mathbf{y}_j\}_{j=t+1}^{k}$.

Since g_0 and σ_0 are known in advance, recalling the notation in Section 10.2.2, we can substitute the profile samples, $\mathbf{z}_j$, $g_0 = 0$, and $\sigma_0^2 = 1$, into the preceding expressions. This will result in the following final expression of $lr_{t,k}$,

$$lr_{t,k} = -2(l_0 - l_1) = \sum_{j=t+1}^{k}\mathbf{z}_j^{\otimes} - (k-t)n\left[\ln\left(\frac{1}{(k-t)n}\sum_{j=t+1}^{k}(\mathbf{z}_j - \mathbf{W}\bar{\mathbf{z}}_{t,k})^{\otimes}\right) + 1\right],$$

where $\bar{\mathbf{z}}_{t,k} = \frac{1}{k-t}\sum_{j=t+1}^{k}\mathbf{z}_j$. Using the classical binary segment procedure leads to the change-point estimator (2.7).

REFERENCES

Fan, J. and Gijbels, I. (1996) Local Polynomial Modeling and Its Applications. Chapman and Hall, London.

Fan, J., Zhang, C., and Zhang, J. (2001). Generalized likelihood ratio statistics and Wilks phenomenon. *The Annals of Statistics*, **29**(1), 153–193.

Guerre, E. and Lavergne, P. (2005) Data-driven rate-optimal specification testing in regression models. *The Annals of Statistics*, **33**(2), 840–870.

Hall, P. and Marron, J. S. (1990) On variance estimation in nonparametric regression. *Biometrika*, **77**(2), 415–419.

Hart, J. D. (1997) Nonparametric Smoothing and Lack-of-Fit Tests. Springer, New York.

Horowitz, J. L. and Spokoiny, V. G. (2001) An adaptive, rate-optimal test of a parametric mean-regression model against a nonparametric alternative. *Econometrica*, **69**(3), 599–631.

Imhof, J. P. (1961) Computing the distribution of quadratic forms in normal variables. *Biometrika*, **48**(3), 419–426.

Jensen, W. A., Birch, J. B. and Woodall, W. H. (2008) Monitoring correlation within linear profiles using mixed models. *Journal of Quality Technology*, **40**(2), 167–183.

Kim, K., Mahmoud, M. A. and Woodall, W. H. (2003) On the monitoring of linear profiles. *Journal of Quality Technology*, **35**(3), 317–328.

Laird, N. M. and Ware, J. H. (1982) Random effects models for longitudinal data. *Biometrics*, **38**(4), 963–974.

Lucas, J. M. and Saccucci, M. S. (1990) Exponentially weighted moving average control scheme: properties and enhancements. *Technometrics*, **32**(1), 1–29.

Qiu, P. and Zou, C. (2010) Control hart for monitoring nonparametric profiles with arbitrary design. *Statistica Sinica*, **20**(4), 1655–1682.

Qiu, P., Zou, C., and Wang, Z. (2010) Nonparametric profile monitoring by mixed effects modeling (with discussions). *Technometrics*, **52**(3), 265–277.

Walker, E. and Wright, S. (2002) Comparing curves using additive models. *Journal of Quality Technology*, **34**(1), 118–129.

Williams, J. D., Birch, J. B., Woodall, W. H., and Ferry, N. M. (2007a) Statistical monitoring of heteroscedastic dose-response profiles from high-throughput screening. *Journal of Agricultural, Biological and Environmental Statistics*, **12**(2), 216–235.

Williams, J. D., Woodall, W. H. and Birch, J. B. (2007b) Statistical monitoring of nonlinear product and process quality profiles. *Quality and Reliability Engineering International*, **23**(8), 925–941.

Wu, H. and Zhang, J. (2002) Local polynomial mixed-effects models for longitudinal data. *Journal of the American Statistical Association*, **97**(3), 883–897.

Zou, C., Qiu, P., and Hawkins, D. (2009) Nonparametric control chart for monitoring profiles using the change point formulation and adaptive smoothing. *Statistica Sinica*, **19**(3), 1337–1357.

Zou, C., Tsung, F., and Wang, Z. (2007) Monitoring general linear profiles using multivariate EWMA schemes. *Technometrics*, **49**(3), 395–408.

Zou, C., Tsung, F., and Wang, Z. (2008) Monitoring profiles based on nonparametric regression methods. *Technometrics*, **50**(4), 512–526.

Zou, C., Zhang, Y., and Wang, Z. (2006) Control chart based on change-point model for monitoring linear profiles. *IIE Transactions*, **38** (12), 1093–1103.

Index

Statistical Analysis of Profile Monitoring, First Edition. Edited by Rassoul Noorossana, Abbas Saghaei and Amirhossein Amiri.

Geometric specification 18, 217, 218, 224, 233

WILEY SERIES IN PROBABILITY AND STATISTICS
ESTABLISHED BY WALTER A. SHEWHART AND SAMUEL S. WILKS

The ***Wiley Series in Probability and Statistics*** is well established and authoritative. It covers many topics of current research interest in both pure and applied statistics and probability theory. Written by leading statisticians and institutions, the titles span both state-of-the-art developments in the field and classical methods.

Reflecting the wide range of current research in statistics, the series encompasses applied, methodological and theoretical statistics, ranging from applications and new techniques made possible by advances in computerized practice to rigorous treatment of theoretical approaches.

This series provides essential and invaluable reading for all statisticians, whether in academia, industry, government, or research.

† ABRAHAM and LEDOLTER • Statistical Methods for Forecasting
AGRESTI • Analysis of Ordinal Categorical Data, *Second Edition*
AGRESTI • An Introduction to Categorical Data Analysis, *Second Edition*
AGRESTI • Categorical Data Analysis, *Second Edition*
ALTMAN, GILL, and McDONALD • Numerical Issues in Statistical Computing for the Social Scientist
AMARATUNGA and CABRERA • Exploration and Analysis of DNA Microarray and Protein Array Data
ANDĚL • Mathematics of Chance
ANDERSON • An Introduction to Multivariate Statistical Analysis, *Third Edition*
* ANDERSON • The Statistical Analysis of Time Series
ANDERSON, AUQUIER, HAUCK, OAKES, VANDAELE, and WEISBERG • Statistical Methods for Comparative Studies
ANDERSON and LOYNES • The Teaching of Practical Statistics
ARMITAGE and DAVID (editors) • Advances in Biometry
ARNOLD, BALAKRISHNAN, and NAGARAJA • Records
* ARTHANARI and DODGE • Mathematical Programming in Statistics
* BAILEY • The Elements of Stochastic Processes with Applications to the Natural Sciences
BALAKRISHNAN and KOUTRAS • Runs and Scans with Applications
BALAKRISHNAN and NG • Precedence-Type Tests and Applications
BARNETT • Comparative Statistical Inference, *Third Edition*
BARNETT • Environmental Statistics
BARNETT and LEWIS • Outliers in Statistical Data, *Third Edition*
BARTOSZYNSKI and NIEWIADOMSKA-BUGAJ • Probability and Statistical Inference
BASILEVSKY • Statistical Factor Analysis and Related Methods: Theory and Applications
BASU and RIGDON • Statistical Methods for the Reliability of Repairable Systems
BATES and WATTS • Nonlinear Regression Analysis and Its Applications

†Now available in a lower priced paperback edition in the Wiley Classics Library.
*Now available in a lower priced paperback edition in the Wiley—Interscience Paperback Series.

BECHHOFER, SANTNER, and GOLDSMAN • Design and Analysis of Experiments for Statistical Selection, Screening, and Multiple Comparisons
BELSLEY • Conditioning Diagnostics: Collinearity and Weak Data in Regression
† BELSLEY, KUH, and WELSCH • Regression Diagnostics: Identifying Influential Data and Sources of Collinearity
BENDAT and PIERSOL • Random Data: Analysis and Measurement Procedures, *Fourth Edition*
BERRY, CHALONER, and GEWEKE • Bayesian Analysis in Statistics and Econometrics: Essays in Honor of Arnold Zellner
BERNARDO and SMITH • Bayesian Theory
BHAT and MILLER • Elements of Applied Stochastic Processes, *Third Edition*
BHATTACHARYA and WAYMIRE • Stochastic Processes with Applications
BILLINGSLEY • Convergence of Probability Measures, *Second Edition*
BILLINGSLEY • Probability and Measure, *Third Edition*
BIRKES and DODGE • Alternative Methods of Regression
BISGAARD and KULAHCI • Time Series Analysis and Forecasting by Example
BISWAS, DATTA, FINE, and SEGAL • Statistical Advances in the Biomedical Sciences: Clinical Trials, Epidemiology, Survival Analysis, and Bioinformatics
BLISCHKE AND MURTHY (editors) • Case Studies in Reliability and Maintenance
BLISCHKE AND MURTHY • Reliability: Modeling, Prediction, and Optimization
BLOOMFIELD • Fourier Analysis of Time Series: An Introduction, *Second Edition*
BOLLEN • Structural Equations with Latent Variables
BOLLEN and CURRAN • Latent Curve Models: A Structural Equation Perspective
BOROVKOV • Ergodicity and Stability of Stochastic Processes
BOULEAU • Numerical Methods for Stochastic Processes
BOX • Bayesian Inference in Statistical Analysis
BOX • R. A. Fisher, the Life of a Scientist
BOX and DRAPER • Response Surfaces, Mixtures, and Ridge Analyses, *Second Edition*
* BOX and DRAPER • Evolutionary Operation: A Statistical Method for Process Improvement
BOX and FRIENDS • Improving Almost Anything, *Revised Edition*
BOX, HUNTER, and HUNTER • Statistics for Experimenters: Design, Innovation, and Discovery, *Second Editon*
BOX, JENKINS, and REINSEL • Time Series Analysis: Forcasting and Control, *Fourth Edition*
BOX, LUCEÑO, and PANIAGUA-QUIÑONES • Statistical Control by Monitoring and Adjustment, *Second Edition*
BRANDIMARTE • Numerical Methods in Finance: A MATLAB-Based Introduction
† BROWN and HOLLANDER • Statistics: A Biomedical Introduction
BRUNNER, DOMHOF, and LANGER • Nonparametric Analysis of Longitudinal Data in Factorial Experiments
BUCKLEW • Large Deviation Techniques in Decision, Simulation, and Estimation
CAIROLI and DALANG • Sequential Stochastic Optimization
CASTILLO, HADI, BALAKRISHNAN, and SARABIA • Extreme Value and Related Models with Applications in Engineering and Science
CHAN • Time Series: Applications to Finance with R and S-Plus®, *Second Edition*
CHARALAMBIDES • Combinatorial Methods in Discrete Distributions
CHATTERJEE and HADI • Regression Analysis by Example, *Fourth Edition*
CHATTERJEE and HADI • Sensitivity Analysis in Linear Regression
CHERNICK • Bootstrap Methods: A Guide for Practitioners and Researchers, *Second Edition*
CHERNICK and FRIIS • Introductory Biostatistics for the Health Sciences
CHILÈS and DELFINER • Geostatistics: Modeling Spatial Uncertainty

*Now available in a lower priced paperback edition in the Wiley Classics Library.
†Now available in a lower priced paperback edition in the Wiley–Interscience Paperback Series.

CHOW and LIU • Design and Analysis of Clinical Trials: Concepts and Methodologies, *Second Edition*
CLARKE • Linear Models: The Theory and Application of Analysis of Variance
CLARKE and DISNEY • Probability and Random Processes: A First Course withApplications, *Second Edition*
* COCHRAN and COX • Experimental Designs, *Second Edition*
COLLINS and LANZA • Latent Class and Latent Transition Analysis: With Applications in the Social, Behavioral, and Health Sciences
CONGDON • Applied Bayesian Modelling
CONGDON • Bayesian Models for Categorical Data
CONGDON • Bayesian Statistical Modelling
CONOVER • Practical Nonparametric Statistics, *Third Edition*
COOK • Regression Graphics
COOK and WEISBERG • Applied Regression Including Computing and Graphics
COOK and WEISBERG • An Introduction to Regression Graphics
CORNELL • Experiments with Mixtures, Designs, Models, and the Analysis of Mixture Data, *Third Edition*
COVER and THOMAS • Elements of Information Theory
COX • A Handbook of Introductory Statistical Methods
* COX • Planning of Experiments
CRESSIE • Statistics for Spatial Data, *Revised Edition*
CRESSIE and WIKLE • Statistics for Spatio-Temporal Data
CSÖRGŐ and HORVÁTH • Limit Theorems in Change Point Analysis
DANIEL • Applications of Statistics to Industrial Experimentation
DANIEL • Biostatistics: A Foundation for Analysis in the Health Sciences, *Eighth Edition*
* DANIEL • Fitting Equations to Data: Computer Analysis of Multifactor Data, *Second Edition*
DASU and JOHNSON • Exploratory Data Mining and Data Cleaning
DAVID and NAGARAJA • Order Statistics, *Third Edition*
* DEGROOT, FIENBERG, and KADANE • Statistics and the Law
DEL CASTILLO • Statistical Process Adjustment for Quality Control
DEMARIS • Regression with Social Data: Modeling Continuous and Limited Response Variables
DEMIDENKO • Mixed Models: Theory and Applications
DENISON, HOLMES, MALLICK and SMITH • Bayesian Methods for Nonlinear Classification and Regression
DETTE and STUDDEN • The Theory of Canonical Moments with Applications in Statistics, Probability, and Analysis
DEY and MUKERJEE • Fractional Factorial Plans
DILLON and GOLDSTEIN • Multivariate Analysis: Methods and Applications
DODGE • Alternative Methods of Regression
* DODGE and ROMIG • Sampling Inspection Tables, *Second Edition*
* DOOB • Stochastic Processes
DOWDY, WEARDEN, and CHILKO • Statistics for Research, *Third Edition*
DRAPER and SMITH • Applied Regression Analysis, *Third Edition*
DRYDEN and MARDIA • Statistical Shape Analysis
DUDEWICZ and MISHRA • Modern Mathematical Statistics
DUNN and CLARK • Basic Statistics: A Primer for the Biomedical Sciences, *Third Edition*
DUPUIS and ELLIS • A Weak Convergence Approach to the Theory of Large Deviations
EDLER and KITSOS • Recent Advances in Quantitative Methods in Cancer and Human Health Risk Assessment
* ELANDT-JOHNSON and JOHNSON • Survival Models and Data Analysis

*Now available in a lower priced paperback edition in the Wiley Classics Library.
†Now available in a lower priced paperback edition in the Wiley–Interscience Paperback Series.

ENDERS • Applied Econometric Time Series
† ETHIER and KURTZ • Markov Processes: Characterization and Convergence
EVANS, HASTINGS, and PEACOCK • Statistical Distributions, *Third Edition*
FELLER • An Introduction to Probability Theory and Its Applications, Volume I, *Third Edition,* Revised; Volume II, *Second Edition*
FISHER and VAN BELLE • Biostatistics: A Methodology for the Health Sciences
FITZMAURICE, LAIRD, and WARE • Applied Longitudinal Analysis
* FLEISS • The Design and Analysis of Clinical Experiments
FLEISS • Statistical Methods for Rates and Proportions, *Third Edition*
† FLEMING and HARRINGTON • Counting Processes and Survival Analysis
FUJIKOSHI, ULYANOV, and SHIMIZU • Multivariate Statistics: High-Dimensional and Large-Sample Approximations
FULLER • Introduction to Statistical Time Series, *Second Edition*
† FULLER • Measurement Error Models
GALLANT • Nonlinear Statistical Models
GEISSER • Modes of Parametric Statistical Inference
GELMAN and MENG • Applied Bayesian Modeling and Causal Inference from Incomplete-Data Perspectives
GEWEKE • Contemporary Bayesian Econometrics and Statistics
GHOSH, MUKHOPADHYAY, and SEN • Sequential Estimation
GIESBRECHT and GUMPERTZ • Planning, Construction, and Statistical Analysis of Comparative Experiments
GIFI • Nonlinear Multivariate Analysis
GIVENS and HOETING • Computational Statistics
GLASSERMAN and YAO • Monotone Structure in Discrete-Event Systems
GNANADESIKAN • Methods for Statistical Data Analysis of Multivariate Observations, *Second Edition*
GOLDSTEIN and LEWIS • Assessment: Problems, Development, and Statistical Issues
GREENWOOD and NIKULIN • A Guide to Chi-Squared Testing
GROSS, SHORTLE, THOMPSON, and HARRIS • Fundamentals of Queueing Theory, *Fourth Edition*
GROSS, SHORTLE, THOMPSON, and HARRIS • Solutions Manual to Accompany Fundamentals of Queueing Theory, *Fourth Edition*
* HAHN and SHAPIRO • Statistical Models in Engineering
HAHN and MEEKER • Statistical Intervals: A Guide for Practitioners
HALD • A History of Probability and Statistics and their Applications Before 1750
HALD • A History of Mathematical Statistics from 1750 to 1930
† HAMPEL • Robust Statistics: The Approach Based on Influence Functions
HANNAN and DEISTLER • The Statistical Theory of Linear Systems
HARTUNG, KNAPP, and SINHA • Statistical Meta-Analysis with Applications
HEIBERGER • Computation for the Analysis of Designed Experiments
HEDAYAT and SINHA • Design and Inference in Finite Population Sampling
HEDEKER and GIBBONS • Longitudinal Data Analysis
HELLER • MACSYMA for Statisticians
HINKELMANN and KEMPTHORNE • Design and Analysis of Experiments, Volume 1: Introduction to Experimental Design, *Second Edition*
HINKELMANN and KEMPTHORNE • Design and Analysis of Experiments, Volume 2: Advanced Experimental Design
HOAGLIN, MOSTELLER, and TUKEY • Fundamentals of Exploratory Analysis of Variance
* HOAGLIN, MOSTELLER, and TUKEY • Exploring Data Tables, Trends and Shapes
* HOAGLIN, MOSTELLER, and TUKEY • Understanding Robust and Exploratory Data Analysis

*Now available in a lower priced paperback edition in the Wiley Classics Library.
†Now available in a lower priced paperback edition in the Wiley–Interscience Paperback Series.

HOCHBERG and TAMHANE • Multiple Comparison Procedures
HOCKING • Methods and Applications of Linear Models: Regression and the Analysis of Variance, *Second Edition*
HOEL • Introduction to Mathematical Statistics, *Fifth Edition*
HOGG and KLUGMAN • Loss Distributions
HOLLANDER and WOLFE • Nonparametric Statistical Methods, *Second Edition*
HOSMER and LEMESHOW • Applied Logistic Regression, *Second Edition*
HOSMER, LEMESHOW, and MAY • Applied Survival Analysis: Regression Modeling of Time-to-Event Data, *Second Edition*
HUBER • Data Analysis: What Can Be Learned From the Past 50 Years
† HUBER and RONCHETTI • Robust Statistics, *Second Edition*
HUBERTY • Applied Discriminant Analysis
HUBERTY • Applied Discriminant Analysis *Second Edition*
HUNT and KENNEDY • Financial Derivatives in Theory and Practice, *Revised Edition*
HURD and MIAMEE • Periodically Correlated Random Sequences: Spectral Theory and Practice
HUSKOVA, BERAN, and DUPAC • Collected Works of Jaroslav Hajek—with Commentary
HUZURBAZAR • Flowgraph Models for Multistate Time-to-Event Data
IMAN and CONOVER • A Modern Approach to Statistics
† JACKSON • A User's Guide to Principle Components
JOHN • Statistical Methods in Engineering and Quality Assurance
JOHNSON • Multivariate Statistical Simulation
JOHNSON and BALAKRISHNAN • Advances in the Theory and Practice of Statistics: A Volume in Honor of Samuel Kotz
JOHNSON and BHATTACHARYYA • Statistics: Principles and Methods, *Fifth Edition*
JOHNSON and KOTZ • Distributions in Statistics
JOHNSON and KOTZ (editors) • Leading Personalities in Statistical Sciences: From the Seventeenth Century to the Present
JOHNSON, KOTZ, and BALAKRISHNAN • Continuous Univariate Distributions,Volume 1, *Second Edition*
JOHNSON, KOTZ, and BALAKRISHNAN • Continuous Univariate Distributions, Volume 2, *Second Edition*
JOHNSON, KOTZ, and BALAKRISHNAN • Discrete Multivariate Distributions
JOHNSON, KEMP, and KOTZ • Univariate Discrete Distributions, *Third Edition*
JUDGE, GRIFFITHS, HILL, LÜTKEPOHL, and LEE • The Theory and Practice of Econometrics, *Second Edition*
JUREČKOVÁ and SEN • Robust Statistical Procedures: Aymptotics and Interrelations
JUREK and MASON • Operator-Limit Distributions in Probability Theory
KADANE • Bayesian Methods and Ethics in a Clinical Trial Design
KADANE AND SCHUM • A Probabilistic Analysis of the Sacco and Vanzetti Evidence
KALBFLEISCH and PRENTICE • The Statistical Analysis of Failure Time Data, *Second Edition*
KARIYA and KURATA • Generalized Least Squares
KASS and VOS • Geometrical Foundations of Asymptotic Inference
† KAUFMAN and ROUSSEEUW • Finding Groups in Data: An Introduction to Cluster Analysis
KEDEM and FOKIANOS • Regression Models for Time Series Analysis
KENDALL, BARDEN, CARNE, and LE • Shape and Shape Theory
KHURI • Advanced Calculus with Applications in Statistics, *Second Edition*
KHURI, MATHEW, and SINHA • Statistical Tests for Mixed Linear Models
KLEIBER and KOTZ • Statistical Size Distributions in Economics and Actuarial Sciences
KLEMELÄ • Smoothing of Multivariate Data: Density Estimation and Visualization

*Now available in a lower priced paperback edition in the Wiley Classics Library.
†Now available in a lower priced paperback edition in the Wiley–Interscience Paperback Series.

KLUGMAN, PANJER, and WILLMOT • Loss Models: From Data to Decisions, *Third Edition*
KLUGMAN, PANJER, and WILLMOT • Solutions Manual to Accompany Loss Models: From Data to Decisions, *Third Edition*
KOTZ, BALAKRISHNAN, and JOHNSON • Continuous Multivariate Distributions, Volume 1, *Second Edition*
KOVALENKO, KUZNETZOV, and PEGG • Mathematical Theory of Reliability of Time-Dependent Systems with Practical Applications
KOWALSKI and TU • Modern Applied U-Statistics
KRISHNAMOORTHY and MATHEW • Statistical Tolerance Regions: Theory, Applications, and Computation
KROESE, TAIMRE, and BOTEV • Handbook of Monte Carlo Methods
KROONENBERG • Applied Multiway Data Analysis
KVAM and VIDAKOVIC • Nonparametric Statistics with Applications to Science and Engineering
LACHIN • Biostatistical Methods: The Assessment of Relative Risks, *Second Edition*
LAD • Operational Subjective Statistical Methods: A Mathematical, Philosophical, and Historical Introduction
LAMPERTI • Probability: A Survey of the Mathematical Theory, *Second Edition*
LANGE, RYAN, BILLARD, BRILLINGER, CONQUEST, and GREENHOUSE • Case Studies in Biometry
LARSON • Introduction to Probability Theory and Statistical Inference, *Third Edition*
LAWLESS • Statistical Models and Methods for Lifetime Data, *Second Edition*
LAWSON • Statistical Methods in Spatial Epidemiology
LE • Applied Categorical Data Analysis
LE • Applied Survival Analysis
LEE and WANG • Statistical Methods for Survival Data Analysis, *Third Edition*
LEPAGE and BILLARD • Exploring the Limits of Bootstrap
LEYLAND and GOLDSTEIN (editors) • Multilevel Modelling of Health Statistics
LIAO • Statistical Group Comparison
LINDVALL • Lectures on the Coupling Method
LIN • Introductory Stochastic Analysis for Finance and Insurance
LINHART and ZUCCHINI • Model Selection
LITTLE and RUBIN • Statistical Analysis with Missing Data, *Second Edition*
LLOYD • The Statistical Analysis of Categorical Data
LOWEN and TEICH • Fractal-Based Point Processes
MAGNUS and NEUDECKER • Matrix Differential Calculus with Applications in Statistics and Econometrics, *Revised Edition*
MALLER and ZHOU • Survival Analysis with Long Term Survivors
MALLOWS • Design, Data, and Analysis by Some Friends of Cuthbert Daniel
MANN, SCHAFER, and SINGPURWALLA • Methods for Statistical Analysis of Reliability and Life Data
MANTON, WOODBURY, and TOLLEY • Statistical Applications Using Fuzzy Sets
MARCHETTE • Random Graphs for Statistical Pattern Recognition
MARDIA and JUPP • Directional Statistics
MASON, GUNST, and HESS • Statistical Design and Analysis of Experiments with Applications to Engineering and Science, *Second Edition*
McCULLOCH, SEARLE, and NEUHAUS • Generalized, Linear, and Mixed Models, *Second Edition*
McFADDEN • Management of Data in Clinical Trials, *Second Edition*
* McLACHLAN • Discriminant Analysis and Statistical Pattern Recognition
McLACHLAN, DO, and AMBROISE • Analyzing Microarray Gene Expression Data
McLACHLAN and KRISHNAN • The EM Algorithm and Extensions, *Second Edition*

*Now available in a lower priced paperback edition in the Wiley Classics Library.
†Now available in a lower priced paperback edition in the Wiley–Interscience Paperback Series.

McLACHLAN and PEEL • Finite Mixture Models
McNEIL • Epidemiological Research Methods
MEEKER and ESCOBAR • Statistical Methods for Reliability Data
MEERSCHAERT and SCHEFFLER • Limit Distributions for Sums of Independent Random Vectors: Heavy Tails in Theory and Practice
MICKEY, DUNN, and CLARK • Applied Statistics: Analysis of Variance and Regression, *Third Edition*
* MILLER • Survival Analysis, *Second Edition*
MONTGOMERY, JENNINGS, and KULAHCI • Introduction to Time Series Analysis and Forecasting
MONTGOMERY, PECK, and VINING • Introduction to Linear Regression Analysis, *Fourth Edition*
MORGENTHALER and TUKEY • Configural Polysampling: A Route to Practical Robustness
MUIRHEAD • Aspects of Multivariate Statistical Theory
MULLER and STOYAN • Comparison Methods for Stochastic Models and Risks
MURRAY • X-STAT 2.0 Statistical Experimentation, Design Data Analysis, and Nonlinear Optimization
MURTHY, XIE, and JIANG • Weibull Models
MYERS, MONTGOMERY, and ANDERSON-COOK • Response Surface Methodology: Process and Product Optimization Using Designed Experiments, *Third Edition*
MYERS, MONTGOMERY, VINING, and ROBINSON • Generalized Linear Models. With Applications in Engineering and the Sciences, *Second Edition*
† NELSON • Accelerated Testing, Statistical Models, Test Plans, and Data Analyses
† NELSON • Applied Life Data Analysis
NEWMAN • Biostatistical Methods in Epidemiology
OCHI • Applied Probability and Stochastic Processes in Engineering and Physical Sciences
OKABE, BOOTS, SUGIHARA, and CHIU • Spatial Tesselations: Concepts and Applications of Voronoi Diagrams, *Second Edition*
OLIVER and SMITH • Influence Diagrams, Belief Nets and Decision Analysis
PALTA • Quantitative Methods in Population Health: Extensions of Ordinary Regressions
PANJER • Operational Risk: Modeling and Analytics
PANKRATZ • Forecasting with Dynamic Regression Models
PANKRATZ • Forecasting with Univariate Box-Jenkins Models: Concepts and Cases
* PARZEN • Modern Probability Theory and Its Applications
PEÑA, TIAO, and TSAY • A Course in Time Series Analysis
PIANTADOSI • Clinical Trials: A Methodologic Perspective
PORT • Theoretical Probability for Applications
POURAHMADI • Foundations of Time Series Analysis and Prediction Theory
POWELL • Approximate Dynamic Programming: Solving the Curses of Dimensionality
PRESS • Bayesian Statistics: Principles, Models, and Applications
PRESS • Subjective and Objective Bayesian Statistics, *Second Edition*
PRESS and TANUR • The Subjectivity of Scientists and the Bayesian Approach
PUKELSHEIM • Optimal Experimental Design
PURI, VILAPLANA, and WERTZ • New Perspectives in Theoretical and Applied Statistics
† PUTERMAN • Markov Decision Processes: Discrete Stochastic Dynamic Programming
QIU • Image Processing and Jump Regression Analysis
* RAO • Linear Statistical Inference and Its Applications, *Second Edition*
RAUSAND and HØYLAND • System Reliability Theory: Models, Statistical Methods, and Applications, *Second Edition*
RENCHER • Linear Models in Statistics
RENCHER • Methods of Multivariate Analysis, *Second Edition*

*Now available in a lower priced paperback edition in the Wiley Classics Library.
†Now available in a lower priced paperback edition in the Wiley–Interscience Paperback Series.

RENCHER • Multivariate Statistical Inference with Applications
* RIPLEY • Spatial Statistics
* RIPLEY • Stochastic Simulation
ROBINSON • Practical Strategies for Experimenting
ROHATGI and SALEH • An Introduction to Probability and Statistics, *Second Edition*
ROLSKI, SCHMIDLI, SCHMIDT, and TEUGELS • Stochastic Processes for Insurance and Finance
ROSENBERGER and LACHIN • Randomization in Clinical Trials: Theory and Practice
ROSS • Introduction to Probability and Statistics for Engineers and Scientists
ROSSI, ALLENBY, and MCCULLOCH • Bayesian Statistics and Marketing
† ROUSSEEUW and LEROY • Robust Regression and Outlier Detection
* RUBIN • Multiple Imputation for Nonresponse in Surveys
RUBINSTEIN and KROESE • Simulation and the Monte Carlo Method, *Second Edition*
RUBINSTEIN and MELAMED • Modern Simulation and Modeling
RYAN • Modern Engineering Statistics
RYAN • Modern Experimental Design
RYAN • Modern Regression Methods, *Second Edition*
RYAN • Statistical Methods for Quality Improvement, *Second Edition*
SALEH • Theory of Preliminary Test and Stein-Type Estimation with Applications
* SCHEFFE • The Analysis of Variance
SCHIMEK • Smoothing and Regression: Approaches, Computation, and Application
SCHOTT • Matrix Analysis for Statistics, *Second Edition*
SCHOUTENS • Levy Processes in Finance: Pricing Financial Derivatives
SCHUSS • Theory and Applications of Stochastic Differential Equations
SCOTT • Multivariate Density Estimation: Theory, Practice, and Visualization
† SEARLE • Linear Models for Unbalanced Data
† SEARLE • Matrix Algebra Useful for Statistics
† SEARLE, CASELLA, and McCULLOCH • Variance Components
SEARLE and WILLETT • Matrix Algebra for Applied Economics
SEBER • A Matrix Handbook For Statisticians
† SEBER • Multivariate Observations
SEBER and LEE • Linear Regression Analysis, *Second Edition*
† SEBER and WILD • Nonlinear Regression
SENNOTT • Stochastic Dynamic Programming and the Control of Queueing Systems
* SERFLING • Approximation Theorems of Mathematical Statistics
SHAFER and VOVK • Probability and Finance: It's Only a Game!
SILVAPULLE and SEN • Constrained Statistical Inference: Inequality, Order, and Shape Restrictions
SMALL and MCLEISH • Hilbert Space Methods in Probability and Statistical Inference
SRIVASTAVA • Methods of Multivariate Statistics
STAPLETON • Linear Statistical Models, *Second Edition*
STAPLETON • Models for Probability and Statistical Inference: Theory and Applications
STAUDTE and SHEATHER • Robust Estimation and Testing
STOYAN, KENDALL, and MECKE • Stochastic Geometry and Its Applications, *Second Edition*
STOYAN and STOYAN • Fractals, Random Shapes and Point Fields: Methods of Geometrical Statistics
STREET and BURGESS • The Construction of Optimal Stated Choice Experiments: Theory and Methods
STYAN • The Collected Papers of T. W. Anderson: 1943–1985
SUTTON, ABRAMS, JONES, SHELDON, and SONG • Methods for Meta-Analysis in Medical Research
TAKEZAWA • Introduction to Nonparametric Regression

*Now available in a lower priced paperback edition in the Wiley Classics Library.
†Now available in a lower priced paperback edition in the Wiley–Interscience Paperback Series.

TAMHANE • Statistical Analysis of Designed Experiments: Theory and Applications
TANAKA • Time Series Analysis: Nonstationary and Noninvertible Distribution Theory
THOMPSON • Empirical Model Building
THOMPSON • Sampling, *Second Edition*
THOMPSON • Simulation: A Modeler's Approach
THOMPSON and SEBER • Adaptive Sampling
THOMPSON, WILLIAMS, and FINDLAY • Models for Investors in Real World Markets
TIAO, BISGAARD, HILL, PEÑA, and STIGLER (editors) • Box on Quality and Discovery: with Design, Control, and Robustness
TIERNEY • LISP-STAT: An Object-Oriented Environment for Statistical Computing and Dynamic Graphics
TSAY • Analysis of Financial Time Series, *Third Edition*
UPTON and FINGLETON • Spatial Data Analysis by Example, Volume II: Categorical and Directional Data
† VAN BELLE • Statistical Rules of Thumb, *Second Edition*
VAN BELLE, FISHER, HEAGERTY, and LUMLEY • Biostatistics: A Methodology for the Health Sciences, *Second Edition*
VESTRUP • The Theory of Measures and Integration
VIDAKOVIC • Statistical Modeling by Wavelets
VINOD and REAGLE • Preparing for the Worst: Incorporating Downside Risk in Stock Market Investments
WALLER and GOTWAY • Applied Spatial Statistics for Public Health Data
WEERAHANDI • Generalized Inference in Repeated Measures: Exact Methods in MANOVA and Mixed Models
WEISBERG • Applied Linear Regression, *Third Edition*
WEISBERG • Bias and Causation: Models and Judgment for Valid Comparisons
WELSH • Aspects of Statistical Inference
WESTFALL and YOUNG • Resampling-Based Multiple Testing: Examples and Methods for p-Value Adjustment
WHITTAKER • Graphical Models in Applied Multivariate Statistics
WINKER • Optimization Heuristics in Economics: Applications of Threshold Accepting
WONNACOTT and WONNACOTT • Econometrics, *Second Edition*
WOODING • Planning Pharmaceutical Clinical Trials: Basic Statistical Principles
WOODWORTH • Biostatistics: A Bayesian Introduction
WOOLSON and CLARKE • Statistical Methods for the Analysis of Biomedical Data, *Second Edition*
WU and HAMADA • Experiments: Planning, Analysis, and Parameter Design Optimization, *Second Edition*
WU and ZHANG • Nonparametric Regression Methods for Longitudinal Data Analysis
YANG • The Construction Theory of Denumerable Markov Processes
YOUNG, VALERO-MORA, and FRIENDLY • Visual Statistics: Seeing Data with Dynamic Interactive Graphics
ZACKS • Stage-Wise Adaptive Designs
ZELTERMAN • Discrete Distributions—Applications in the Health Sciences
* ZELLNER • An Introduction to Bayesian Inference in Econometrics
ZHOU, OBUCHOWSKI, and MCCLISH • Statistical Methods in Diagnostic Medicine, *Second Edition*

*Now available in a lower priced paperback edition in the Wiley Classics Library.
†Now available in a lower priced paperback edition in the Wiley–Interscience Paperback Series.

www.ingramcontent.com/pod-product-compliance
Lightning Source LLC
LaVergne TN
LVHW020057220926
848484LV00002BB/612